AQUATIC MICROBIOLOGY

THE SOCIETY FOR APPLIED BACTERIOLOGY
SYMPOSIUM SERIES NO. 6

AQUATIC MICROBIOLOGY

Edited by

F. A. SKINNER

AND

J. M. SHEWAN

1977

ACADEMIC PRESS

LONDON . NEW YORK . SAN FRANCISCO
A Subsidiary of Harcourt Brace Jovanovich, Publishers

ACADEMIC PRESS INC. (LONDON) LTD
24-28 OVAL ROAD
LONDON N.W.1

U.S. Edition published by
ACADEMIC PRESS INC.
111 FIFTH AVENUE
NEW YORK, NEW YORK 10003

Library of Congress Catalog Card Number 77-85109
ISBN: 0-12-648030-3

Printed in Great Britain by
Whitstable Litho Ltd., Whitstable, Kent

Contributors

AVRIL E. ANSON, *Department of Biological Sciences, Hatherly Laboratories, Prince of Wales Road, University of Exeter, Exeter EX4 4PS, Devon, England*

P. A. AYRES, *Ministry of Agriculture, Fisheries and Food, Fisheries Laboratory, Burnham-on-Crouch, Essex, England*

C. M. BROWN, *Department of Biological Sciences, University of Dundee, Dundee DD1 4HN, Scotland*

N. CHRISTOFI, *Department of Biological Sciences, University of Dundee, Dundee DD1 4HN, Scotland*

C. R. CURDS, *Department of Zoology, British Museum (Natural History), Cromwell Road, London SW7 5BD, England*

D. R. CULLIMORE, *Regina Water Research Institute, University of Regina, Regina, Saskatchewan, Canada*

M. J. DAFT, *Department of Biological Sciences, University of Dundee, Dundee DD1 4HN, Scotland*

I. B. DUNCAN, *Department of Agriculture and Fisheries for Scotland, Marine Laboratory, Victoria Road, Aberdeen, Scotland*

D. M. GIBSON, *Torry Research Station, 135 Abbey Road, Aberdeen AB9 8DG, Scotland*

MARGARET S. HENDRIE, *Torry Research Station, 135 Abbey Road, Aberdeen AB9 8DG, Scotland*

R. A. HERBERT, *Department of Biological Sciences, University of Dundee, Dundee DD1 4HN, Scotland*

P. HIRSCH, *Institut für Allgemeine Mikrobiologie der Universität, 2300 Kiel, Germany*

G. HOBBS, *Torry Research Station, 135 Abbey Road, Aberdeen AB9 8DG, Scotland*

N. C. HOUSTON, *Torry Research Station, 135 Abbey Road, Aberdeen AB9 8DG, Scotland*

M. HUTCHINSON, *Tynes Laboratory of the South West Water Authority, Upton Pyne, Exeter, Devon, England*

H. W. JANNASCH, *Woods Hole Oceanographic Institution, Woods Hole, Massachusetts 02543, USA*

J G. JONES, *Freshwater Biological Association, Ferry House, Ambleside, Cumbria LA22 0LP, England*

ANNETTE E. McCANN, *Canadian Water Resources Engineering Corporation, Saskatchewan, Canada*

D. H. McCARTHY, *Ministry of Agriculture, Fisheries and Food, Fish Diseases Laboratory, The Nothe, Weymouth, Dorset, England*

M. MÜLLER, *Institut für Allgemeine Mikrobiologie der Universität, 2300 Kiel, Germany*

A. L. S. MUNRO, *Department of Agriculture and Fisheries for Scotland, Marine Laboratory, Victoria Road, Aberdeen, Scotland*

J. W. RIDGWAY, *Distribution Division, Water Research Centre, Medmenham Laboratory, P.O. Box 16, Henley Road, Medmenham, Marlow SL7 2HD, Buckinghamshire, England*

H. SCHLESNER, *Institut für Allgemeine Mikrobiologie der Universität, 2300 Kiel, Germany*

F. SINADA, *Department of Biological Sciences, University of Dundee, Dundee DD1 4HN, Scotland*

S. O. STANLEY, *Dunstaffnage Marine Laboratory, Oban, Argyll, Scotland*

W. D. P. STEWART, *Department of Biological Sciences, University of Dundee, Dundee DD1 4HN, Scotland*

G. C. WARE, *Department of Bacteriology, Medical School, University of Bristol, Bristol BS8 1TD, Avon, England*

Preface

FOR MANY years aquatic microbiology has been the Cinderella of this area of the science; indeed almost all of its practitioners, even on a world scale, could be counted on the fingers of one's hand. The remarkable change that has taken place within the last decade or so in this field is now well recognized and the Committee of the Society for Applied Bacteriology thought it appropriate that some of the more interesting aspects of the subject should be discussed at the Summer Conference at the University of Lancaster, 6–8 July 1976. The papers presented at the Symposium on 'Aquatic Microbiology', organized by Vera G. Collins of the Freshwater Biological Association, Ambleside, comprise this volume.

The more fundamental aspects of fresh water and marine microbiology receive ample consideration in the first seven chapters; these include a review of interactions between protozoa and other microbes and a description of unusual budding aquatic bacteria. More practical aspects of the subject are covered by papers on the maintenance of satisfactory drinking water supplies, pollution of estuaries by sewage and fish diseases. This wide range of topics should give the general reader as well as the expert a good idea of the exciting and challenging work now being pursued in the field of aquatic microbiology.

Included in the text are abstracts of papers on aquatic microbiology given in the open sessions of the Summer Conference, since many of these excellent papers indicate other areas in the subject where important advances are being made.

F. A. SKINNER
Rothamsted Experimental Station
Harpenden AL5 2JQ
Hertfordshire
England

J. M. SHEWAN
Ministry of Agriculture, Fisheries and Food
Torry Research Station
135 Abbey Road
Aberdeen AB9 8DG
Scotland

Contents

The Study of Aquatic Microbial Communities

J. G. JONES

Freshwater Biological Association,
Ambleside, Cumbria, England

CONTENTS

1. Introduction

THE COMMUNITIES of aquatic micro-organisms may be divided into three broad categories, the plankton, the haptobenthos and the herpobenthos. The plankton comprises those organisms which float in the water, the haptobenthos those which grow on solid substrata immersed in the water (e.g. stone, macrophytes, glass slides), and the herpobenthos those which grow in or on easily penetrable mud. The nomenclature of the benthic communities is that recommended by Hutchinson (1975) and although the division contains a degree of subjectivity, the term haptobenthos was considered preferable to 'aufwuchs' or periphyton and the many subdivisions spawned from the latter.

The structure of the community, the proportion of algae, bacteria, protozoa and fungi present will vary with site and season. This variability in the three communities makes a significant contribution to the problems encountered in obtaining reliable estimates of population size and activity in the natural environment. The organizers of this symposium considered that some discussion on methodology and techniques was required. Accordingly, this contribution attempts to provide a general summary of current methodology applicable to the three communities in marine and fresh water habitats. Several recent publications and meetings have been concerned specifically with microbiological techniques in ecology (Sorokin & Kadota 1972; Rosswall 1973; Vollenweider 1974; Proceedings of the Society for General Microbiology 1975). To avoid repetition I have tried to take a general view of the methods, and to discuss in more detail the problems of studying whole communities, how their activities might be partitioned and the contribution of each microbial component assessed. The methods for measuring growth rates (reviewed by Brock 1971), the study of viruses and the details of microbial kinetics are omitted since they will be discussed at this Symposium by experts in these subject areas.

2. Sampling the Microbial Community

The method of sampling and the size of the sampler will be governed by the information required. Gross changes in the phytoplankton in surface waters may be followed satisfactorily with a tube sampler (Lund *et al.* 1958) consisting of a length of translucent vinyl tubing *ca.* 2.5 cm i.d. with cord attached to a weighted end, which is lowered gently into the water. The upper end is then closed and the weighted end raised to the surface, thus trapping an integrated water sample to the depth required. If more details of vertical distribution patterns in planktonic populations are required then a Friedinger, Universal or Van Dorn water bottle is often more suitable. This may be of any volume or length but practical considerations of handling the samples in a small boat usually limits the sampler volume to between 1 and 5 l, and its length to between 30 cm and 1 m. A desirable feature in such a sampler is that it should allow unimpeded flow of water through the main compartment as it is lowered to the desired depth where the lids may be closed by a 'messenger'. Such a device thus traps a water sample, the volume and dimensions of which are both known. If a sterile sampling device is required, which may be filled by air displacement or by preliminary evacuation, then the original location of the water which fills the sampler cannot be determined with certainty.

If detailed knowledge of the fine structure of vertical stratification is required (some indication of which is shown in Fig. 1) then a series of small sample vessels separated by relatively short distances mounted on a single frame and designed to be filled simultaneously or consecutively can provide the required information. Such a simple device was described by Baker (1970).

Heaney (1974) devised a more sophisticated apparatus which is easier to use in a boat. Some information on sites of particular interest in the water column may be obtained by *in situ* measurements of temperature and turbidity such as those presented in Fig. 1 obtained from *x/y* plots of signals from a thermistor and a Sauberer field transparency meter (Philipp Schenk, Vienna) attached to a pressure transducer.

It is very difficult to obtain undisturbed samples of sediment or mud in lakes. The sampler must be designed to allow easy penetration of the surface mud, with minimum compaction of the sample and without disturbing the surface, which might release reducing substances or nutrients from the subsurface layers,

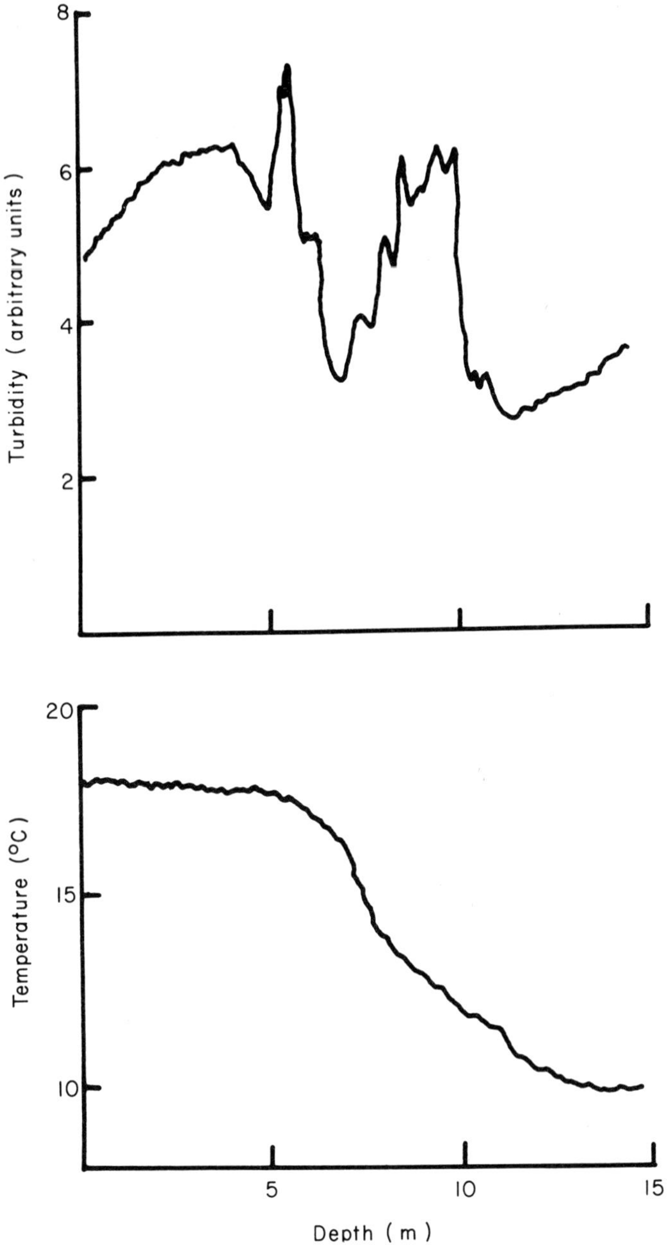

Fig. 1. Measurements *in situ* of temperature and turbidity in Esthwaite Water, a eutrophic lake, on September 1975.

or cause downward 'streaking' of material or organisms from the surface. Some degree of success might be achieved in shallow waters where the coring device, usually a perspex tube, may be driven into the mud by hand. In deeper waters, however, some degree of sample disturbance must be accepted, the effects of which might be partially overcome by allowing the sample to re-stabilize at environmental temperature in the dark for about 24 h. The Jenkin surface mud sampler has been used with some success on soft organic sediments in the English Lake District (Mortimer 1971). It is easily adapted to take subsamples for investigation of vertical distribution patterns in the mud (Goulder 1971) and for continued subsampling of the overlying water (Jones 1976). The Hargrave (1969) box sampler is somewhat larger and particularly suitable for shallow areas.

The haptobenthic community is virtually impossible to sample without disturbance. It is often diverse, consisting of representatives of microbial groups, many members of which are fragile and cannot be removed easily. The population may be on submerged stone, wood or macrophyte leaves etc. The researcher may remove part of the population by some mechanical means (Douglas 1958), examine the population *in situ* where this is possible (Jones 1974), but where not possible he may be obliged to introduce an artificial substrate such as glass or perspex to be colonized and then examined microscopically. The disadvantage of mechanical removal is that the community structure is destroyed, spatial relationships are difficult to determine except at a very crude level, and individual organisms may be rendered unrecognizable. Direct examination of the substratum is possible only in a limited number of circumstances and although spatial distribution patterns are easily seen, reliable quantitative estimates of population size often require large numbers of samples to be taken. Artificial substrates suffer from the drawback that they are artificial and therefore possibly selective. It is relatively easy to make population estimates on the surface of glass slides but the results may bear no relation to those obtained on the natural substratum. We have had some limited success with modified natural substrata which may be examined directly; for example stone surfaces which are too uneven for direct microscopic examination may be ground down to a flat surface. Colonization by algae and bacteria may then be observed by epifluorescence microscopy using the autofluorescence of chlorophyll *a*, and a suitable application of fluorochrome or an optical brightener for the bacteria. Some of the problems and methods for dealing with macrophyte haptobenthos are discussed by Fry & Humphrey (1976).

It is difficult to generalize about sampling problems since individual requirements are rarely completely satisfied by any recommended method. The methods discussed above are those most frequently reported in the literature. Further details of these and other sampling devices, and the problems associated with the marine habitat are provided by Collins *et al.* (1973).

3. Estimation of Population Size and Biomass

The most frequently used methods for estimating microbial numbers in the aquatic environment are summarized in Table 1. Detailed discussions of the methods are to be found in the references provided and only a brief outline is provided here.

Table 1

Methods for the enumeration of micro-organisms

Organism	Method	Reference
Algae	Sedimentation chamber and other counting chambers (P)	Vollenweider (1974)
	Membrane filtration (P)	Dozier & Richerson (1975)
	Scraping and subsampling into chambers (Ha)	Douglas (1958)
	Stripping with synthetic film (Ha)	Margalef (1949)
	Direct counting by epifluorescence (Ha)	Jones (1974)
	Dilution and epifluorescence counts in chambers (He)	Unpublished
Fungi	Plate counts (P, B)	Jones (1971)
	Baiting techniques quantified by MPN method (P, B)	Gaertner (1968)
	Direct counts of parasitic forms on phytoplankton hosts (P, B)	Jones (1976)
	Fluorescence antibody techniques (P, B)	Schmidt (1973)
Protozoa	Direct counts in chambers or in drops of known volume (P, B)	Goulder (1971, 1972)
Bacteria	Direct counts on membrane filters	
	Acridine epifluorescence (P, B)	Jones & Simon (1975*a*)
	Erythrosine staining (P)	Razumov (1947)
	Unstained material (P)	Jones (1975)
	Viable counts by	
	Spread plate or membrane technique (P, B)	Collins *et al.* (1973)
	MPN technique (P, B)	Harris & Sommers (1968)

P, plankton; B, benthos; Ha, haptobenthos; He, herpobenthos; Cu, cultures; MPN, most probable number.

The available techniques may be divided into six main categories, direct counts, in chambers, on surfaces or on membranes and counts of the viable organism by agar plate, membrane or MPN (most probable number) techniques. The problems associated with direct counts on surfaces have been mentioned in Section 2. Counts of plankton or benthos in chambers are subject to two main errors, the loss of material during transfer and differences in volume each time the chamber is filled. With shallow chambers suitable for counting bacterial populations the user is advised to measure the chamber depth each time it is filled. Direct counts on membrane filters are being used with increasing frequency, particularly for planktonic populations. The attraction of the method is that it provides a fairly rapid and easy method for concentration and display of the relatively sparse planktonic organisms. It is often unsuitable for fragile organisms such as protozoa and some algae and it is suggested that some bacteria may lose their metabolic activity on contact with the membrane (see discussions to several papers in Stevenson & Colwell 1973). If a membrane filter method is considered suitable for the particular organism of interest then care should also be taken to determine its distribution pattern in the membrane if reliable quantitative estimates are to be made. Of the membrane, direct count procedures available, the one which has increasingly been used in the past five years has been enumeration of microbes by epifluorescence microscopy on black membrane filters. Fluorochromes such as acridine dyes and fluorescein isothiocyanate are used to stain the organisms, particularly bacteria. Automation of direct count procedures has included computer-linked scanning electron microscope studies of plankton displayed on polycarbonate membranes. The perfect counting method would allow the microbiologist to count the organisms present and to estimate the proportion which was viable. Unfortunately this is rarely possible, the exceptions usually being the application of an experienced eye to certain eukaryotic populations. A technique which might go some way to meet these demands, at least for eukaryotes and blue-green algae uses the fluorigenic ester, fluorescein diacetate. This is considered to be a good indicator of intact cell membrane structure, the esterase enzymes sited there cleave the substrate and release the highly fluorescent end product (fluorescein) which is trapped within the cell. The method has been used successfully to estimate yeast populations (Paton & Jones 1975).

It is well known that the cultural methods used for viable populations often yield results several orders of magnitude lower than the direct counts. The microbiologist usually chooses between the precision of the plate count and membrane filter methods and the (often) higher but less precise estimates obtained by MPN techniques. There is no need to discuss here the well known problems of selectivity associated with cultural methods except possibly to reiterate that with certain groups (e.g. fungi and actinomycetes) the selectivity may be not only for particular species but also for spores rather than vegetative cells. The novel technique of Jorgensen *et al.* (1973) based on the gelling reaction of bacterial endotoxin with lysate of *Limulus* (the horseshoe crab) amoebocytes may provide some interesting information, particularly where low numbers of bacteria are encountered.

Methods for estimating the biomass of micro-organisms either as a whole community or as components of the whole are summarized in Table 2. The ideal biomass indicator is a substance, uniform in concentration in all living cells, which may be detected with a high degree of precision and sensitivity, is rapidly decomposed on the death of the organism, and is not adsorbed in a detectable state on to mineral or detrital material in the environment. Adenosine triphosphate, ATP, (Table 2) comes closest to fulfilling all these requirements. Its major disadvantages are that although the C to ATP ratios of a wide variety of micro-organisms when tested in culture tended to centre on 250 : 1, values considerably smaller and larger than this have been observed; there is also a need for information on C to ATP ratios of organisms taken from the natural environment. More useful conclusions may be drawn if values for adenosine diphosphate (ADP) and adenosine monophosphate (AMP) are also obtained. The method appears however to provide realistic results for the plankton (Hobbie *et al.* 1972) although there are problems associated with extraction efficiency or inhibition by high salts concentrations when benthic or saline material is examined (Hodson *et al.* 1976). In all instances it is advisable to include internal standards, preferably in the form of pure cultures rather than an ATP solution, to check the overall efficiency of the method.

Table 2
Methods for estimating microbial biomass

Organism	Biomass indicator	Reference
Whole community	ATP (P)	Holm-Hansen & Booth (1966)
	(P, B)	Hodson *et al.* (1976)
	(B)	Karl & LaRock (1975)
	DNA (P)	Holm-Hansen *et al.* (1968)
Algae	Chlorophyll *a* (P, B)	Vollenweider (1974)
	(P)	Loftus & Seliger (1975)
Fungi	Chitin (B)	Willoughby (1976)
		Swift (1973)
Protozoa	Cell volume (P)	Michiels (1974)
	(B)	Fenchel (1975)
	(Cu)	Curds *et al.* (1976)
Bacteria	Muramic acid (He)	Moriarty (1975)
	Cell volume (Cu)	Drake & Tsuchiya (1973)
Photosynthetic bacteria	Bacteriochlorophyll (P)	Trüper & Yentsch (1967)
		Takahashi & Ichimura (1968)

See footnote to Table 1 for explanation of abbreviations.

Two other methods for the estimation of general microbial biomass exist (determination of DNA and measurement of cell volume) but both are more tedious and sometimes less accurate than the ATP method. For certain microbial groups such as protozoa, measurement of cell volume may be the only method available, and where shapes other than simple geometric patterns are encountered, it may be necessary to construct three-dimensional models of the organisms to obtain a volume factor. Use of more sophisticated versions of the Coulter counter on a wide size range of organisms (bacteria to ciliates) in mixed culture (Drake & Tsuchiya 1973) suggest that with care some useful information might be obtained, at least about planktonic communities. Biomass values are usually derived from volume estimates through conversion factors (Winberg

1971) for volume : wet weight : dry weight : cell carbon. Needless to say such factors should be used with extreme caution on results derived from material of natural mixed populations.

The group-specific biomass indicators listed in Table 2 are far from ideal but may provide information of comparative value in the absence of anything else. They fall into two main categories, photosynthetic pigments and cell wall polymers, both of which may survive for some time after the death of the cell. The algal pigment chlorophyll *a* has been the subject of a considerable amount of research both *in vitro* and *in vivo* and has been used extensively to monitor changes in the algal components of the benthos and the plankton. A more recent innovation has been the measurement of fluorescence *in vivo* of chlorophyll *a* in the field and the laboratory, allowing rapid and detailed examination of vertical and horizontal variation in the plankton population. It is evident however, (Loftus & Seliger 1975) that if insufficient attention is paid to light conditions, temperature or the species composition of the population, then significant errors in the estimates may occur.

The cell wall polymers of bacteria and fungi are usually determined as their monomers after hydrolysis or enzymic cleavage. The longevity of these polymers after cell death, the lack of specificity of the substances analysed to the group in question, and low recoveries in the procedures adopted suggest that the methods may be only of limited use.

4. Measurement of Community Metabolic Activity

It is impossible to summarize here all the possible metabolic activities which might be encountered in a microbial community, and how these might be measured. The major reactions of microbial primary producers and decomposers towards oxygen and carbon are summarized in Fig. 2. It can be seen from this that the use of light and dark incubation conditions will allow rough estimates to be made of primary production and mineralization. The most frequently used techniques for measurement of primary production are determination of changes in dissolved O_2 concentration and measurement of CO_2 uptake (Vollenweider 1974). The former are often so small, particularly in hypolimnetic samples, that oxygen electrodes lack the necessary precision for sufficiently reliable estimates to be made. Under these circumstances, where the range on five replicate readings would often need to be no greater than 0.4%, it is necessary to use the Winkler titration, in the form of one of the more recent modifications which allow a high degree of precision (e.g. Talling 1973; Riley *et al.* 1975). Similarly high-precision determinations of CO_2 may be made and these are particularly valuable in studies of mineralization processes (Skirrow 1975; Stainton 1973; Talling 1973). Net changes in O_2 and CO_2 may be measured in enclosed samples in the laboratory or the field. Diurnal and/or seasonal changes in the water body

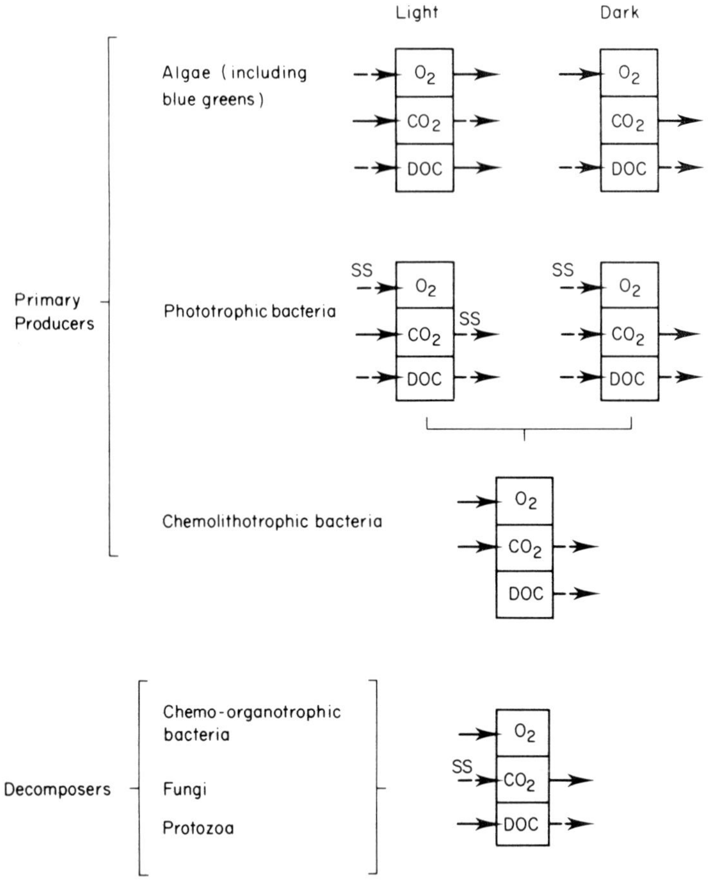

Fig. 2. The activity of community components towards carbon and oxygen. Dotted arrows indicate that the reaction occurs at low rates (e.g. the release of DOC due to cell 'leakiness') or the alternative reaction occurs at a far higher rate. SS indicates that the reaction occurs in some species and DOC is dissolved organic carbon.

itself may be used to estimate production and mineralization processes in communities which have not been enclosed. The problems associated with enclosure of a microbial community are discussed in Section 5.

The use of radioactive tracers, particularly those of high specific activity, results in a higher degree of sensitivity than is possible when measuring changes in O_2 and CO_2 concentrations. This sensitivity may not, of course, be necessary when studying the more active communities such as the benthos where reliable values for O_2 and CO_2 may be obtained in a few hours. In sparse plankton populations, however, prolonged incubation ($\geqslant$24 h) may be necessary for respirometric determinations, and under these circumstances more reliable information may be obtained from short-term incubation with radioactive tracers.

The pros and cons of the two approaches, particularly with reference to primary production measurements, are discussed by Talling & Fogg (1974). The major errors which may occur in the use of $^{14}CO_2$ to measure photosynthetic and chemosynthetic production are associated with the release of labelled organic material from the cells during incubation and during the chemical fixation and filtration processes used to stop the process. As with all experiments in which radioactive tracers are used, controls to check for non-biological adsorption must be set up with care.

Whereas realistic estimates of primary production rates may be obtained by measurement of the substrates and end products involved (CO_2 and O_2) the absence of a unique substrate for the estimation of heterotrophic production has considerably delayed research in this direction. At the time of writing the worker who wishes to measure community heterotrophic activity may attempt it in one of the following ways.

1. Measure dark oxygen consumption and convert using a suitable respiratory quotient – this method may provide useful information on mineralization rates (if ΔO_2 is large enough) but none on assimilation and therefore, production (e.g. Jones 1976). An interesting alternative approach is the measurement of electron transport system (ETS) activity as the reduction of tetrazolium dyes. This has been performed with some success on plankton (Hobbie *et al.* 1972) and benthos (Zimmerman 1975). Kenner & Ahmed 1975*a*, *b*) have suggested that when compared with O_2 uptake measurements the ETS activity might provide a respiratory control ratio which, if constant in the microbial population may be used to calculate the O_2 uptake.
2. Measure the kinetics of uptake of a range of concentrations of specific substrates (Wright & Hobbie 1965) applying a correction for respiratory losses (Hobbie & Crawford 1969) thus allowing calculation of the kinetic parameters V_{max} (the maximum possible rate of uptake or utilization under conditions of enzyme saturation); K_t, the transport constant (a measure of the affinity of the organisms for the substrate and numerically equivalent to the substrate concentration at 0.5 V_{max}); turnover time (the time in hours for the natural substrate concentration, S_n, to be utilized completely) and flux (the utilization rate at S_n which requires that the natural substrate concentration be measured independently) – this method provides information only on the specific substrates used and therefore provides comparative rather than absolute measurements of heterotrophic production. A valuable discussion on the problems associated with this approach is provided by Wright (1973) and on errors associated with the fixation process by Baross *et al.* (1975).

3. Measure assimilation and respiration of trace additions of labelled organic compounds (Williams 1970) – this method parallels that for measurement of primary production by addition of $^{14}CO_2$, in that it is assumed that the added material does not alter the natural substrate concentration to a significant degree. In this way the natural turnover time may be calculated more easily. Because of the small additions at a single concentration, substrates of the highest possible specific activity must be used, but considerable economy in materials and effort allows a wider range of individual substrates, and mixtures of substrates to be tested. In the latter case the relevance of the mixture to natural concentrations, and the possible effects of competetive inhibition (Burnison & Morita 1973) should be considered.
4. Measure the movement of a particular isotopic label through the whole community – the turnover of organic and inorganic carbon in the whole community is studied by labelling it at the primary producer level. The primary producers are either grown in the presence of (or pulse labelled with) $^{14}CO_2$ and then re-introduced to enclosures in the environment. The transport of the label to other organisms and detritus is then followed over suitable time periods. The method may allow more realistic estimates of transfer and production rates to be obtained since 'natural' substrates and communities are being used. Examples of the results obtained using this approach are provided by Wetzel *et al.* (1972) and Saunders (1972*a*). Some of the short-term (diurnal) oscillations in community activity have been demonstrated by this technique (Saunders 1972*b*, 1976).
5. Measure heterotrophic production using ^{35}S – labelled sulphate uptake – this method devised for aerobic planktonic populations by Monheimer (1974) is based on the assumption that carbon and sulphur are taken up in the same proportions as they are found in microbial cells. Monheimer (1974), who first used the method to examine natural populations, argued that the theoretical C to S ratio in phytoplankton was 500 : 1. Any reduction in this ratio in the observed uptake of $^{35}SO_4$ and $^{14}CO_2$ by plankton in the light was therefore due to heterotrophic activity. Jassby (1975) used dark $^{35}SO_4$ uptake, and an assumed C to S ratio of 50 : 1 to estimate bacterial productivity. The methods assume that sulphate is not limiting in the field and is the major source of cell S, that the S content of the cells is constant, and that dark SO_4 uptake is negligible in phytoplankton (this last point was shown to be true by Jassby (1975) when organic C levels and algal heterotrophy were low). The method obviously cannot be used to study communities which develop under anoxic conditions but deserves further investigation in, for example, the epilimnetic plankton. It provides a useful alternative to the estimation of heterotrophic activity by dark $^{14}CO_2$ uptake (Sorokin 1974).

The theoretical basis of these radioactive tracer techniques with a rough guide to the concentrations of natural substrates encountered and of the radioactive substrate added are summarized in Fig. 3. Most of these values are derived from work on the plankton since the kinetics of tracer incorporation into benthic biota are more difficult to estimate reliably (Hall *et al.* 1972; Wood & Chua 1973) and effort is usually confined to the measurement of mineralization rates (Harrison *et al.* 1971).

If information is required on which component of the microbial community is actively utilizing substrates then this may be obtained using autoradiographic techniques. These have been applied successfully to planktonic (Peroni & Lavarello 1975) haptobenthic (Ramsay 1974; Fry & Humphrey 1976) and herpobenthic (Munro & Brock 1968) communities.

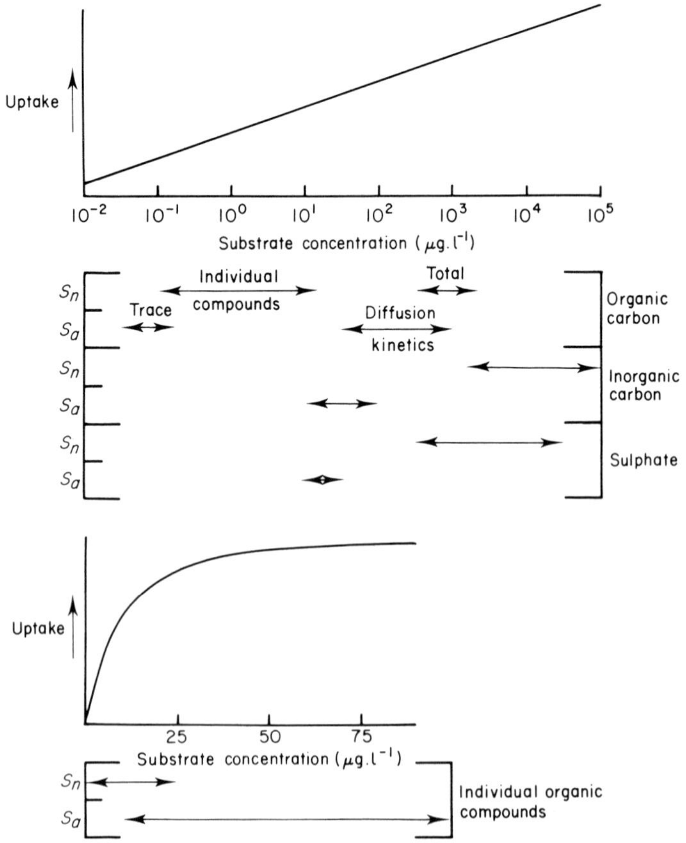

Fig. 3. The assumed relationships between uptake rates of natural (S_n) and added (S_a) substrates in natural waters. ⟷ Indicates the concentration ranges observed in aquatic systems and added in the form of radioactive tracers.

5. The Effect of Enclosure on Community Activity

The removal of a portion of a community to determine its activity (also measurement in the field) often involves enclosure in a sampling bottle or some suitable container for a period of time sufficient to allow a detectable change to occur. The community has therefore been changed and now possesses new boundaries. The effect of this change has been discussed at length in the past, often under the name of the 'bottle effect' which is variously reported as an increased, decreased or unchanged community activity depending on the temperature, incubation period, size of the enclosure and the community being studied. The changes most frequently reported are in respiratory activity and the size of the viable bacterial population, as estimated by plate counts. Generally speaking the herpobenthic communities activity is relatively unchanged by enclosure, particularly if the sediment micro-zones are allowed to become re-established (usually within 24 h) before measurement commences. As shown in Fig. 4 it is possible to measure O_2 uptake in Jenkin surface mud cores, without significant change in the rate, for several days after taking the sample. The final reduction in activity is caused by O_2 limitation rather than the effect of prolonged sample storage. Since it is usually possible to obtain reliable estimates of activity at the sediment–water interface within 3 h it is unlikely that the effect of enclosure will be significant. Further support for this suggestion comes from the fact that the 'bottle effect' has often been considered to be due to the increased surface area made available to the organism by enclosure. In this instance the increase in surface will be negligible compared with that already available in the sediment itself. Removal of sediment

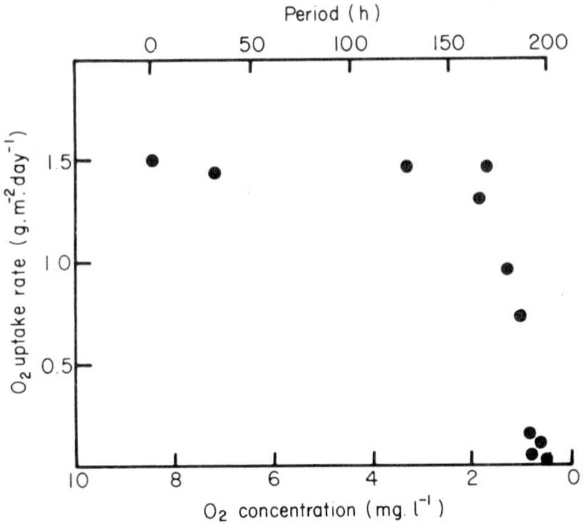

Fig. 4. Dark oxygen uptake in Jenkin surface mud cores, incubated at environmental temperature, plotted against time and oxygen concentration (from Jones 1976).

subsamples and resultant destruction of the chemical, physical and biological community structure, such as occurs on dilution and mixing with water, can, however, result in extremely high apparent metabolic rates (Jones & Simon 1975*b*). A similar effect would be expected if the haptobenthic community was sampled in a destructive manner (e.g. by scraping) the results of which would be of comparative value only. The most realistic values are likely to be obtained by measurement of the activity *in situ*; this has been done successfully with both haptobenthos (Schindler *et al.* 1973) and herpobenthos (Davies 1975). It would appear, however, that removal of portions of the benthic community to the laboratory will not produce serious errors in the estimates obtained as long as conditions of temperature, light intensity and degree of movement of the overlying water resemble those of the original site.

It is with reference to the plankton that much of the discussion on the 'bottle effect' has taken place, due to the belief that the extra surface area made available to this community in a bottle would be more likely to affect its activity. This increase is often discussed in terms of substrate concentration and increased microbial growth at the solid–liquid interface. It is partly to avoid this phenomenon that many workers submit bottles to slow rotation during the incubation period (also, of course, to simulate the turbulence to which the plankton is usually exposed). It is with these possibilities in mind that most aquatic microbiologists keep the period of artificial enclosure to a minimum. If radioactive tracer work is being performed then no real problems arise since most determinations of uptake and mineralization can be performed in 0.5–4 h, well before the population begins to respond to the changed conditions. When respiratory activity is being measured, however, particularly in sparse plankton samples or those from cooler hypolimnetic regions, then a much longer incubation period may be required to obtain a detectable change in O_2 or CO_2. This may be as long as (sometimes longer than) 24 h during which time considerable changes in the community may take place. Observations in the literature suggest that the most likely events to occur with prolonged incubation are a decline in the activity of the algal component and a considerable increase in the bacterial population. It is this latter event which has been the basis of so much of the criticism of the five-day BOD test (which is conducted at 20°C).

Since so many aquatic microbiologists have examined the 'bottle effect', usually to ensure the linearity of a given reaction with time, it would appear to be pertinent to summarize some more recent observations. The effects of incubation temperature and time on the viable and total bacterial population (as estimated by aerobic spread plate counts and the Euchrysine – 2GNX epifluorescence method) are shown in Fig. 5. The results come from a large number of experiments with water samples from a number of lakes and sampling depths, hence the representation of population change as a percentage in an attempt to draw some general conclusions. The percentage increase in the plate counts is

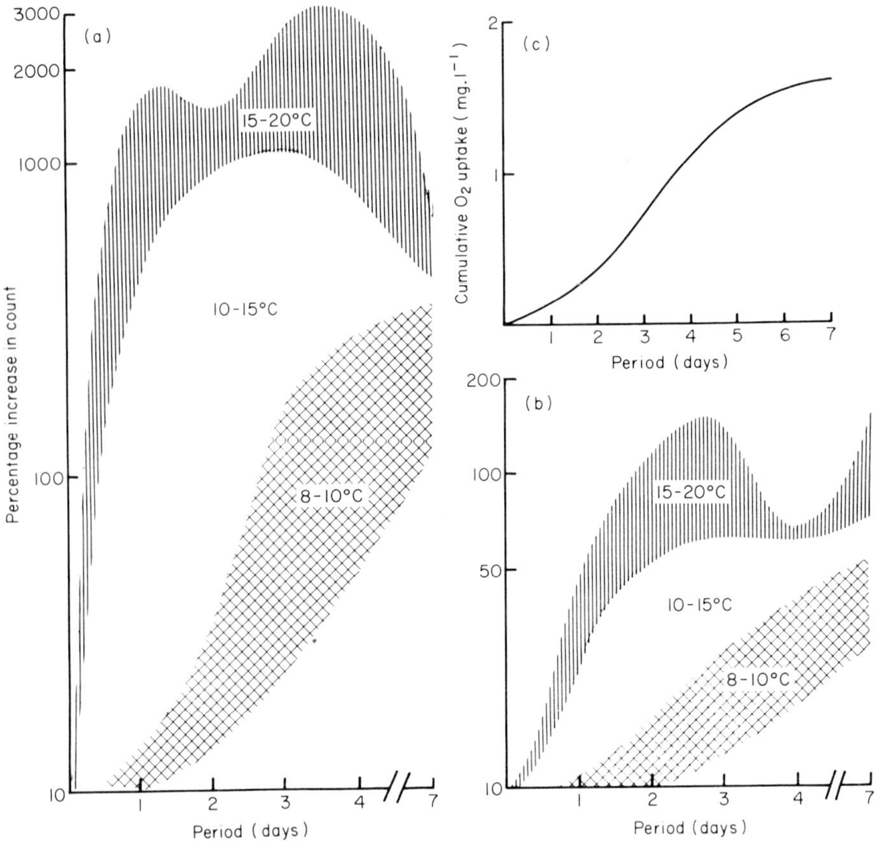

Fig. 5. The effect of temperature and time on estimates of viable (a) and total bacteria (b) and oxygen uptake (c) in stored water samples (from J. G. Jones, G. H. Hall, B. M. Simon & M. R. Reynolds, Freshwater Biological Association).

considerably greater than the direct counts (N.B. the ordinates are presented on a logarithmic scale). The lowest significant increase is considered to be 10% (the usual 95% confidence limit of the direct count) but, as is well known, a 10% increase in the direct count (usually $>10^6$ bacteria.ml^{-1}) is equivalent numerically to a 10,000% increase in the plate count (usually $>10^3$ bacteria.ml^{-1}). It is worth noting that it is often possible to incubate plankton samples at low temperatures for up to 24 h before a significant change in the population is observed. Under such conditions it might, therefore, be possible to obtain realistic estimates of respiration rates from prolonged incubations. Any anomolous results could be checked very quickly and easily by a direct count on the incubated sample. Incubation periods longer than 24 h often result in an unacceptable increase in the bacterial population and its respiration rate (Fig. 5c).

One further observation about Fig. 5 relates to the cyclic population changes sometimes observed, which often depend on the incubation temperature and the concentration of nutrients in the water. Similar cycles in the direct count and activity of heterotrophic bacteria have been observed by Shishkin & Kalinina (1974). Interesting parallel observations have been made on different scales of incubation vessel and time. Saunders (1976), for example, using 3.8 l bottles incubated *in situ* has demonstrated diurnal cyclic phenomena in the bacterial population activity related to phytoplankton photosynthetic activity. Cyclic changes in the species composition of the viable bacterial population may also occur throughout the season. The result of enclosure of large volumes of water (18,000 m^3) in experimental tubes (Fig. 6) placed in a lakeland tarn (Lack & Lund 1974) was that cycles of single bacterial species dominated the plankton on occasions (Jones 1973), possibly related to a slow decrease in nutrient levels in the tubes (Fig. 7). The plankton of the tarn, where the nutrients were replenished by inflowing streams consisted of the normal mixed bacterial populations. It is, therefore, evident that a variety of cyclic phenomena occur in the plankton, the amplitude and frequency of which should be taken into consideration when investigations are undertaken and sampling programmes designed.

Whereas bacterial populations are stimulated by storage the opposite effect is often observed with algal and other microbial populations. Gibson (1975) observed a linear oxygen uptake with time, often extending to 48 h, in a plankton sample dominated by blue-green algae. Further examination of some of his data (Fig. 8a) shows that the overall dark uptake rate falls rapidly to a level slightly greater than half that observed during the initial 36 h (Fig. 8b). The observed fluctuations in the instantaneous rate of oxygen uptake (Fig. 8c) obviously tempts speculation on possible small cyclic changes in the bacterial component superimposed on the greater overall rate caused by the blue-green algae during this period. Although such cyclic phenomena are not uncommon in samples containing blue-green algae the initial drop in respiration rate is only observed with populations which are not dark adapted (C. E. Gibson pers. comm.).

6. Partitioning of Microbial Communities

Although the measurement of total microbial metabolic activity or biomass may present problems, these often pale into insignificance with those encountered when attempts are made to estimate the contributions of the component parts of the microbiota. Although the movement of a particular substrate through a community may be traced autoradiographically, the contribution of the major

Fig. 6. Large experimental enclosures placed in Blelham Tarn (Lack & Lund 1974). Photograph by A. E. Ramsbottom, Freshwater Biological Association.

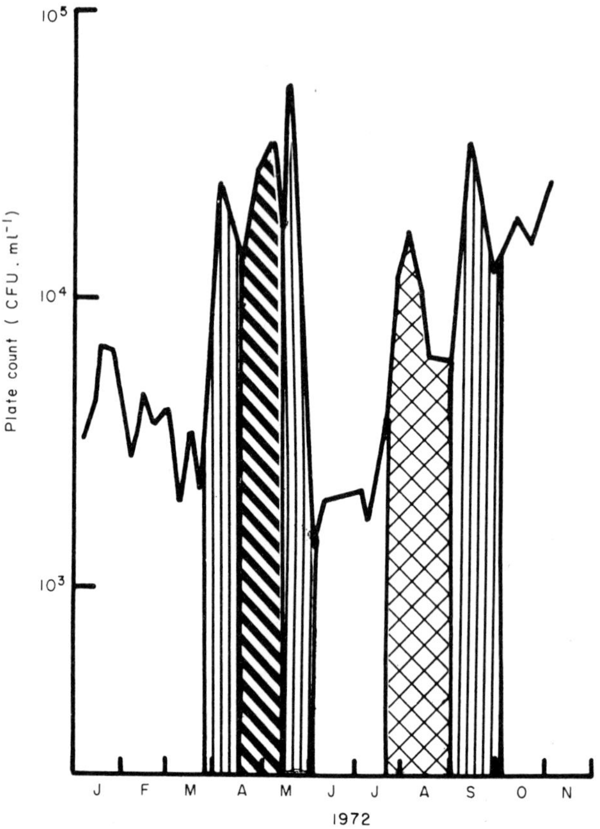

Fig. 7. Plate counts of bacterioplankton in large experimental enclosures. Hatched areas indicate cycles of dominance by different species of bacteria (from Jones 1973).

microbial components to a basic metabolic activity, such as respiration, may be determined only with considerable difficulty.

Physical separation, based on size, is feasible only for the plankton since destruction of the community structure with resultant drastic changes in rates would be experienced if it was attempted on benthos. The size ranges of algae, bacteria, protozoa and fungi overlap to varying degrees in different aquatic environments. Clearly size alone will not allow complete separation of the various community constituents but judicious filtration (e.g. on occasions when the algae in the plankton are all fairly large) can provide useful information on relative biomass or respiratory activities. Membrane filtration under vacuum or pressure can damage micro-organisms from all groups and therefore a very gentle filtration process such as that described by Pomeroy & Johannes (1968) is necessary, in which the filters used may be of nylon, stainless steel or silver mesh

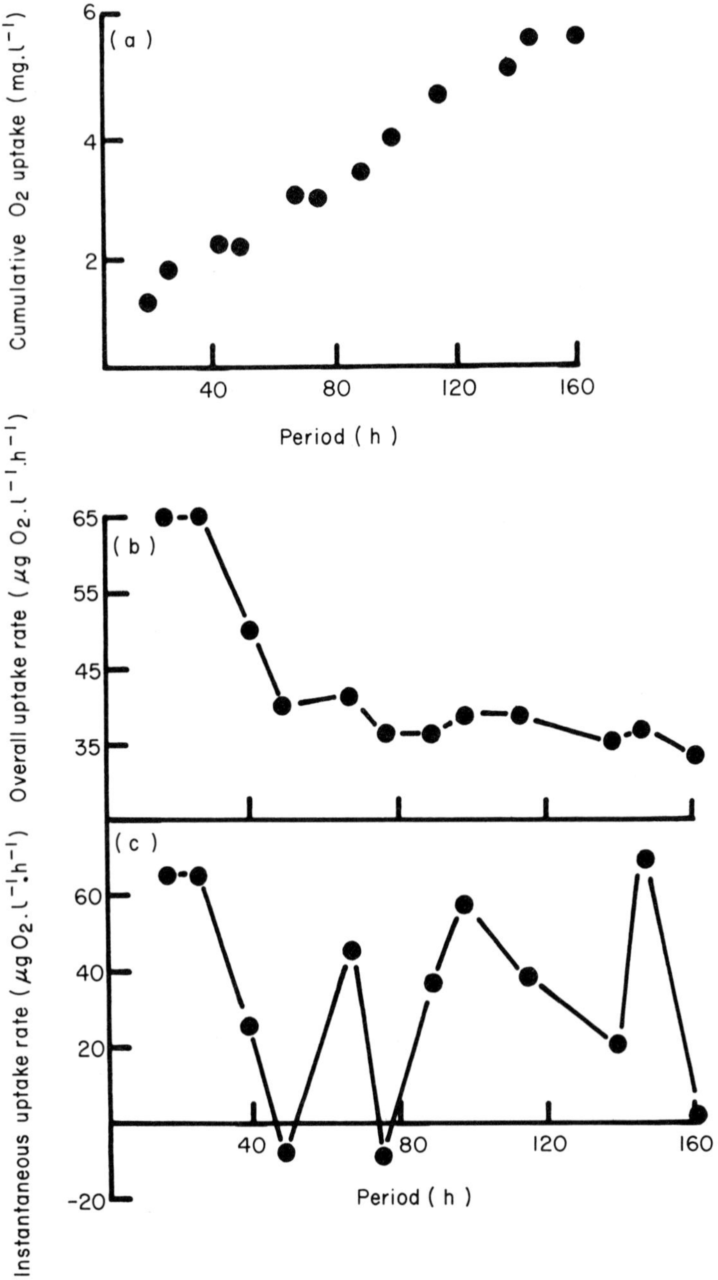

Fig. 8. The effect of storage at 16°C on dark oxygen uptake of a plankton sample dominated by blue-green algae (from Gibson 1975).

net or cellulose ester or polycarbonate membranes. The only other physical method available depends on the destruction of a particular group within the community, followed by a measurement of the change in activity. Coleman & Laurie (1976) have commented on the use of ultrasound for the selective disintegration of ciliates in the rumen. We have had mixed success with the immersion of benthic and planktonic samples in an ultrasonic water bath. Care should be taken to avoid stimulation of the remaining microbiota in samples where large numbers of ciliates are lysed.

The alternative to physical partitioning is the use of inhibitors specific to one or more components of the community. The characteristics of a suitable inhibitor depend on the type of study to be undertaken. If metabolic activity is to be measured then the ideal inhibitor will act rapidly (within a few hours) and be absolutely specific for the target group or groups of organisms. If, on the other hand, long-term population studies are being attempted then the speed of action of the inhibitor is not so critical. The most common uses of inhibitors in the aquatic environment have been: (*a*) to partition bacterial and algal activity or to prevent bacterial growth during the enclosure of plankton; (*b*) to partition bacterial, algal and fungal activity in the benthos; (*c*) to partition biological and chemical oxygen demand in the benthos. A summary of the relevant literature and the assumed activities of the inhibitors is presented in Table 3. The activity of any given inhibitor in a microbial community may be divided into three phases with respect to time.

Phase 1: during which the inhibitor is mixed with the community and establishes contact with the target organism or metabolic site within the organism.

Phase 2: during which the inhibitor is active.

Phase 3: during which the activity of the inhibitor declines possibly due to non-specific adsorption onto benthic particles or the glassware. The population may recover during this phase, occasionally due to the introduction of organisms remote from the site of the inhibitor's activity e.g. from the deeper layers of a benthic sample.

If partitioning experiments are to be successful the measurements must be made during phase 2. Unfortunately much of the published work provides little information on the duration of phase 1 and is reported in such a way as to lead one to suspect that measurements are made immediately after the addition of the inhibitor. The remainder of this section will be devoted to a discussion of some of the work summarized in Table 3 and some of my own observations in relation to the time course of events.

Of the long-term studies with inhibitors, probably some of the most successful have been those of Kaushik & Hynes (1968, 1971) who observed the bacterial and fungal colonization of the fallen leaves in streams. These authors were able to demonstrate the importance of fungi in the conditioning of leaves

Table 3
The use of inhibitors in microbial ecology

Inhibitor (final concentration)	Reported activity	Reference	Community
Long-term studies			
Nystatin (50 mg.l^{-1}) Actidione, Cycloheximide (50 mg.l^{-1})	Inhibit fungi	Kaushik & Hynes (1968, 1971) Flegler *et al.* (1974)	(Ha) (He)
Streptomycin (1.5–30 mg.l^{-1}) Penicillin (1.5–30 mg.l^{-1})	Inhibit bacteria	Kaushik & Hynes (1968, 1971) Flegler *et al.* (1974)	(Ha) (He)
2,4-Dinitrophenol (5 x 10^{-5} M)	Prevent bacterial growth	Golterman (1971, 1972) Ganf (1974)	(P) (P)
3(3,4-dichlorophenyl-1,1-dimethyl urea (5 x 10^{-6} M)	Inhibit algal photosynthesis	Golterman (1971)	(P)
Chloramphenicol	Delay bacterial growth	See text	(P)
Trimethylchlorsilene on glassware	To discourage bacterial growth on glassware	Fry & Staples (1974)	(P)
Germanium dioxide	Inhibit diatom growth by preventing silica incorporation	Dickman (1974)	(Ha)
Short-term studies			
KCN (5 x 10^{-3} M) Formaldehyde (2%) Na azide (0.65 g.l^{-1})	Inhibit biological activity	Liu (1973) Hargrave (1969, 1972*a, b*) See text	(He) (He) (He)
Streptomycin (50 mg.l^{-1}) + neomycin Streptomycin (1000 mg.kg^{-1})	To inhibit bacterial component of community respiration	Hargrave (1969) Smith (1973) Anderson & Domsch (1973)	(He) (He) (Soil)
Actidione (2000 mg.kg^{-1})	To inhibit fungal component	Anderson & Domsch (1973)	(Soil)
Polymixin (240 U.ml^{-1}) Streptomycin (325 U.ml^{-1}) Neomycin (500 mg.l^{-1})	To stress the microbial community	Wood (1973)	(He)
2,4-Dinitrophenol 10^{-3} M	To inhibit phosphate uptake	Yall *et al.* (1970)	(Sewage)
2,4-Dinitrophenol 10^{-4} M	To inhibit acetate uptake	Paerl & Goldman (1972)	(He)

See footnote to Table 1 for explanation of abbreviations.

as a source of food for the zoobenthos. These observations are comparative rather than absolute but by ensuring that the antibiotic concentrations were maintained in the environment the authors were able to prolong phase 2 and demonstrate the interaction of fungi and bacteria in the colonization process.

Both Golterman (1971, 1972) and Ganf (1974) have used dinitrophenol (DNP) to prevent bacterial growth during the measurement of mineralization in the plankton. My experience with this inhibitor (Fig. 9, III) has been that bacteria were able to grow in its presence at the concentrations used, and that

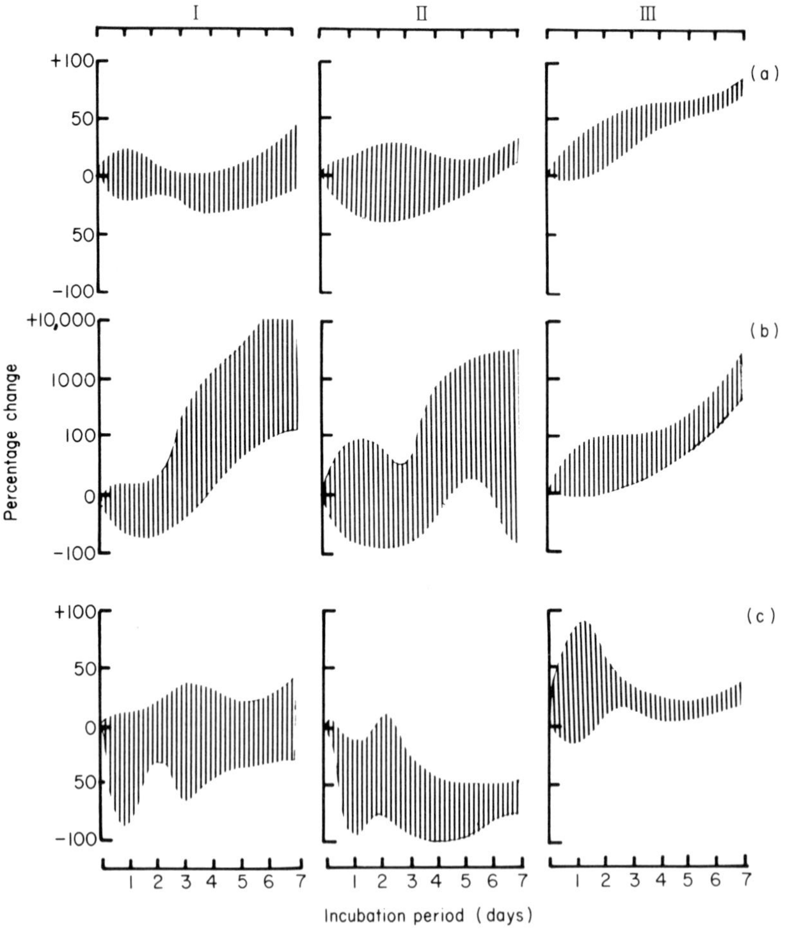

Fig. 9. The effect of inhibitors on total bacterial counts (a), viable bacterial counts (b), and oxygen uptake (c) in stored plankton samples. I: chloramphenicol, 50 mg.l^{-1}; II: streptomycin and neomycin, 50 mg.l^{-1}; III: dinitrophenol, 5 x 10^{-5} M. All samples were incubated at 15°C.

increased bacterial respiration (theoretically possible with such an uncoupling agent) always occurred. This stimulation occurred in both short- and long-term experiments but was less marked in the presence of oxidizable substrate (Table 4). Considerable care should, therefore, be exercised if DNP is to be used, particularly since it may also stimulate algal photosynthesis and respiration

Table 4

The effect of dinitrophenol on oxygen uptake by plankton

Sample O_2 consumed ($mg.l^{-1}$)	O_2 uptake after (h) 6	30	89
Plankton	0.04	0.143	0.306
Plankton + DNP*	0.304	0.302	0.357
Plankton + glucose†	3.036	0.982	1.684
Plankton + glucose + DNP	3.152	1.21	1.758

* Final concentration is 5×10^{-5} M.
† Glucose added is 10 μmoles.

(Mironyuk 1972). It is difficult to see how this inhibitor might be used effectively, as reported, to prevent substrates being taken up by microbial cells.

The use of antibiotics to control bacterioplankton during prolonged incubations is frought with difficulties and the results are so variable (Fig. 9, I and II) that it is advisable to check their effect by some independent test on each occasion. In my experience chloramphenicol in the concentration range of 50 to 75 $mg.l^{-1}$ conferred most stability, in the qualitative and quantitative sense, on the bacterial population.

Partitioning of microbial community activities such as respiration require an inhibitor which acts rapidly and specifically at the functional site of interest. Thus antibiotics such as streptomycin and neomycin which interfere with ribosomal protein synthesis, will only prevent bacterial respiration in the long term i.e. by the slow death of bacterial cells. The ability of these substances to prevent colonial development on solid media is no test of their efficacy as inhibitors in the natural environment. At low concentrations (*ca.* 50 $mg.l^{-1}$) these antibiotics act so slowly that replenishment is required to maintain phase 2 of their activity particularly in benthic samples. Although inhibition may be obtained, considerable caution should be exercised in interpreting this in a quantitative sense, particularly in the light of the variable response noted in Fig. 9 and the more detailed tests conducted by Yetka & Wiebe (1974) who concluded that the antibiotic activity varied so much with the bacterial species and its growth phase that it could not be used for ecological purposes to delineate bacterial respiration in mixed microbial communities. My own experience during investigations of benthic respiration was that streptomycin–neomycin mixtures often caused stimulation in the measured oxygen uptake except at concentrations so high, such as those used by Anderson & Domsch (1973), that other organisms such as algae would be affected. Even at these levels results were variable, and it is interesting that Anderson & Domsch (1973) pretreated their soil with glucose to stimulate biosynthetic activity before the application of the antibiotics. Since no information is available on the duration of the preincubation or phase 1 under their experimental conditions it is difficult to judge how much the community might have changed before the measurements were made. With reference to the addition of antibiotics on benthic communities it is also interesting to note Wood's observations (Wood 1973) where assimilation of labelled glucose decreased but the proportion mineralized increased. This might account for some of the stimulation in O_2 uptake noted above. I have made several attempts to determine the separate contributions of prokaryotes and eukaryotes to the community respiration, using antimycin A in the concentration range 1 to 1000 $\mu g.l^{-1}$. None were successful, largely due, I suspect, to the inability of the inhibitor to enter the eukaryotic cells.

A wide variety of metabolic poisons have been used in attempts to partition benthic oxygen demand into its biological and chemical components. Of these formaldehyde has been the most popular, but this inhibitor suffers from the disadvantage that it interferes with the Winkler determination for oxygen, which must be used if very low uptake rates occur. It has been my experience that the activity of formaldehyde, at least at the commonly used 2% final concentration, is somewhat variable, particularly in the duration of phase 2. It is not uncommon to observe motile bacteria in sediments after 24 h contact with the inhibitor suggesting that phase 3 has been reached. Similar results were obtained with $HgCl_2$ and $CuCl_2$ at 200–2000 $mg.l^{-1}$ (these poisons also interfere with the Winkler titration). The speed of action and the specificity of cyanide makes it a useful inhibitor for such studies, but considerable laboratory precautions would have to be taken if it was to be used in conjunction with the Winkler method which involves additions of concentrated mineral acid. Liu (1973) used cyanide in a Gilson respirometer with a 30 min pretreatment period (phase 1). If sediment is to be removed in such a way as to disturb it, an increase in chemical oxygen demand might be anticipated, and a period of stabilization (24 h is usually sufficient) is advisable. We have recently tested azide (final concentration 1–1.5 $g.l^{-1}$) as an inhibitor of biological oxygen demand in sediment cores, and found a 92 to 97% inhibition of uptake after a phase 1 pretreatment of 1.5 h. Given the variability in chemical demand between cores it would appear that azide was an effective inhibitor of biological uptake. Further experiments with bacteria (usually the most difficult group to inhibit) cast some doubt on this claim. The bacterial populations of mud and water samples were enriched and their respiration measured in a Gilson respirometer. Six trials were run and the results with azide, 5×10^{-3} M KCN and 2% formaldehyde were remarkably consistent producing 60–80%, 70–90% and 92–98% inhibition of O_2 uptake, respectively. These results do not appear to agree with those obtained from intact cores unless one accepts that either the azide was more effective on the natural population or the bacterial contribution to the community respiration is relatively small (⩽30%). The speed and effectiveness of formaldehyde is ample evidence to suggest that it is effective, particularly in short-term studies.

An alternative to the use of inhibitors, which we are currently investigating for benthic communities, involves the analysis of many samples of widely ranging activities. The respirometric and population data are then subjected to a multivariate analysis so that the variance of the former might be partitioned into its component parts.

7. Summary

It would not be possible in a contribution such as this to provide an exhaustive review of current methodology but I have tried to provide some indication of the range of methods in use and which might provide useful information in the future. The largest gap in current knowledge is the absence of a reliable method for the estimation of heterotrophic production. A variety of methods are available which can provide comparative information, but absolute values are at present only obtained from systems where radioactive tracers have been introduced at the primary producer level. Thus the activity and exchange within the community may be measured, as well as estimates of total community activity from respirometric data. Although most total community activities may be measured the methods available for partitioning these activities should be used with caution and preferably accompanied by an independent check of their effectiveness.

8. Acknowledgements

I am grateful to Mrs M. R. Reynolds and Mr B. M. Simon for their technical assistance in obtaining some of the original data presented in this contribution and to Mr M. J. Nield for help with the field work. Mr G. H. Hall kindly provided some of the data on the 'bottle effect', Mr A. E. Ramsbottom the photograph of the Blelham tubes and Dr C. E. Gibson gave permission for use of his data on oxygen uptake by planktonic blue-green algae. Dr V. G. Collins kindly agreed to read the manuscript.

9. References

ANDERSON, J. P. E. & DOMSCH, K. H. 1973 Quantification of bacterial and fungal contributions to soil respiration. *Archiv für Mikrobiologie* **93**, 113–127.

BAKER, A. L. 1970 An inexpensive microsampler. *Limnology and Oceanography* **15**, 158–160.
BAROSS, J. A., HANUS, F. J., GRIFFITHS, R. P. & MORITA, R. Y. 1975 Nature of incorporated ^{14}C-labelled material retained by sulfuric acid fixed bacteria in pure cultures and in natural aquatic populations. *Journal of the Fisheries Research Board of Canada* **32**, 1876–1879.
BROCK, T. D. 1971 Microbial growth rates in nature. *Bacteriological Reviews* **35**, 39–58.
BURNISON, B. K. & MORITA, R. Y. 1973 Competetive inhibition for amino acid uptake by the indigenous microflora of Upper Klamath Lake. *Applied Microbiology* **25**, 103–106.
COLEMAN, G. S. & LAURIE, J. I. 1976 Investigations on the metabolism of rumen ciliate protozoa and their closely associated bacteria. In *Techniques for the Study of Mixed Populations*, eds Lovelock, D. W. & Davies, R. Society for Applied Bacteriology Technical Series No. 11. London & New York: Academic Press.
COLLINS, V. G., JONES, J. G., HENDRIE, M. S., SHEWAN, J. M., WYNN-WILLIAMS, D. D. & RHODES, M. E. 1973 Sampling and estimation of bacterial populations in the aquatic environment. In *Sampling – Microbiological Monitoring of Environments*, eds Board, R. G. & Lovelock, D. W. Society for Applied Bacteriology Technical Series No. 7. London & New York: Academic Press.
CURDS, C. R., ROBERTS, D. M., SMITH, I. C. H. & BAZIN, M. J. 1976 The use of continuous cultures and electronic sizing devices to study the growth of two species of ciliated protozoa. In *Techniques for the Study of Mixed Populations*, eds Lovelock, D. W. & Davies, R. Society for Applied Bacteriology Technical Series No. 11. London & New York: Academic Press.
DAVIES, J. M. 1975 Energy flow through the benthos in a Scottish sea loch. *Marine Biology* **31**, 353–362.
DICKMAN, M. 1974 Changes in periphyton community structure following diatom inhibition. *Oikos* **25**, 187–193.
DOUGLAS, B. 1958 The ecology of the attached diatoms and other algae in a small stony stream. *Journal of Ecology* **46**, 295–322.
DOZIER, B. J. & RICHERSON, P. J. 1975 An improved membrane filter method for the enumeration of phytoplankton. *Verhandlung der Internationalen Vereinigung für theoretische und angewandte Limnologie* **19**, 1524–1529.
DRAKE, J. F. & TSUCHIYA, H. M. 1973 Differential counting in mixed cultures with Coulter counters. *Applied Microbiology* **26**, 9–13.
FENCHEL, T. 1975 The quantitative importance of benthic microfauna of an arctic tundra pond. *Hydrobiologia* **46**, 445–464.
FLEGLER, S. L., McNABB, C. D. & FIELDS, W. G. 1974 Antibiotic treatment of lake sediments to determine the effect of fungi on decomposition. *Water Research* **8**, 307–310.
FRY, J. C. & HUMPHREY, N. C. B. 1976 Techniques for the study of bacteria epiphytic on aquatic macrophytes. In *Techniques for the Study of Mixed Populations*, eds Lovelock, D. W. & Davies, R. Society for Applied Bacteriology Technical Series No. 11. London & New York: Academic Press.
FRY, J. C. & STAPLES, D. G. 1974 The occurrence and role of *Bdello vibrio bacteriovorus* in a polluted river. *Water Research* **8**, 1029–1035.
GAERTNER, A. 1968 Eine Methode des quantitativen Nachweises niederer, mit Pollen köderbarer Pilze im Meerwasser und im Sediment. *Veröffentlichungen des Instituts für Meeresforschung in Bremerhaven, Sonderbund* **3**, 75–91.
GANF, G. G. 1974 Rates of oxygen uptake by the planktonic community of a shallow equatorial lake (Lake George, Uganda) *Oecologia* **15**, 17–32.
GIBSON, C. E. 1975 A field and laboratory study of oxygen uptake by planktonic blue-green algae. *Journal of Ecology* **35**, 867–879.
GOLTERMAN, H. L. 1971 The determination of mineralization losses in correlation with the estimation of net primary production with the oxygen method and chemical inhibitors. *Freshwater Biology* **1**, 249–256.

GOLTERMAN, H. L. 1972 The role of phytoplankton in detritus formation. *Memorie dell' Istituto Italiano di Idrobiologia Dott Marco de Marchi* **29** Suppl., 89–103.

GOULDER, R. 1971 Vertical distribution of some ciliated protozoa in two freshwater sediments. *Oikos* **22**, 199–203.

GOULDER, R. 1972 The vertical distribution of some ciliated Protozoa in the plankton of a eutrophic pond during summer stratification. *Freshwater Biology* **2**, 163–176.

HALL, K. J., KLEIBER, P. M. & YESAKI, I. 1972 Heterotrophic uptake of organic solutes by microorganisms in the sediment. *Memorie dell' Istituto Italiano di Idrobiologia Dott Marco de Marchi* **29** Suppl., 441–471.

HARGRAVE, B. T. 1969 Epibenthic algal production and community respiration in the sediments of Marion Lake. *Journal of the Fisheries Research Board of Canada* **26**, 2003–2026.

HARGRAVE, B. T. 1972*a* Aerobic decomposition of sediment and detritus as a function of particle surface area and organic content. *Limnology and Oceanography* **17**, 583–596.

HARGRAVE, B. T. 1972*b* Oxidation-reduction potentials, oxygen concentration and oxygen uptake of profundal sediments in a eutrophic lake. *Oikos* **23**, 166–177.

HARRIS, R. F. & SOMMERS, L. E. 1968 Plate dilution frequency technique for assay of microbial ecology. *Applied Microbiology* **16**, 330–334.

HARRISON, M. J., WRIGHT, R. T. & MORITA, R. Y. 1971 Method for measuring mineralization in lake sediments. *Applied Microbiology* **21**, 698–702.

HEANEY, S. I. 1974 A pneumatically operated water sampler for close intervals of depth. *Freshwater Biology* **4**, 103–106.

HOBBIE, J. E. & CRAWFORD, C. C. 1969 Respiration corrections for bacterial uptake of dissolved organic compounds in natural waters. *Limnology and Oceanography* **14**, 528–533.

HOBBIE, J. E., HOLM-HANSEN, O., PACKARD, T. T., POMEROY, L. R., SHELDON, R. W., THOMAS, J. P. & WIEBE, W. J. 1972 A study of the distribution and activity of micro-organisms in ocean water. *Limnology and Oceanography* **17**, 544–555.

HODSON, R. E., HOLM-HANSEN, O. & AZAM, F. 1976 Improved methodology for ATP determination in marine environments. *Marine Biology* **34**, 143–150.

HOLM-HANSEN, O. & BOOTH, C. R. 1966 The measurement of adenosine triphosphate in the ocean and its ecological significance. *Limnology and Oceanography* **11**, 510–519.

HOLM-HANSEN, O., SUTCLIFFE, W. H. & SHARP, J. 1968 Measurement of deoxyribonucleic acid in the ocean and its ecological significance. *Limnology and Oceanography* **13**, 507–514.

HUTCHINSON, G. E. 1975 *A Treatise on Limnology, III Limnological Botany.* New York & Chichester: Wiley.

JASSBY, A. D. 1975 Dark sulfate uptake and bacterial productivity in a subalpine lake. *Ecology* **56**, 627–636.

JONES, E. B. G. 1971 Aquatic Fungi. In *Methods in Microbiology*, Vol. 4, ed. Booth, C. London & New York: Academic Press.

JONES, J. G. 1973 Studies on freshwater bacteria: the effect of enclosure in large experimental tubes. *Journal of Applied Bacteriology* **36**, 445–456.

JONES, J. G. 1974 A method for observation and enumeration of epilithic algae directly on the surface of stones. *Oecologia* **16**, 1–8.

JONES, J. G. 1975 Some observations on the occurrence of the iron bacterium *Leptothrix ochracea* in fresh water, including reference to large experimental enclosures. *Journal of Applied Bacteriology* **39**, 63–72.

JONES, J. G. 1976 The microbiology and decomposition of seston in open water and experimental enclosures in a productive lake. *Journal of Ecology* **14**, 241–278.

JONES, J. G. & SIMON, B. M. 1975*a* An investigation of errors in direct counts of aquatic bacteria by epifluorescence microscopy, with reference to a new method for dyeing membrane filters. *Journal of Applied Bacteriology* **39**, 317–329.

JONES, J. G. & SIMON, B. M. 1975*b* Some observations on the fluorimetric determination of glucose in freshwater. *Limnology and Oceanography* **20**, 882–887.

JORGENSEN, J. H., CARVAJAL, H. F., CHIPPS, B. E. & SMITH, R. F. 1973 Rapid detection of Gram-negative bacteriuria by use of the *Limulus* endotoxin assay. *Applied Microbiology* **26**, 38–42.

KARL, D. M. & LAROCK, P. A. 1975 Adenosine triphosphate measurements in soil and marine sediments. *Journal of Fisheries Research Board of Canada* **32**, 599–607.

KAUSHIK, N. K. & HYNES, H. B. N. 1968 Experimental study on the role of autumn-shed leaves in aquatic environments. *Journal of Ecology* **56**, 229–243.

KAUSHIK, N. K. & HYNES, H. B. N. 1971 The fate of dead leaves that fall into streams. *Archiv für Hydrobiologie* **68**, 465–515.

KENNER, R. A. & AHMED, S. I. 1975*a* Measurements of electron transport activities in marine phytoplankton. *Marine Biology* **33**, 119–127.

KENNER, R. A. & AHMED, S. I. 1975*b* Correlation between oxygen utilization and electron transport activity in marine phytoplankton. *Marine Biology* **33**, 129–133.

LACK, T. J. & LUND, J. W. G. 1974 Observations and experiments on the phytoplankton of Blelham Tarn, English Lake District 1. The experimental tubes. *Freshwater Biology* **4**, 399–415.

LIU, D. 1973 Application of the manometric technique in the study of sediment oxygen depletion. *Canadian Research and Development* **6**, 35–37.

LOFTUS, M. E. & SELIGER, H. H. 1975 Some limitations of *in vivo* fluorescence technique. *Chesapeake Science* **16**, 79–92.

LUND, J. W. G., KIPLING, C. & LE CREN, E. D. 1958 The inverted microscope method of estimating algal numbers and the statistical basis of estimation by counting. *Hydrobiologia* **11**, 143–170.

MARGALEF, R. 1949 A new limnological method for the investigation of thin-layered epilithic communities. *Hydrobiologia* **1**, 215–216.

MICHIELS, M. 1974 Biomass determination of some freshwater ciliates. *Biologisch Jaarboek – Koninklijk Natuurwetenschappelijk Genootschap* **42**, 132–136.

MIRONYUK, V. I. 1972 Effect of 2,4-dinitrophenol on content of pigments and oxygen metabolism in *Dunaliella salina* Teod. and *D. minuta* Larche. *Ukrayins'kyi botanichnyi zhurnal* **29**, 559–565.

MONHEIMER, R. H. 1974 Sulfate uptake as a measure of planktonic microbial production in freshwater ecosystems. *Canadian Journal of Microbiology* **20**, 825–831.

MORIARTY, D. J. W. 1975 A method for estimating the biomass of bacteria in aquatic sediments and its application to trophic studies. *Oecologia* **20**, 219–229.

MORTIMER, C. H. 1971 Chemical exchanges between sediments and water in the Great Lakes – speculations on probable regulatory mechanisms. *Limnology and Oceanography* **16**, 387–404.

MUNRO, A. L. S. & BROCK, T. D. 1968 Distinction between bacterial and algal utilization of soluble substances in the sea. *Journal of General Microbiology* **51**, 35–42.

PAERL, H. W. & GOLDMAN, C. R. 1972 Heterotrophic assays in the detection of water masses at Lake Tahoe, California. *Limnology and Oceanography* **17**, 145–148.

PATON, A. M. & JONES, S. M. 1975 The observation and enumeration of micro-organisms in fluids using membrane filtration and incident fluorescence microscopy. *Journal of Applied Bacteriology* **38**, 199–200.

PERONI, C. & LAVARELLO, O. 1975 Microbial activities as a function of water depth in the Ligurian Sea: an autoradiographic study. *Marine Biology* **30**, 37–50.

POMEROY, L. R. & JOHANNES, R. E. 1968 Occurrence and respiration of ultraplankton in the upper 500 meters of the ocean. *Deep Sea Research* **15**, 381–391.

PROCEEDINGS OF THE SOCIETY FOR GENERAL MICROBIOLOGY 1975 Recent advances in methodology in the study of microbial ecology. Abstracts of papers presented at the Symposium of the Microbial Ecology Group, 63–66.

RAMSAY, A. J. 1974 The use of autoradiography to determine the proportion of bacteria metabolising in an aquatic habitat. *Journal of General Microbiology* **80**, 363–373.

RAZUMOV, A. S. 1947 *Methods of Microbiological Studies of Water* Moscow: W.O.D.G.E.O.

RILEY, J. P., ROBERTSON, D. E., DUTTON, J. W. R., MITCHELL, N. T. & WILLIAMS, P. J. Le B. 1975 Analytical chemistry of sea water. In *Chemical Oceanography*, 2nd Edn, eds Riley, J. P. & Skirrow, G. **2**, 193–514. London & New York: Academic Press.

ROSSWALL, T. (eds) 1973 *Modern Methods in the Study of Microbial Ecology*. Bulletin No. 17 from the Ecological Research Committee. Stockholm: Swedish Natural Science Research Council.

SAUNDERS, G. W. 1972*a* The transformation of artificial detritus in lake water. Memorie *dell'Istituto Italiano di Idrobiologia Dott Marco de Marchi* **29** Suppl., 261–288.

SAUNDERS, G. W. 1972*b* The kinetics of extracellular release of soluble organic matter by plankton. *Verhandlung der Internationalen Vereinigung für theoretische und angewandte Limnologie* **18**, 140–146.

SAUNDERS, G. W. 1976 Microbial activity in freshwater in relation to organic materials. *In The Role of Terrestrial and Aquatic Organisms in Decomposition Processes.* 17th Symposium, British Ecological Society. Oxford: Blackwell Scientific Publications.

SCHINDLER, D. W., FROST, V. E. & SCHMIDT, R. V. 1973 Production of epilithiphyton in two lakes of the Experimental Lakes Area, north western Ontario. *Journal of the Fisheries Research Board of Canada* **30**, 1511–1524.

SCHMIDT, E. L. 1973 Fluorescent antibody techniques for the study of microbial ecology. In *Modern Methods in the Study of Microbial Ecology*, ed Rosswall, T. Bulletin No. 17 from the Ecological Research Committee. Stockholm: Swedish Natural Science Research Council.

SHISHKIN, B. A. & KALININA, A. A. 1974 Reproduction of planktonic bacteria in bottles with filtered water. *Hydrobiological Journal* **10**, 11–17.

SKIRROW, G. 1975 *Chemical Oceanography*, eds Riley, J. P. & Skirrow, G. Vol. 2. London & New York: Academic Press, pp. 1–192.

SMITH, K. L. 1973 Respiration of a sublittoral community. *Ecology* **54**, 1065–1075.

SOROKIN, Y. I. 1974 Dark assimilation of CO_2 In *A Manual on Methods for Measuring Primary Production in Aquatic Environments*, ed. Vollenweider, R. A. Oxford: Blackwell Scientific Publications.

SOROKIN, Y. I. & KADOTA, H. (eds) 1972 *Techniques for the Assessment of Microbial Production and Decomposition in Fresh Waters.* International Biological Programme Handbook No. 23. Oxford: Blackwell Scientific Publications.

STAINTON, M. P. 1973 A syringe gas-stripping procedure for gas chromatographic determination of dissolved inorganic and organic carbon in fresh water and carbonates in sediments. *Journal of the Fisheries Research Board of Canada* **30**, 1441–1445.

STEVENSON, L. H. & COLWELL, R. R. (eds) 1973 *Estuarine Microbial Ecology* Columbia: University of South Carolina Press.

SWIFT, M. J. 1973 The estimation of mycelial biomass by determination of the hexosamine content of wood tissue decayed by fungi. *Soil Biology and Biochemistry* **5**, 321–332.

TAKAHASHI, M. & ICHIMURA, S. 1968 Vertical distribution and organic matter production of photosynthetic sulfur bacteria in Japanese lakes. *Limnology and Oceanography* **13**, 644–655.

TALLING, J. F. 1973 The application of some electrochemical methods to the measurement of photosynthesis and respiration in freshwaters. *Freshwater Biology* **3**, 335–362.

TALLING, J. F. & FOGG, G. E. 1974 Methods for measuring production rates. Possible limitations and artificial modifications. In *A Manual on Methods for Measuring Primary Production in Aquatic Environments*, ed. Vollenweider, R. A. Oxford: Blackwell Scientific Publications.

TRÜPER, H. G. & YENTSCH, C. S. 1967 Use of glass fibre filters for the rapid preparation of *in vivo* absorption spectra of photosynthetic bacteria. *Journal of Bacteriology* **94**, 1255–1256.

VOLLENWEIDER, R. A. (Ed.) 1974 *A Manual on Methods for Measuring Primary Production in Aquatic Environments.* International Biological Programme Handbook No. 12, 2nd Edn. Oxford: Blackwell Scientific Publications.

WETZEL, R. G., RICH, P. H., MILLER, M. C. & ALLEN, H. L. 1972 Metabolism of dissolved and particulate detrital carbon in a temperate hard-water lake. *Memorie dell' Istituto Italiano di Idrobiologia Dott Marco di Marchi* **29** Suppl., 185–243.

WILLIAMS, P. J. Le B. 1970 Heterotrophic utilization of dissolved organic compounds in the sea 1. Size distribution of population and relation between respiration and incorporation of growth substrates. *Journal of the Marine Biological Association of the United Kingdom* **50**, 859–870.

WILLOUGHBY, L. G. 1976 Methods for studying micro-organisms on decaying leaves and wood in Fresh water. In *Techniques for the Study of Mixed Cultures*, eds Lovelock, D. W. & Davies, R. Society for Applied Bacteriology Technical Series No. 11. London & New York: Academic Press.

WINBERG, G. G. (ed.) 1971 *Symbols, Units and Conversion Factors in Studies of Fresh Water Productivity*. London: International Biological Programme.

WOOD, L. W. 1973 Pollution stress and respiration of glucose by natural microbial communities of the sediment and water of Toronto Harbor. *Proceedings of the Conference of Great Lakes Research* **16**, 204–213.

WOOD, L. W. & CHUA, K. E. 1973 Glucose flux at the sediment-water interface of Toronto Harbor, Lake Ontario, with reference to pollution stress. *Canadian Journal of Microbiology* **19**, 413–420.

WRIGHT, R. T. 1973 Some difficulties in using ^{14}C organic solutes to measure heterotrophic bacterial activity. In *Estuarine Microbial Ecology*, eds Stevenson, L. H. & Colwell, R. R. Columbia: University of South Carolina Press.

WRIGHT, R. T. & HOBBIE, J. E. 1965 Uptake of organic solutes in lake water. *Limnology and Oceanography* **10**, 22–28.

YALL, I., BOUGHTON, W. H., KNUDSON, R. C. & SINCLAIR, N. A. 1970 Biological uptake of phosphorus by activated sludge. *Applied Microbiology* **20**, 145–150.

YETKA, J. E. & WIEBE, W. J. 1974 Ecological applications of antibiotics as respiratory inhibitors of bacterial populations. *Applied Microbiology* **28**, 1033–1039.

ZIMMERMAN, A. P. 1975 Electron transport analysis as an indicator of biological oxidations in freshwater sediments. *Verhandlung der Internationalen Vereinigung für theoretische und angewandte Limnologie* **19**, 1518–1523.

Primary Production and Microbial Activity in Scottish Fresh Water Habitats

W. D. P. STEWART, F. SINADA, N. CHRISTOFI AND M. J. DAFT

Department of Biological Sciences,
University of Dundee, Dundee,
Scotland

CONTENTS

1. Introduction

A DIVERSITY of micro-organisms occurs in fresh water lakes. These range from phytoplankton organisms which are important sources of energy input in the water column, to other autotrophs such as the photosynthetic and the sulphur-oxidizing bacteria which occur predominantly in or near the sediments, and to the obligate heterotrophs which are important in geochemical and energy cycling. This paper provides some original data on how the micro-organisms, particularly the phytoplankton, vary with change in the water quality of three lochs in the east of Scotland and on how the ultimate status of each body of water is determined largely by the microbial processes which go on within it. The data presented have some bearing on the management of fresh water ecosystems for amenity, recreational and water supply purposes.

2. The Study Area

The three lochs investigated were Balgavies Loch, Forfar Loch and Loch of the Lowes which lie 20–50 km north of Dundee. These are shallow, usually well-mixed lochs, which increase in eutrophication in the order: Lowes → Balgavies → Forfar. Some physical and chemical characteristics of the lochs are summarized in Table 1 and more details of them and their catchments are presented elsewhere (Stewart *et al.* 1975, 1976).

3. The Phytoplankton

The lochs are rich in primary producers with the total number of planktonic species of algae exceeding 160 species. Several trends are noticeable, however, as the eutrophic status of the water column increases from Lowes → Balgavies → Forfar. First, species diversity decreases markedly in the poly-eutrophic Forfar where only 50 species are present, compared with 105 in Balgavies and 122 in Lowes (Fig. 1). Second, only four algal classes: Chlorophyceae, Euglenophyceae, Bacillariophyceae and Cryptophyceae are represented in Forfar whereas in Lowes and Balgavies 7 algal classes are represented, thus: Chlorophyceae, Euglenophyceae, Dinophyceae, Chrysophyceae, Bacillariophyceae, Cryptophyceae and Cyanophyceae.

Third, the algal classes which predominate, in terms of species, vary in the three lochs (Fig. 2). In Lowes there are four major classes: Chlorophyceae, Dinophyceae, Bacillariophyceae and Cyanophyceae, while in Balgavies, Dinophyceaen algae are rare and Chlorophyceae, Bacillariophyceae and Cyanophyceae predominate. In Forfar the two main classes are the Chlorophyceae and Bacillariophyceae. Heterocystous species of Cyanophyceae are more abundant in Lowes than in Balgavies which is dominated by *Microcystis flos-aquae.*

Table 1

Mean values of various chemical parameters in Lowes, Balgavies and Forfar Lochs during March 1974 to March 1975

Factors	Lowes	Balgavies	Forfar
Secchi disc visibility (cm)	305	144	132
Oxygen ($mg.l^{-1}$)	12	13	12
Total alkalinity (mg $CaCO_3 .l^{-1}$)	20	94	143
Nitrate-nitrogen ($mg.l^{-1}$)	0.1	1.6	3.5
Nitrite-nitrogen ($\mu g.l^{-1}$)	11	33	220
Ammonium-nitrogen ($\mu g.l^{-1}$)	35	68	485
Phosphate-phosphorus (μg.l)	64	146	2620
Silicate ($mg.l^{-1}$)	0.9	3.3	2.8

The data are the means of samples taken at approximately two-week intervals from six depths throughout the water column (apart from the Secchi disc readings). The following methods were used: oxygen, Winkler's method (Mackereth 1963); alkalinity (American Public Health Association 1965); nitrate (Wood *et al.* 1967); nitrite (Bendschneider & Robinson 1952); ammonium-nitrogen (Chaney & Marbach 1962); phosphate (American Public Health Association 1965); silicate (Golterman & Clymo 1969).

Fourth, there is a reduction in the total number of dominant algal species with increase in eutrophication. Thus, in Lowes 10 different algae dominated at certain times of the year and were mainly diatoms; in Balgavies 10 species dominated and were diatoms and blue-green algae whereas in Forfar the flora

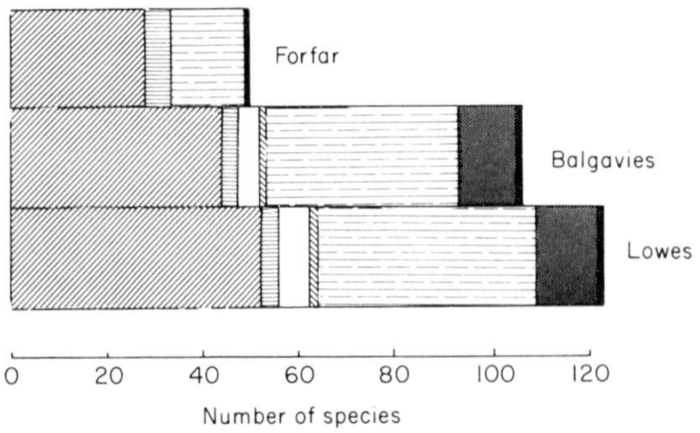

Fig. 1. The mean abundance of total species of micro-algae in Lowes, Balgavies and Forfar Lochs, and the distribution of such species among different algal classes during March 1974 to March 1975. ▨, Chlorophyceae; ▧, Chrysophyceae; ⊟, Bacillariophyceae; □, Dinophyceae; ⊟, Euglenophyceae; ▩, Cyanophyceae; ■, Cryptophyceae.

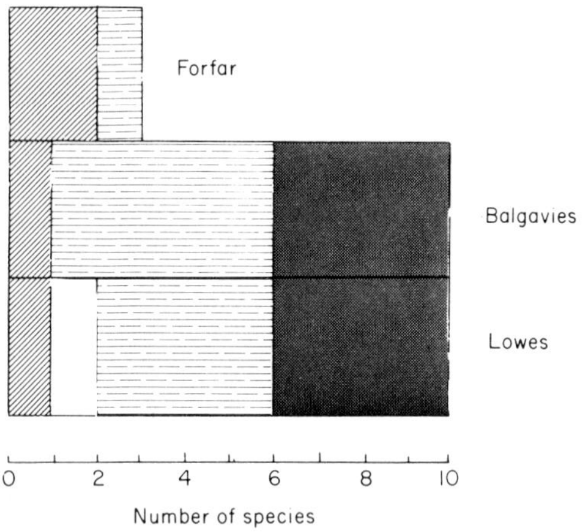

Fig. 2. The mean abundance of dominant species of micro-algae in Lowes, Balgavies and Forfar Lochs, and the distribution of such species in the different algal classes during March 1974 to March 1975. ▨, Chlorophyceae; ⊟, Bacillariophyceae; □, Dinophyceae; ▩, Cyanophyceae.

was dominated only by strains or species of the chlorophyte *Scenedesmus*, with two other green algae, *Ankistrodesmus falcatus* and *Schroederia setigera*, and by one diatom, *Stephanodiscus hantzschii*. The abundance of *Steph. hantzschii* in the polyeutrophic Forfar Loch and its dominance there, and in Loch Leven (Bailey-Watts 1974), suggests that its presence may provide a simple indication of highly eutrophic waters. In the spring of 1976 appreciable numbers of *Steph. hantzschii* (up to 16.5×10^3 cells.ml^{-1} on 23/2/76) were also recorded for the first time over a seven-year period in Balgavies.

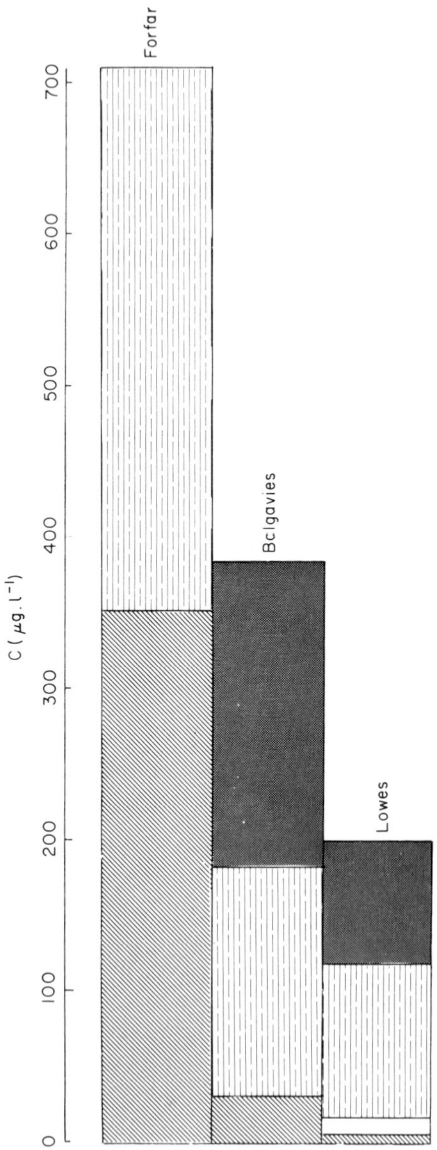

Fig. 3. The mean contribution of different algal classes (C) towards the mean total biomasses of micro-algae occurring in Lowes, Balgavies and Forfar Lochs during March 1974 to March 1975. ▨, Chlorophyceae; ⊟, Bacillariophyceae; □, Dinophyceae; ▩, Cyanophyceae.

Fifth, the data in Fig. 3 show that with increase in eutrophication not only does total algal biomass increase but also in the polyeutrophic Forfar Loch, the biomass of Chlorophyceae is almost as great as the total biomass of algae in Balgavies where the biomass is dominated by Cyanophyceae and it exceeds that in Lowes where the biomass is dominated by Bacillariophyceae.

Sixth, Fig. 4 provides data on the increase in chlorophyll *a* which occurs with

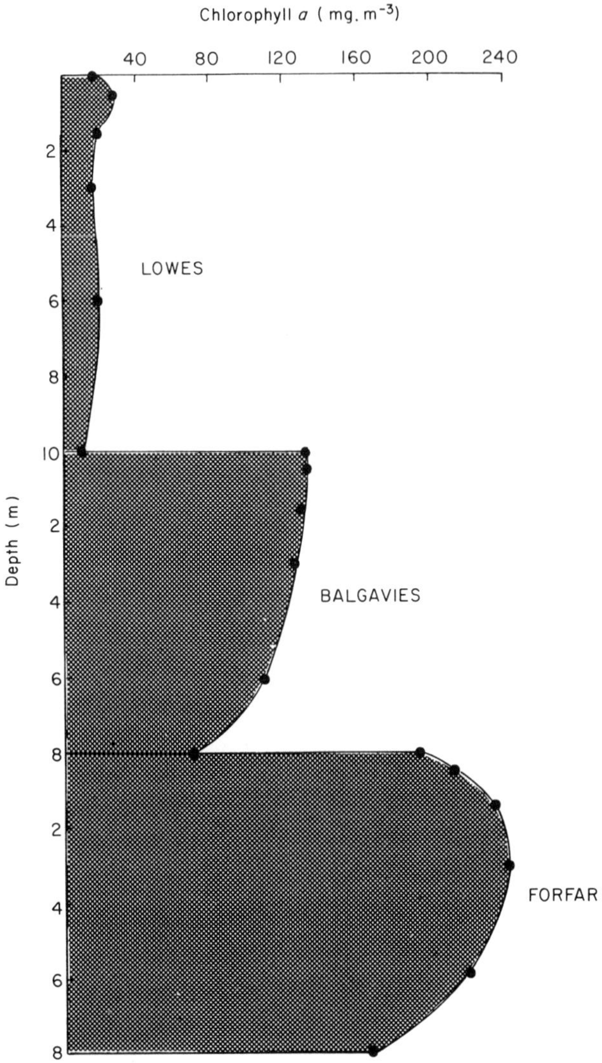

Fig. 4. The mean vertical distribution of micro-algae on a chlorophyll *a* basis in Lowes, Balgavies and Forfar Lochs during March 1974 to March 1975.

increase in eutrophication. It is seen that not only do the lochs remain fairly well mixed but that the total phytoplankton biomass of Balgavies is four times greater than that in Lowes, and is approximately half that in the polyeutrophic Forfar. Maximum biomass values, which are often a more useful guide to the extent of eutrophication, especially in lochs such as Forfar where very large fluctuations in biomass occur for uncertain reasons, are: Lowes 34 mg.m^{-3} on 28/8/74; Balgavies, 158 mg.m^{-3} on 17/7/74 and Forfar, 419 mg.m^{-3} on 27/8/75.

In sum, therefore, with increase in eutrophication there is an increase in plant biomass, but with decrease in species diversity and in the number of dominant species, and with the replacement of dinoflagellates and blue-green algae by Chlorophyceae.

4. Aquatic Macrophytes

In addition to phytoplankton, aquatic macrophytes may be important energy sources for the growth of heterotrophic micro-organisms. In our lochs a variety of submerged and emergent macrophytes occur, the most abundant being the emergent *Phragmites australis.* This reed, according to Westlake (1963), is one of the most productive macrophytes of temperate regions. Figure 5 shows the data obtained on harvesting shoot material per unit area each month during 1975. The growth pattern is rather similar in all three lochs with increases in biomass first occurring in March and reaching a maximum in August. The Lowes plants,

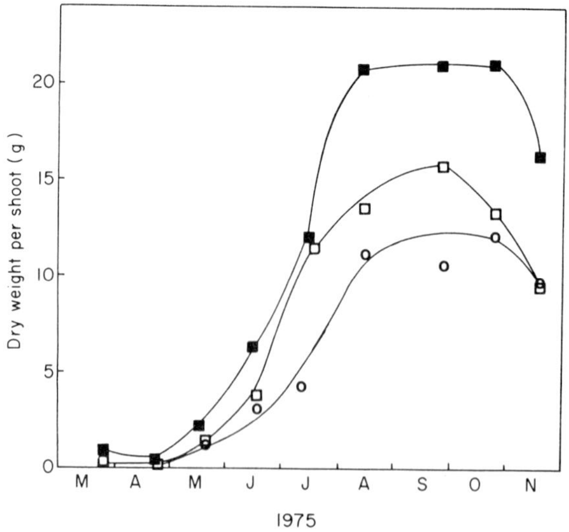

Fig. 5. The seasonal pattern of mean shoot growth of *Phragmites australis* in Lowes, Balgavies and Forfar Lochs during March 1975 to November 1975. □, Balgavies; ■, Forfar; ○, Lowes.

however, increased in biomass slightly later than in the other two lochs. This contrasts with the phytoplankton biomass in Lowes which developed earlier than did that of the other two lochs. The area covered by *Phragmites* increased from Lowes → Balgavies → Forfar (Table 2) and with increasing eutrophication more total nutrients accumulated in *Phragmites*. Calculations on the average amounts of carbon accumulated per annum in *Phragmites* shoot material and in the phytoplankton show that over the year *Phragmites* carbon values, as percentages of the *Phragmites* plus phytoplankton carbon were 45% in Lowes, 90% in Balgavies and 98% in Forfar (Table 3).

Table 2

The mean areas covered by Phragmites australis *in Lowes, Balgavies and Forfar Lochs during 1975, and the maximum amounts of phosphorus, nitrogen and carbon accumulated in the aerial parts of such plants*

Loch	*Phragmites* cover (ha)	Phosphorus (kg)	Nitrogen (kg)	Carbon (kg)
Lowes	0.84	8	68	2600
Balgavies	0.89	24	188	5400
Forfar	2.05	165	1425	37,200

Phragmites cover was estimated using aerial photography coupled with quadrat analyses and the harvesting of the shoot material present each month in average 1 m quadrats. The material was dried and then milled prior to its chemical analyses. Phosphorus was assayed, after acid digestion, according to the American Public Health Association (1965); nitrogen and carbon were analysed using a Hewlett-Packard Model 185B CHN analyser.

Table 3

The mean carbon bound in the phytoplankton compared with that in the aerial parts of Phragmites australis *in 1975*

Loch	Carbon (kg)		Carbon in phytoplankton as percentage of carbon in *Phragmites* shoot material plus carbon in phytoplankton
	Phytoplankton	Macrophytes	
Lowes	1444	1176	55.1
Balgavies	273	2403	10.2
Forfar	313	15,022	2.1

5. Micro-organisms Affecting the Primary Producers

Heterotrophic bacteria, which are dependent on fixed carbon compounds for growth and metabolism, are abundant in all three lochs, numbers being highest in Forfar and in the sediment–water interface. These micro-organisms affect the algae in numerous ways and three main groups will be considered here: those which stimulate algal growth; those which inhibit algal growth and those involved in nutrient cycling.

(a) *Bacteria which stimulate algal growth*

Bacteria may stimulate algal growths in various ways. These range from the very specific effects caused by the production of essential growth factors, to more general effects such as those in which the bacterial flora may alter the pCO_2 and pO_2 and in this way affect algal photosynthesis, photorespiration and ultimately growth (see Tolbert 1974). The possible importance of bacteria as producers of organic factors such as vitamins, growth substances and morphological regulators has been considered by Provasoli & Carlucci (1974). In general vitamin B_{12} is the vitamin most required by phytoplankton, being essential for the growth of many Euglenophyceae, Cryptophyceae, Dinophyceae and Chrysophyceae, which show a predominance of auxotrophy. The Bacillariophyceae are intermediate in their vitamin requirements and in general the Chlorophyceae, Xanthophyceae, and particularly the Cyanophyceae, seldom have vitamin requirements.

We have examined the levels of dissolved vitamin B_{12} in Lowes, Balgavies and Forfar using the *Lactobacillus leichmannii* assay of Guttman (1963). Typical data on the abundance of B_{12} with depth in the water column of Balgavies are presented in Fig. 6. It is seen that the levels remain fairly similar in the top 4 m but increase at lower depths to a maximum near the sediment water interface.

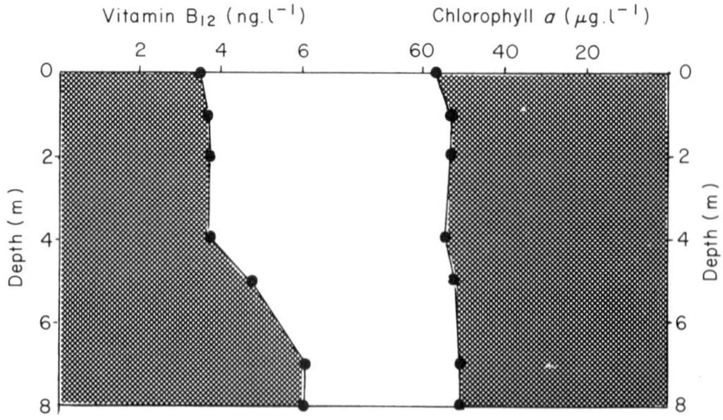

Fig. 6. The vertical distribution of soluble vitamin B_{12} and its correlation with particulate chlorophyll in Balgavies Loch on 8/5/76. Vitamin B_{12} was measured using the *L. leichmannii* assay. Each value is the mean of triplicate determinations. Chlorophyll *a* concentrations were measured according to Strickland & Parsons (1968).

Total chlorophyll *a* concentrations, in contrast, remain evenly mixed throughout the water column. The mean levels of vitamin B_{12} present in the three lochs over the period November 1974 to May 1976 (Table 4) are lowest in Lowes, increase in Balgavies, and are highest in Forfar. In Lowes, vitamin B_{12} is

Table 4

*The mean concentrations of vitamin B_{12} (*Lactobacillus leichmannii* assay) in the surface waters of Lowes, Balgavies and Forfar Lochs during November 1974 to May 1976*

Loch	Vitamin B_{12} ($ng.l^{-1}$)	
Lowes	1.1	(0–8.3)
Balgavies	13.6	(1.9–27.3)
Forfar	19.0	(2.0–28.3)

The samples were collected every 2 weeks over the period from the top 0.5 m of the water column at a standard sampling site near the middle of each loch. Measurements were carried out using the *Lactobacillus leichmannii* assay according to Guttman (1963). The values in parentheses show the ranges of values obtained.

frequently undetectable in winter, and in Forfar maximum levels occur near the outflow from the sewage works (Fig. 7). These results compare with values of *ca.* 0–15 ng $B_{12}.l^{-1}$ (*Euglena gracilis* assay) observed by Daisley (1969) in the English lakes. There, vitamin B_{12} levels tended to increase with increase in the eutrophic state of the water. In Lake Sagami in Japan, Ohwada & Taga (1972) observed soluble levels of vitamin B_{12} (*L. leichmannii* assay) of 0–12 $ng.l^{-1}$, with the pattern changing with season and with algal biomass.

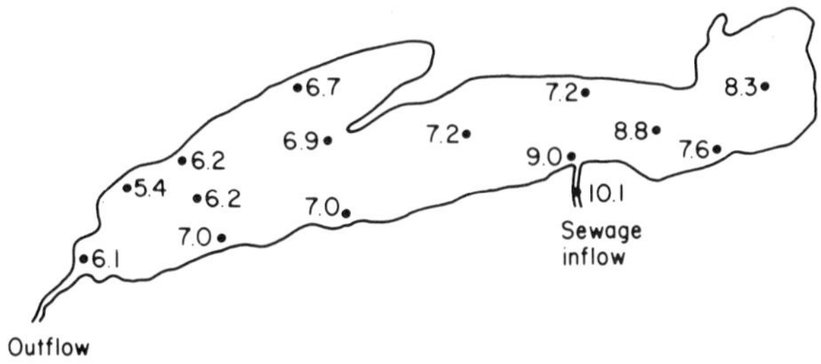

Fig. 7. The pattern of distribution of vitamin B_{12} in the top 0.5 m of Forfar Loch on 4/4/76. Note that the highest levels of vitamin B_{12} are found in the inflow water from the sewage works. Values are expressed as ng vitamin $B_{12}.l^{-1}$.

On assaying representative bacterial isolates for vitamin B_{12} production it was found (Table 5) that the total numbers of B_{12} producers increased from Lowes → Balgavies → Forfar, although as a percentage of the total bacteria present the opposite tended to occur. In all three lochs, therefore, heterotrophic bacteria appear to be responsible in part at least for the levels of B_{12} found. The blue-green algae, which are also important sources of vitamin B_{12} (Brown *et al.* 1956) could also contribute in Lowes and Balgavies, but not in Forfar where they were absent during the study period, but where input from the sewage works may be important.

Table 5

The abundance of total heterotrophic bacteria and vitamin B_{12}-producing bacteria in water samples from Lowes, Balgavies and Forfar Lochs in March 1976

Loch	Total bacteria ml^{-1}	Total B_{12}-producing bacteria ml^{-1}	B_{12} producers as percentage of total bacteria
Lowes	5.6×10^3	1.6×10^3	29
Balgavies	3.0×10^4	9.3×10^3	31
Forfar	4.1×10^5	8.2×10^4	20

The values obtained are for the top 0.5 m of the water column from 12 sites in each loch which were sampled on 2/3/76 (Lowes), 4/3/76 (Balgavies) and 25/3/76 (Forfar). Total bacteria were measured by dilution counts on the medium of Collins (1963); vitamin B_{12}-producers were estimated according to Guttman (1963).

The ecological significance of the measured levels of soluble vitamins in bioassays is uncertain since the values measured simply reflect the extent to which production exceeds consumption in the natural water sample taken, and since the optimum concentration for one organism may not be optimum for another (see Provasoli & Carlucci 1974). Nevertheless, it is notable that B_{12} levels increase with increase in the abundance of B_{12}-producing bacteria from Lowes → Balgavies → Forfar and that this is accompanied by a change in the flora with the blue-green algae, which, in general, do not require B_{12}, disappearing in Forfar, where small rapidly growing green algae and diatoms which may require vitamin B_{12}, predominate. It is also possible, on the other hand, that the algae present in Forfar may themselves be responsible for producing much of the measured B_{12}, or that the sewage outlet is the major source of B_{12}.

(b) *Bacteria and other agents which inhibit algal growth*

Within recent years it has become appreciated that a variety of microbial agents including fungi, viruses and bacteria may attack and cause the death of healthy, as well as unhealthy, algal cells. The characteristics of many of these agents are summarized in recent reviews (Stewart & Daft 1976, 1977). These, and algal-grazing zooplankton, may be responsible for some of the rapid disappearances of algal blooms frequently observed in fresh waters.

The bacteria found to be particularly important in our lochs in causing algal lysis are the non-fruiting myxobacteria with high guanine + cytosine ratios (65–69%), although a variety of other bacteria also exist. These non-fruiting myxobacteria rapidly lyse Chlorophyceae and Cyanophyceae and have been detected in all three lochs with numbers increasing from Lowes → Balgavies → Forfar. In Forfar the sewage works appears to be an important source of such bacteria with numbers decreasing away from the sewage works inflow (Daft *et al.* 1975). However, in all three lochs there is a direct correlation between the abundance of phytoplankton and the abundance of the bacteria with both being highest in spring, summer and autumn and lowest in winter.

Bacteria of the CP-1 type (Daft & Stewart 1971, 1973; Daft *et al.* 1975) are not obligate parasites in that they can be grown free from other organisms. Earlier (Daft & Stewart 1973) we provided casitone as an organic source for growth, but this can be replaced by a mixture of glutamic acid, 3 mM; histidine, 3 mM; asparagine, 6 mM; cysteine, 5 mM; tryptophan, 5 mM; arginine, 5 mM and sodium acetate (0.01%, w/v). In nature, the bacteria and algae probably co-exist under most conditions, as they may be shown to do in continuous culture where only slight inverse fluctuations between algal and bacterial biomasses can be observed (Fig. 8). However, when an imbalance occurs e.g. by environmental conditions stimulating bacterial growth, or inhibiting algal growth, large scale lysis of algal growths may occur (see Daft *et al.* 1975).

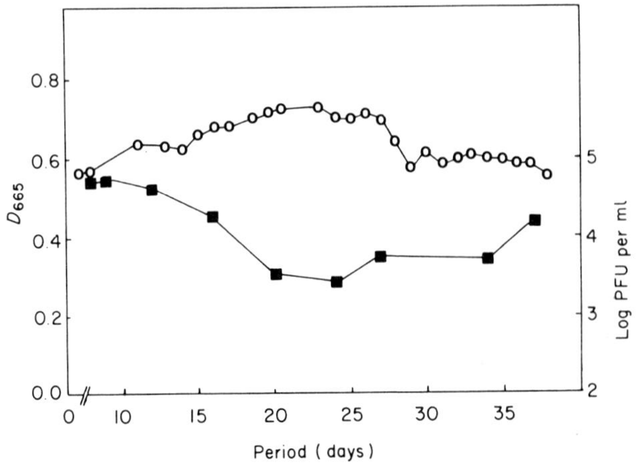

Fig. 8. The relationship between biomass of *Anacystis nidulans* (○) and myxobacterium CP-1 (■) grown in continuous culture at 26°C and 3000 lux over a 38-day period. Each value is the mean of duplicate determinations.

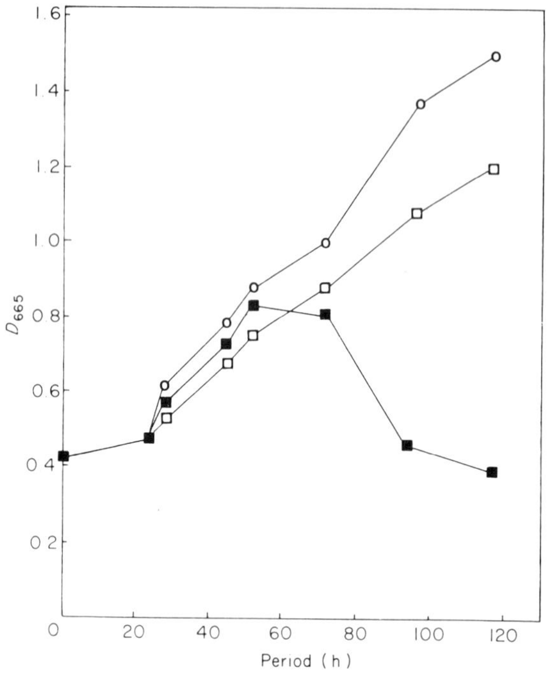

Fig. 9. A comparison of the rate of lysis of *Nostoc ellipsosporum* inoculated, at 0 time, with myxobacterium CP-1 in static (■) and shake (80 r.min^{-1}) (□) culture. ○, Is the control series without added CP-1. The experiment was carried out in 100 ml aliquots of suspension in 250 ml flasks; the CP-1 inoculum added at 0 time was 10 cells.ml^{-1}, the temperature was 26°C and the light intensity continuous at 3000 lux.

Bacteria of the CP-1 type require a contact mechanism for lysis and lysis occurs more rapidly in static cultures than in shake cultures of algae (Fig. 9). The bacteria which attach end-on (see Daft & Stewart 1973) produce at least three types of enzyme: (1) a lysozyme-like material which breaks down the L_2 layer of the algal cell wall (Fig. 10a); (2) proteolytic enzymes, as can be demonstrated by their ability to break down gelatine, either in petri dishes, or in a

Fig. 10(a) Electron micrograph of a section of *Anabaena cylindrica* showing the L_2 layer which is susceptible to attack by myxobacterium CP-1. The four wall layers L_1 – L_4 are shown (x 80,000).

(b) The effect of adding a drop of suspension of myxobacterium CP-1 to a colour negative film. Note clear central area due to the proteolytic activity of CP-1 on the gelatine film. There is no clearing when medium without CP-1, or when an autoclaved suspension of bacterium is added (x 3).

(c) The ability of myxobacterium CP-1 to hydrolyse starch. The bacterium was grown on starch-containing medium, which was subsequently flooded with iodine. Note the clear area surrounding the bacterial colonies, indicating that the bacteria have hydrolysed the starch (x 0.8).

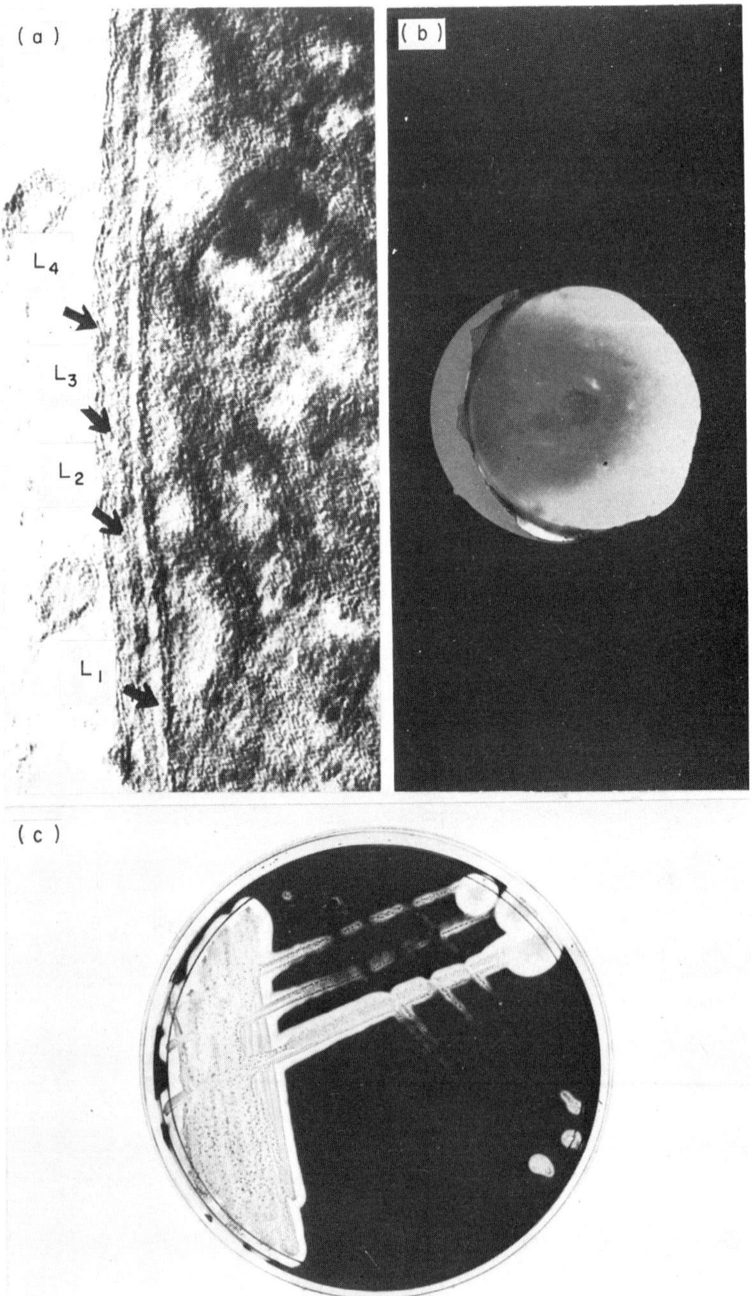
(a)
L4
L3
L2
L1
(b)
(c)

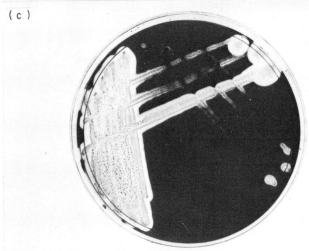

photographic emulsion (Fig. 10b); (3) starch-hydrolysing enzymes (Fig. 10c). The end result of such an attack is the complete rupture of the algal cells and a loss of soluble organic material into the environment (see below). CP-1 represents only one type of algal-lysing bacterium which we have isolated from Scottish fresh waters, and it is probable that many others still remain to be discovered.

(c) *Micro-organisms involved in nutrient cycling*

In fresh water lakes, as in other habitats, bacteria play a key role in nutrient cycling (see e.g. Svensson & Soderlund 1976). Our studies have been concerned primarily with those micro-organisms involved in the cycling of nitrogen, the levels of which increase from Lowes → Balgavies → Forfar (Table 1), and some of our preliminary findings are reported here.

Table 6

Nitrogenase activity (acetylene reduction assay) by heterocystous algae and by dominant bloom-forming blue-green algae in Lowes and Balgavies Lochs

Alga	Lowes	Balgavies
Anabaena circinalis	+	+
An. flos-aquae	+	+
An. spiroides	+	+
An. sp.	+	+
Aphanizomenon flos-aquae	nt	+
Microcystis flos-aquae	nt	–
Coelosphaerium naegelianum	–	nt
Rivularia sp.	nt	+

Acetylene reduction was measured in uni-algal populations over a 60 min period at loch temperatures. +, rates in excess of 1 μmol C_2H_4.mg chl^{-1}.h^{-1} recorded; –, no acetylene reduction detected; nt, not tested since the alga was not abundant. The *Rivularia* species tested occurred in the benthos.

(i) *Nitrogen-fixing micro-organisms*

The most obvious N_2-fixing agents in our lakes are the blue-green algae which occur in the phytoplankton of Lowes and Balgavies. We have examined uni-algal populations of all the heterocystous blue-green algae present in Lowes and Balgavies, and the dominant non-heterocystous forms, for a capacity to reduce acetylene to ethylene (Stewart *et al.* 1967). The data (Table 6) show that all heterocystous species tested reduced acetylene to ethylene whereas those without heterocysts, including the common bloom-forming *Microcystis* species, did not. All assays were carried out under aerobic conditions which are characteristic of those well-mixed lakes. No prolonged dark nitrogenase activity, characteristic of heterotrophic N_2-fixing bacteria, was detected in alga-free samples of the water column, indicating that heterotrophic N_2-fixing agents were unimportant there.

Sediment nitrogenase activity by non-photosynthetic organisms did, however, occur in all three lochs where, in particular, it was associated with the presence of aquatic macrophytes. This general finding is in accord with the report of Sylvester-Bradley (1976) who isolated acetylene-reducing bacteria resembling *Spirillum lipoferum* from the roots of *Potamogeton filiformis* in Loch Leven. In our lochs activity was associated, in particular, with *Nuphar lutea*, *Juncus effusus*, *Phrag. australis*, *Phalaris arundinaceae* and *Polygonium amphibium*. In these studies, which will be reported in detail elsewhere (M. Ogan & W. D. P. Stewart, *in preparation*), the assays were carried out at field temperatures using only short (<4 h) incubation periods. Activity was usually higher under anaerobic or microaerobic conditions, than in well-oxygenated systems (see e.g. Table 7), and was usually associated with the rhizomes and/or roots, except in

Table 7

Nitrogenase activity (acetylene reduction) associated with various fresh water macrophytes in Balgavies Loch

Plant	Material	Gas phase	C_2H_4 (nmol.g fresh $wt^{-1}.h^{-1}$)
Polygonium amphibium	Roots/rhizomes	Aerobic	7
		Anaerobic	25
Phalaris arundinaceae	Roots/rhizomes	Aerobic	20
		Anaerobic	18
Phragmites australis	Roots	Aerobic	15
		Anaerobic	20
Nuphar lutea	Petioles	Aerobic	1.3
		Anaerobic	1.4
Juncus effusus	Roots	Aerobic	45
		Anaerobic	70

The values given are the maxima observed on incubating samples at field temperatures for 3–4 h. Each value is the mean of triplicate determinations.

the case of the water lily *Nuphar lutea* where highest activities were associated with the petioles. Several N_2-fixing heterotrophic bacteria have been isolated from such systems including species of *Azotobacter* and possibly *Spirillum*. The observed rates of acetylene reduction vary with season and there is also a wide variation in rates from one N_2-fixing plant association to another, but average values differ little from those reported for *Potamogeton filiformis* by Sylvester-Bradley (1976).

The amounts of nitrogen fixed by such heterotrophic bacteria are unlikely to be important in the nitrogen economy of the lochs as a whole, especially in

Balgavies and Forfar, where there are large inputs of combined nitrogen from other sources. In Lowes, on the other hand, biological N_2-fixation could provide a significant input of new nitrogen, not only from heterotrophic bacteria and blue-green algae, but also from the *Alnus glutinosa* trees which occur round parts of this loch. Another source of bacterial N-input which requires consideration is that due to N_2-fixing methane-oxidizing bacteria. These organisms which cannot be assayed for nitrogenase activity using the acetylene reduction assay since they metabolise ethylene (Flett *et al.* 1975; De Bont & Mulder 1976) are likely to be present, particularly in Balgavies and Forfar where large amounts of methane are produced from the sediment in summer.

Table 8

The seasonal variation in the abundance of ammonifying bacteria in the top 5.0 cm of the sediment of Balgavies Loch during February 1975 to March 1976

Sampling date	Bacteria (number.g dry wt^{-1} x 10^{-4})
14/2/75	200
13/3/75	96
15/4/75	96
15/5/75	18
16/6/75	1
14/7/75	96
18/8/75	960
18/9/75	219
16/10/75	20
17/11/75	200
16/12/75	447
22/1/76	960
23/2/76	250
22/3/76	250

(ii) *Ammonifying bacteria*

The abundance of ammonifying bacteria in the three lochs was determined by the MPN technique using the peptone–water medium of Meiklejohn (1965). Ammonifying bacteria are present in high numbers in all three lochs in the water column, but especially so at the sediment–water interface and in the sediment (2.72×10^6 g.dry wt^{-1}). Data obtained on their seasonal variation in the top 0–5 cm of Balgavies Loch sediment show (Table 8) that numbers are low in spring, increase to a peak in August then decrease once more in October before increasing in winter to a January peak which is as high as the summer peak in August. The pattern of seasonal variation in the number of ammonifying bacteria in the sediment correlates rather closely in Balgavies with the levels of interstitial NH_4^+ present. Thus, NH_4^+–N levels are as low as 1 $mg.l^{-1}$ in April, increase to a peak of 12 $mg.l^{-1}$ in August, decrease to 5 $mg.l^{-1}$ in October and increase to a second peak of almost 12 $mg.l^{-1}$ in November, after which a decline occurs.

(iii) *Nitrifying bacteria*

Tests for the presence of phase I nitrifiers (those able to convert $NH_4^+ \rightarrow NO_2^-$), phase II nitrifiers (those able to convert $NO_2^- \rightarrow NO_3^-$) and heterotrophic nitrifiers were carried out in all the three lochs using modifications of the techniques of Gode & Overbeck (1972). Data on their vertical distribution in Balgavies show (Table 9) that nitrifying bacteria were present both in the water column and in the sediment and that autotrophic nitrifiers, particularly phase I types, were much more abundant than heterotrophic nitrifiers. Data on seasonal variation in the abundance of nitrifying bacteria in the sediment water interface of Balgavies show (Table 10) that during the period February 1975 to March 1976, phase I autotrophic nitrifiers exceeded phase II autotrophs which throughout were present in much higher numbers than were heterotrophic nitrifiers. However, the seasonal patterns for all three groups were very similar with numbers being high in winter, decreasing in April, increasing during early summer and decreasing during August and September before peaking in the winter.

The quantitative significance of autotrophic nitrification has been assessed by determining the rates of production of $NO_3^- + NO_2^-$ in samples incubated in the dark at loch temperatures in the presence and absence of N-Serve (Dow Chemical Co.). This compound, which inhibits the conversion of $NH_4^+ \rightarrow NO_2^-$ by autotrophic nitrifying bacteria (see Goring 1962; Campbell & Aleem 1965) enabled us to distinguish the extent of autotrophic nitrification and heterotrophic nitrification (which was negligible). Autotrophic nitrification in the sediment–water interface during June to December 1975 showed a peak activity in July [1100 $\mu g\ (NO_2^- + NO_3^-) - N.l^{-1}.day^{-1}$] followed by a decrease to a minimum in October [10 $\mu g\ (NO_2^- + NO_3^-) - N.l^{-1}.day^{-1}$] and followed by a second peak [250 $\mu g\ (NO_2^- + NO_3^-) - N.l^{-1}.day^{-1}$] in November, before rates again fell. Both the July and November peaks in activity correlated directly with peaks in the numbers of autotrophic nitrifying bacteria at these times.

(iv) *Denitrifying bacteria*

Tests have been made for the presence of heterotrophic and autotrophic denitrifying bacteria in all three lochs. Heterotrophic denitrifiers were tested for by the MPN method using the medium of Collins (1963) and examining the composition of the gas produced in Durham tubes by gas chromatography. Trautwein medium (Rodina 1972) and the medium of Baalsrud & Baalsrud (1954) were used in tests for autotrophic denitrifiers. Denitrifying bacteria were detected in all three lochs with numbers increasing from Lowes → Balgavies →

Table 9

The mean abundance of nitrifying bacteria in Balgavies Loch during February 1975 to March 1976

Type	Depth in water column (m)			Sediment
	0.5	3	8	
Phase I autotrophic nitrifiers ($NH_4^+ \rightarrow NO_2^-$)	60 (0–250)	209 (0–2500)	54 (0–45)	1.24×10^5 (2.5×10^3 – 1.24×10^5)
Phase II autotrophic nitrifiers ($NO_2^- \rightarrow NO_3^-$)	32 (0–250)	35 (0–250)	16 (0–45)	6.16×10^4 (4.5×10^2 – 4.5×10^5)
Heterotrophic nitrifiers	35 (0–100)	18 (0–100)	28 (0–100)	2×10^4 (1×10^3 – 1×10^5)

The data are the means for samples collected monthly during 1975. Autotrophic and heterotrophic nitrifiers were determined according to Gode & Overbeck (1972). The water column data are expressed as bacteria.ml^{-1}; the sediment (top 5 cm) data are expressed as bacteria.g dry wt^{-1}.

Forfar and with autotrophic denitrifiers being much less common than heterotrophic denitrifiers. In all three lochs numbers were lowest in the water column and highest in the top 5 cm of the sediment. The heterotrophic denitrifying bacteria were almost all Gram negative rods which formed colourless colonies on nutrient agar and about 50% produced fluorescein under suitable conditions. They showed little saccharolytic activity but most were highly proteolytic, and catalase and oxidase positive. Most appeared to be pseudomonads but vibrios were also common.

Table 10

Seasonal variation in the numbers of nitrifying bacteria in the top 5.0 cm of the sediment of Balgavies Loch during February 1975 to March 1976

Date	Bacteria (number.g dry wt^{-1})		
	Phase I autotrophic nitrifiers (x 10^{-3})	Phase II autotrophic nitrifiers (x 10^{-3})	Heterotrophic nitrifiers (x 10^{-3})
14/2/75	45	25	10
13/3/75	15	2.5	100
15/4/75	4.5	2.5	10
15/5/75	9	4.5	10
16/6/75	25	4.5	1
14/7/75	200	4.5	1
18/8/75	2.5	0.45	1
18/9/75	2.5	2.5	10
16/10/75	115	4.5	1
17/11/75	75	4.5	100
16/12/75	450	450	10
22/1/76	95	250	10
23/2/76	250	95	10
22/3/76	250	11.5	10

Data on the seasonal variation in the numbers of heterotrophic denitrifiers in the top 5.0 cm of the Balgavies sediment (Table 11) show that numbers are low in spring, increase to a maximum in summer, decrease in autumn and increase to a second peak in winter. The average number of heterotrophic denitrifiers (8×10^7.g dry wt^{-1}) greatly outnumber the autotrophic denitrifiers (mainly *Thiobacillus denitrificans*) (7×10^5.g dry wt^{-1}) in these sediments. In general, the seasonal pattern in the rate of potential denitrification in the sediment, measured as $^{15}N_2$ production from added $Na^{15}NO_3$, almost parallels the observed seasonal variation in the abundance of denitrifying bacteria (Table 11). Additional information on denitrification in these lochs has been presented elsewhere (Stewart *et al.* 1975, 1976).

Table 11

Seasonal variation in the numbers of denitrifying bacteria in the top 5 cm of the sediment of Balgavies Loch

Month	Bacteria (number.g dry wt^{-1})
February	1.26
March	2.0
April	1.26×10^2
May	3.98×10^2
June	1.59×10^4
July	2.5×10^6
August	3.98×10^5
September	3.98×10^5
October	2.5×10^3
November	3.2×10^3
December	6.3×10^4
January	6.3×10^4

6. The Primary Producers and Bacterial Activity

The data presented above indicate that increasing eutrophication is accompanied in our lochs by increases both in the biomasses of algae and of certain emergent macrophytes such as *Phragmites.* The amounts of nutrients which thus accumulate in such primary producers may be considerable, as shown by the data in Table 2 and this may result in a lowering of nutrient levels in the waters, particularly during the summer when rapid plant growth occurs. In addition, however, the primary producers also serve as major sources of carbon and energy which, when released, play a most important role in sustaining heterotrophic microbial activity within the lochs. Certainly the abundance of primary producers in the three lochs, the levels of dissolved organic carbon which the lochs contain (Fig. 11) and the microbial biomasses present all show a similar trend, although in the case of Forfar the input of organic matter and bacteria, in particular, from the sewage works complicates the picture.

The release of fixed carbon from the primary producers may occur at any time as a result of extracellular production from normal healthy cells (see Hellebust 1974; Fogg 1975) or by cell autolysis as a result of adverse environmental conditions, microbial attack, grazing by protozoa etc. An example of how soluble organic carbon may be released into the environment as a result of algal autolysis is seen from laboratory experiments in which the algal-lysing bacterium CP-1 was added to healthy cells of *Anabaena cylindrica.* The results (Fig. 12) show that the level of soluble organic carbon in the environment increased markedly as algal autolysis occurred. In nature, this

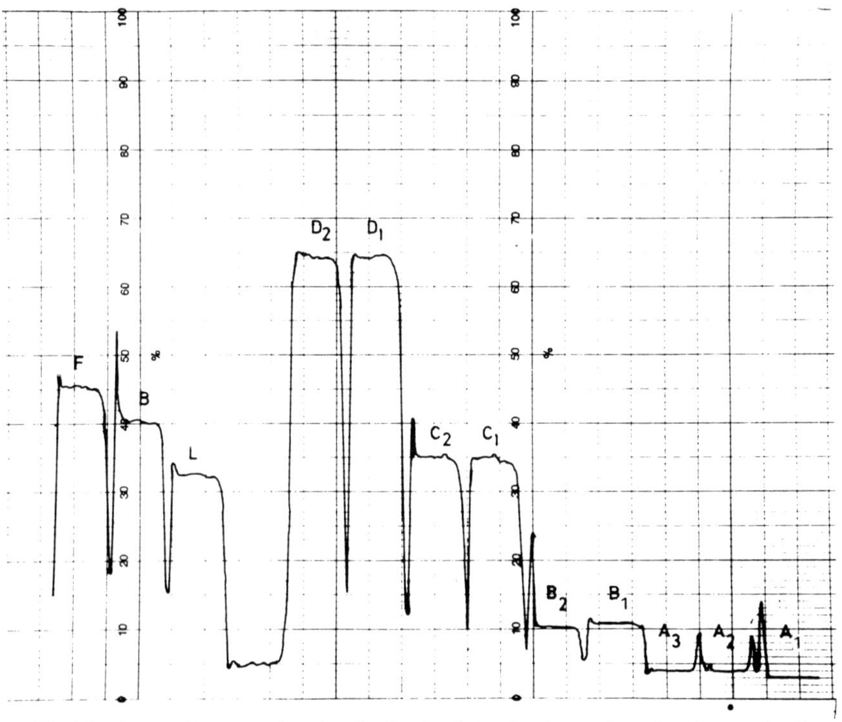

Fig. 11. A recorder trace showing the levels of dissolved organic carbon in samples from the top 0.5 m of the water column at mid-point positions in Lowes, Balgavies and Forfar Lochs on 14/4/76. A_1, A_2, A_3, blanks; B_1, B_2, standards containing 1.25 mg.l^{-1} of organic carbon (as potassium hydrogen phthalate); C_1, C_2, standards containing 4 mg.l^{-1} of organic carbon; D_1, D_2, standards containing 8 mg.l^{-1} of organic carbon; L, soluble organic carbon in Lowes; B, soluble organic carbon in Balgavies; F, soluble organic carbon in Forfar.

extracellular organic carbon would become available to various heterotrophic micro-organisms. Similarly in the case of macrophytes, the production of extracellular carbon from healthy cells may occur, but with maximum release occurring in autumn, winter and early spring as a result of breakdown of decaying plant material. In sum, therefore, there is evidence that with increasing eutrophication there are increased levels of primary producers, which in turn release organic carbon which may, directly or indirectly, affect photosynthetic, photoheterotrophic and heterotrophic microbial processes.

The presence of additional organic carbon in a natural population may affect algal growth in at least three ways. First, it may enable certain algae to grow photoheterotrophically and thus sustain rapid growth at light intensities which are insufficiently high to sustain rapid photoautotrophic growth. This could account in part at least for the high phytoplankton densities found in Forfar Loch compared with the other two lochs. Second, through stimulating bacterial

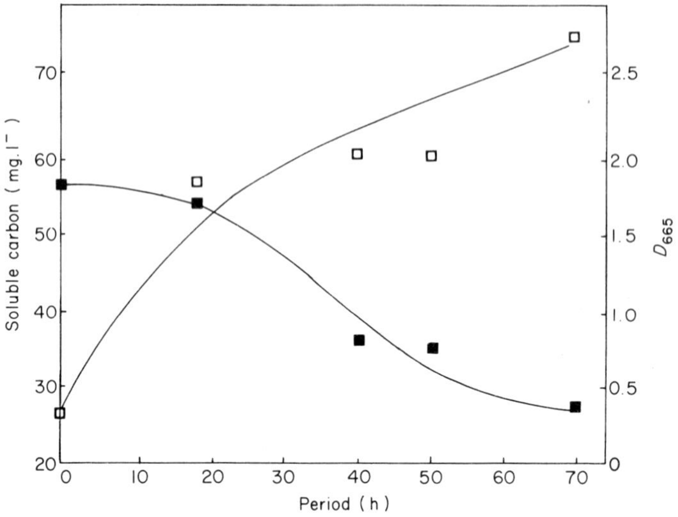

Fig. 12. The effect of adding myxobacterium CP-1 to a batch culture of *A. cylindrica.* Note that algal lysis occurs and that this is accompanied by an increase in soluble organic carbon. The maximum levels of organic carbon produced by CP-1 and *A. cylindrica* when grown separately were 12 and 35 $mg.l^{-1}$, respectively. The experiment was carried out at 26°C at a continuous light intensity of 3000 lux. ■, chlorophyll *a* concentration; □, dissolved organic carbon.

activity it may result in increased levels of vitamin B_{12} and thus in the growth of small, rapidly growing B_{12}-requiring algae. Third, by stimulating respiration by associated bacteria, oxygen tensions may be lowered, and thus algal photorespiration decreased. The first of the above three suggestions is likely to be more important than the other two.

The availability of soluble organic carbon also affects the activity of various heterotrophic micro-organisms, among which will be certain of those involved in nitrogen cycling, for example ammonifying bacteria and denitrifying bacteria. The Balgavies data, for example, show that there is a distinct temporal pattern of activity in the loch and that increased plant growth resulting from increased nutrient levels may be of primary importance in determining the extent of various processes involved in a train of events leading to nitrogen losses from the ecosystem by microbial denitrification. Thus the autumn breakdown of algal and macrophyte material is followed in November by peaks in NH_4^+, probably as a result of heterotrophic ammonification, and in nitrification (mainly autotrophic) and there then follows, in December, a peak in denitrification.

In any body of water, therefore, the various biological processes are intimately interlinked and large-scale man-made interference with any one process, be it nutrient input, primary production, nitrification, denitrification etc. is likely to lead to an imbalance until a new equilibrium is re-established.

Specifically, our studies suggest that the primary producers are of key importance in regulating microbial nitrogen cycling in fresh waters and that there are possible dangers in reducing them selectively, for example by adding algicides or by cutting down the macrophytes, since this may remove a source of fixed carbon, so necessary for the removal of nitrate from the water by microbial activity.

7. Acknowledgements

This study was made possible through research support to W.D.P.S. from the Natural Environment Research Council. We thank E. May, Y. B. Ho, T. Preston and M. Ogan who were responsible for obtaining some of the nutrient cycling data and Miss Gail Alexander for technical assistance.

8. References

AMERICAN PUBLIC HEALTH ASSOCIATION 1965 *Standard Methods for the Examination of Water and Waste Water.* 12th edn New York: American Public Health Association.

BAALSRUD, K. & BAALSRUD, K. S. 1954 Studies on *Thiobacillus denitrificans. Archiv für Mikrobiologie* **20**, 34–62.

BAILEY-WATTS, A. E. 1974 Algal plankton of Loch Leven, Kinross. *Proceedings of the Royal Society of Edinburgh B.* **74**, 135–156.

BENDSCHNEIDER, K. & ROBINSON, R. J. 1952 A new spectrophotometric determination of nitrite in sea water. *Journal of Marine Research* **11**, 87–96.

BROWN, F., CUTHBERTSON, W. F. J. & FOGG, G. E. 1956 Vitamin B_{12} activity of *Chlorella vulgaris* Beij. and *Anabaena cylindrica. Nature, London* **177**, 188.

CAMPBELL, N. E. R. & ALEEM, M. J. H. 1965 The effect of 2-chloro-6 (trichloromethyl) pyridine on the chemoautotrophic metabolism of nitrifying bacteria. 1. Ammonia and hydroxylamine oxidation by *Nitrosomonas. Antonie van Leeuwenhoek* **31**, 124–136.

CHANEY, A. L. & MARBACH, E. P. 1962 Modified reagents for the determination of urea and ammonia. *Clinical Chemistry* **8**, 130–132.

COLLINS, V. G. 1963 The distribution and ecology of bacteria in fresh waters. *Proceedings of the Society for Water Treatment and Examination* **12**, 40–67.

DAFT, M. J. & STEWART, W. D. P. 1971 Bacterial pathogens of freshwater blue-green algae. *New Phytologist* **70**, 819–829.

DAFT, M. J. & STEWART, W. D. P. 1973 Light and electron microscope observations on algal lysis by bacterium CP-1. *New Phytologist* **72**, 799–808.

DAFT, M. J., MCCORD, S. & STEWART, W. D. P. 1975 Ecological studies on algal-lysing bacteria in fresh waters. *Freshwater Biology* **5**, 577–596.

DAISLEY, K. W. 1969 Monthly survey of vitamin B_{12} concentrations in some waters of the English Lake District. *Limnology and Oceanography* **14**, 224–228.

DE BONT, J. A. M. & MULDER, E. G. 1976 Invalidity of the acetylene-reduction assay in alkane-utilizing nitrogen-fixing bacteria. *Applied and Environmental Microbiology* **31**, 640–647.

FLETT, R. J. J., RUDD, J. W. M. & HAMILTON, R. D. 1975 Acetylene-reduction assays for nitrogen fixation in freshwaters: a note of caution. *Applied and Environmental Microbiology* **29**, 580–583.

FOGG, G. E. 1975 Biochemical pathways in unicellular plants. In *Photosynthesis and Productivity in Different Environments,* ed. Cooper, J. P. International Biological Programme, Vol. 3. Cambridge: Cambridge University Press.

GODE, P. & OVERBECK, J. 1972 Studies on heterotrophic nitrification in a lake. *Zeitschrift für Allgemeine Mikrobiologie* **12**, 567–574.

GOLTERMAN, H. L. & CLYMO, R. S. 1969 *Chemical Analysis of Freshwaters.* International Biological Programme Handbook No. 8. Oxford: Blackwell Scientific Publications.

GORING, C. A. 1962 Control of nitrification by 2-chloro-6-(trichloromethyl) pyridine. *Soil Science* **93**, 211.

GUTTMAN, H. N. 1963 Vitamin B_{12} and congeners. In: *Analytical Microbiology*, ed. Kavanagh, F. New York & London: Academic Press.

HELLEBUST, J. A. 1974 Extracellular Products. In: *Algal Physiology and Biochemistry*, ed. Stewart, W. D. P. Oxford: Blackwell Scientific Publications.

MACKERETH, F. J. H. 1963 Some methods of water analysis for limnologists. *Freshwater Biological Association Scientific Publication No. 21.*

MEIKLEJOHN, J. 1965 Microbiological studies on large termite mounds. *Rhodesia, Zambia and Malawi Journal of Agricultural Research* **3**, 67–79.

OHWADA, K. & TAGA, N. 1972 Seasonal cycles of vitamin B_{12}, thiamine and biotin in Lake Sagami. Patterns of their distribution and ecological significance. *Internationale Revue der gesamten Hydrobiologie und Hydrographie* **58**, 851–871.

PROVASOLI, L. & CARLUCCI, A. F. 1974 Vitamins and growth regulators. In *Algal Physiology and Biochemistry*, ed. Stewart, W. D. P. Oxford: Blackwell Scientific Publications.

RODINA, A. J. 1972 *Methods in Aquatic Microbiology*. London: Butterworths.

STEWART, W. D. P. & DAFT, M. J. 1976 Algal lysing agents of freshwater habitats. In *Microbiology in Agriculture, Fisheries and Food.* Society for Applied Bacteriology Symposium Series No. 4, ed. Skinner, F. A. & Carr, J. G. London & New York: Academic Press.

STEWART, W. D. P. & DAFT, M. J. 1977 Microbial pathogens of cyanophycean blooms. In *Advances in Aquatic Microbiology* **1**, 177–219.

STEWART, W. D. P., FITZGERALD, G. P. & BURRIS, R. H. 1967 *In situ* studies on N_2 fixation using the acetylene reduction technique. *Proceedings of the National Academy of Sciences of the United States of America* **58**, 2071–2078.

STEWART, W. D. P., TUCKWELL, S. B. & MAY, E. 1975 Eutrophication and algal growths in Scottish fresh water lochs. Symposium of the British Ecological Society No. 15. Oxford: Blackwell Scientific Publications, pp. 57–80.

STEWART, W. D. P., MAY, E. & TUCKWELL, S. B. 1976 Nitrogen and phosphorus from agricultural land and urbanization and their fate in shallow freshwater lochs. In *Agriculture and Water Quality.* Ministry of Agriculture, Fisheries & Food Technical Bulletin No. 32.

STRICKLAND, J. D. H. & PARSONS, T. R. 1968 A *MSA*, 2nd Edn. Bulletin of the Fisheries Research Board of Canada No. 125.

SVENSSON, B. H. & SODERLUND, R. (eds) 1976 Nitrogen, phosphorus and sulphur – global cycles. SCOPE Report No. 7. Stockholm: Ecology Bulletin.

SYLVESTER-BRADLEY, R. 1976 Isolation of acetylene-reducing spirilla from the roots of *Potamogeton filiformis* from Loch Leven (Kinross). *Journal of General Microbiology* **97**, 129–132.

TOLBERT, N. E. 1974 Photorespiration. In *Algal Physiology and Biochemistry*, ed. Stewart, W. D. P. Oxford: Blackwell Scientific Publications.

WESTLAKE, D. F. 1963 Comparisons of plant productivity. *Biological Reviews* **38**, 385–425.

WOOD, E. D., ARMSTRONG, F. A. J. & RICHARDS, F. A. 1967 The determination of nitrate in sea water by cadmium-copper reduction to nitrite. *Journal of the Marine Biological Association of the United Kingdom* **47**, 23–31.

Growth Kinetics of Aquatic Bacteria

H. W. JANNASCH

Woods Hole Oceanographic Institution,
Woods Hole, Massachusetts, USA

CONTENTS

1. Introduction

GROWTH KINETICS and population dynamics of micro-organisms are well defined terms with respect to pure culture work, especially in the areas of metabolic regulation and genetics. The basic measurements are direct (microscopic) or cultural cell numbers, the latter in the presence of well-known growth requirements, or the more commonly used determinations of various cell constituents.

Studying kinetics of complex natural populations of aquatic bacteria is another matter. The goal is the same, namely to understand the response of the population to changes of specific environmental conditions. However, the presence of almost any concievable species or metabolic type makes the analysis of events difficult, in most cases impossible. In addition, the use of defined and simple combinations of substrates in convenient concentrations defeat the ecological purpose of such studies.

The usefulness of kinetic approaches in aquatic microbiology, therefore, is limited to a number of studies based on certain compromises. In cases where pure cultures or mixtures of a few strains were used, the data are reproducible and interpretable, but their direct ecological significance is limited. On the other hand, integrating measurements on the response of complex natural populations usually lack reproducibility and interpretability for obvious reasons, but the results may be of direct ecological interest as for instance, in the study of microbial blooms. Assay-type approaches may also be classified under kinetic studies. In the absence of a suitable chemical determination, the concentration of a specific substance, e.g. a vitamin, may be assessed by the growth response or the yield of the appropriate indicator organism As long as the growth of such an organism, a prokaryote in the exponential phase, is limited by one known factor or substrate, it can be described by two parameters: μ_{max}, the maximum growth rate and K_s, the substrate affinity (or half saturation) constant. If the dilution rate of a continuous culture system, meeting the technical requirements of a chemostat, is smaller than μ_{max}, a steady state will establish. Any further change of a simulated 'environmental' factor, e.g. temperature, will disturb the existing steady state and, in most cases, lead to the establishment of a new steady state. Thus, the effect of a particular environmental change will be expressed as a change of the population density. The indiscriminate dilution of cells in a continuous culture system, as opposed to a batch culture system, is, in principle, the simple experimental operation which makes growth kinetics an amenable tool in many disciplines of microbiology.

The basic definitions and mathematical derivations of growth parameters were ably presented by Herbert *et al.* in 1956, and have often been referred to subsequently in the literature, so, for the sake of brevity, will be excluded from this discussion as much as possible.

On the basis of a few selected continuous culture experiments, from the author's own laboratory, the present article attempts to discuss concisely the principles, usefulness and limitations of kinetic approaches with respect to their ecological significance. To a certain degree, the discussion represents an updated excerpt from a more detailed review (Jannasch & Mateles 1974).

2. Selective Continuous Cultures

Selective culturing or enrichment techniques have been the classical tool and the initial step in studying the existence and role of specific metabolic groups of micro-organisms in a given natural environment. By adding a particular substrate to a natural population, or by increasing the concentration of a nutrient or energy source, those micro-organisms engaged specifically in its transformation will increase in numbers and become amenable to subsequent isolations for pure culture studies. In closed culture systems the artificially-induced kinetic processes during these population changes are highly complex. They are not only difficult to analyse but also depend heavily on a pronounced metabolic specificity of the organism to be enriched.

Both of these drawbacks are overcome in open culture systems where the continuous dilution of the growing mixed population prevents the accumulation of the unsuccessful competitors and waste products. Small growth advantages, which are not detectable in short-lived closed culture systems, are amplified in prolonged continuous culture runs. As a consequence, a pronounced metabolic substrate specificity of the organism to be enriched is not a necessity. In fact, organisms competing for the 'same' substate, but growing at a different rate, can be separated by continuous culture enrichment procedures. A study, specifically aimed at microbial chemostat enrichments from off-shore sea water, resulted reproducibly in a large variety of species as individual successful competitors dependent, in each case, on the particular dilution rate and substrate (lactate) concentration applied (Jannasch 1967).

This new principle of selective culturing of micro-organisms in open systems is not restricted to the nature of growth limiting nutrients or energy sources but is also applicable to their specific concentration as well as to physicochemical factors such as pH, redox potential, temperature and light (with phototrophic micro-organisms).

3. Competition Studies

From the ecological point of view, the common enrichment culture principles are based on growth competition. The kinetics of these processes in natural population are too complex for simple analyses. Their basic principles, however, have been studied in continuous culture using mixtures of strains with known growth parameters.

Theoretically, the outcome of the competition for the same limiting substrate is predictable if no other interaction between the two species occurs. On these lines, the fate of a contaminant or a mutant in a pure culture has been discussed early in the developments of continuous culture techniques (Powell 1958). More recently, predictable cases of successful and unsuccessful competition have been experimentally verified in several independent studies. Cultures of aquatic bacteria in mixtures of two species have been shown to result in mutual displacement dependent upon the concentration of limiting substrate (Jannasch 1967, 1968; Ishida & Kadota 1974; Kuenen *et al.* 1976), temperature (Harder & Veldkamp 1971), or light intensity in photosynthetic purple bacteria (Van Gemerden & Jannasch 1971). More detailed studies in this area have been reviewed by Veldkamp & Jannasch (1972), Meers (1973) and Jannasch & Mateles (1974).

As an example (Fig. 1), mutual displacement of *Escherichia coli* and *Spirillum* sp. has been demonstrated (Jannasch 1968) in a lactate-limited chemostat culture. The growth parameters obtained in pure culture (the maximal growth rate, μ_{max}; and the half saturation constant, K_s) are both lower in the *Spirillum* sp. than in *E. coli*. Thus, in mixed culture, *E. coli* could be predicted, and was shown, to be displaced when the chemostat was run at relatively low dilution rates and/or low concentrations of the limiting substate. Vice versa, the *Spirillum* sp. turned out to be the unsuccessful competitor when the chemostat was run at 'high' dilution rates and substrate concentrations relative to the growth parameters of the two organisms.

From an ecological point of view, the *Spirillum* sp. appears to be better

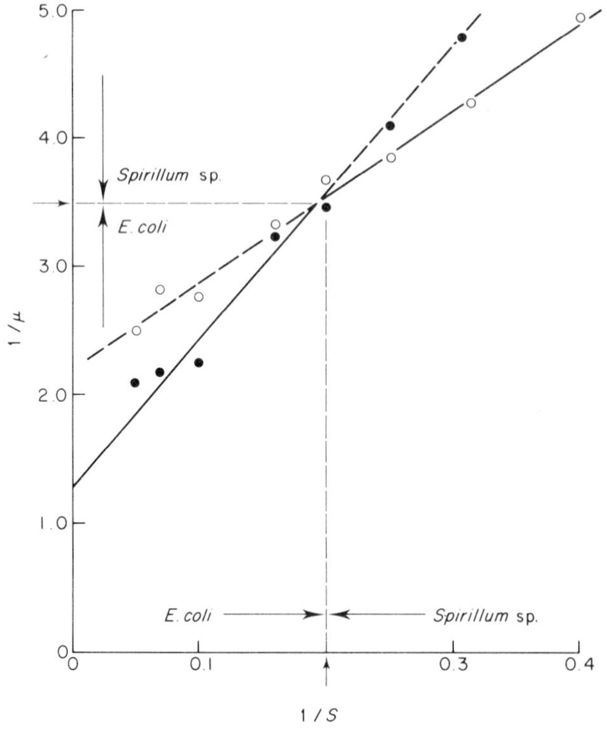

Fig. 1. Reciprocal plot of specific growth rate (μ in h) versus concentration of limiting substrate (S in mg of lactate.l $^{-1}$) demonstrating the range of successful competition between *E. coli* (•) and *Spirillum* sp. (○) (from Jannasch 1968).

adapted to the conditions of the aquatic habitat than *E. coli*. The two organisms were isolated from inshore sea water where lactate had been found in trace quantities, probably as an intermediate product during decomposition of seaweed beds. *Escherichia coli* is found as a consequence of local sewage run-offs. The persistence of *E. coli* in these waters will depend on its viability in the absence of growth and on die-off and grazing rates by protozoa and filter feeders. These factors could be taken as being equivalent to the removal of cells from a chemostat culture by dilution. However, most of these mechanisms for removing cells from a population are highly selective in contrast to the indiscriminate removal of cells from a chemostat by dilution. This is the major reason why results of competition studies in continuous culture cannot be directly applied to natural conditions. While natural exclusion of one species by another in the chemostat is an artifact of the experimental conditions, the data can only be used as an indirect criterion for growth competition.

Graphically, plots of limiting substrate concentration versus growth rate, or of the equivalent reciprocals, show the characteristic cross-over of two competitors (Jannasch 1967; Harder & Veldkamp 1971). Theoretically the two organisms competing for the same substrate will both remain in steady state at the particular substrate concentration indicated by the cross-over point (Fig. 1). A number of other stable steady states of mixed populations have been observed, however, which do not depend simply on equal growth rates but are based on metabolic interrelationships. In an early example, Powell (1958) using a mixed culture of *Bacterium cloacae* and *Pseudomonas pyocyanea*, established the role of the 'primary' organism, limited by an external factor such as a specific substrate, and the role of the 'secondary' organism, limited by a metabolic product of the former. If the potential growth rate of the secondary organism is faster than that of the primary, its dependence on the latter will result in a stable steady state of the two species.

Other cases of interdependence between species in continuous culture were studied by Contois & Yango (1964) e.g. ammonia limited *Dictyostelium discoideum* feeding on *Klebsiella* sp. and *E. coli* maintaining a constant phage titre. Paynter & Bungay (1969) and Horne (1970) became engaged in an interesting controversy concerning the stability of bacteria–phage systems in continuous culture with respect to the accumulations of mutants. A growth stimulating metabolite was assumed to be effective in competing populations of *Bacillus megaterium* and *Torula utilis.* In these experiments, Meers & Tempest (1968) point out the role of initial population density as a factor in the outcome of the competition.

Of special kinetic as well as ecological interest are *oscillations* between the populations of two species as an alternative to well-defined steady state conditions. Contois & Yango (1964) grew two mutants of *Klebsiella* which differed in their ability to form lactic acid and in their pH tolerance. In continuous culture, sustained oscillations of the pH as well as the total population density resulted. Yeoh *et al.* (1968) described a mixed chemostat culture of *Proteus vulgaris* and *Bacillus polymyxa* and explained resulting oscillations as being caused by a combined effect of vitamin and inhibitor production. Sustained as well as damped oscillations have been observed in more complex prey–predator systems containing bacteria and protozoa and are dealt with by C. R. Curds in Chapter 4.

4. Kinetic Growth Rate Determinations

Kinetic studies of individual bacterial species in continuous culture offer the possibility of measuring their growth rate in a natural water sample and in the presence of the unknown limiting substrate. A test strain inoculated in a filter-sterilized water sample would theoretically attain a steady state if the (unknown) maximum growth rate would be equal to or higher than the experimental dilution rate given.

This has never been achieved in a series of experiments run with in-shore sea water (Jannasch 1969). Invariably a decrease of the initial population density of the test strain occurred (as measured by plate counts) at retention times of 16 h or less. If the following equation describing transient state conditions:

$$x = x_0 \, e^{(\mu - D)t} .$$

(x, population density; μ, growth rate; D, dilution rate; t, time) is solved for μ:

$$\mu = D + \left(-\frac{1}{t} \ln x/x_0 \right) ,$$

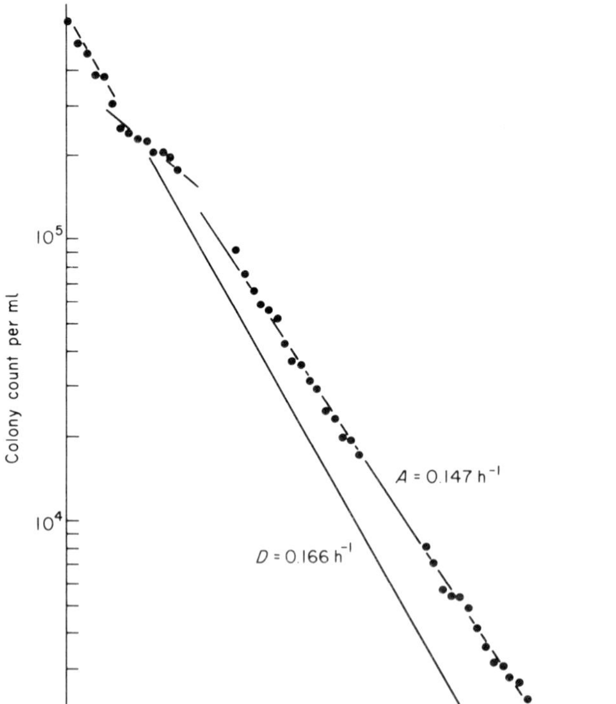

Fig. 2. Washout rate (A) of *Achromobacter* sp. in filter-sterilized sea water as compared to the dilution rate (D). The calculated growth rate is 0.019 h^{-1} (from Jannasch 1969).

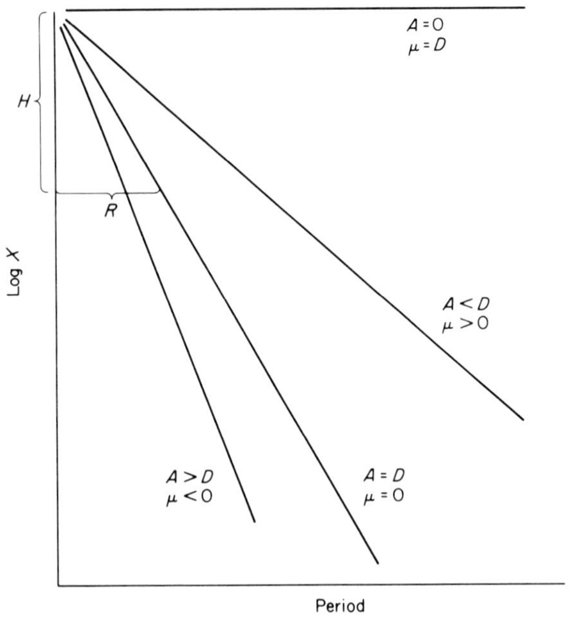

Fig. 3. Population density (x) of a test strain versus time in a chemostat (schematic): A, washout rate; D, dilution rate; μ, growth rate; H, halving of population density by dilution in the absence of growth; R, retention time (from Jannasch 1969).

the expression put in parenthesis represents the washout rate (A) of cells from the chemostat. If this rate is constant for a given period of time, the growth rate μ can be simply calculated. Figure 4 shows an example where the generation time ($1/\mu$) was calculated at 53 h.

Postgate & Hunter (1962) observed generation times of up to nearly 100 h. At that stage of the growing population, however, 60% of the cells were non-viable as established by a slide culture technique (Postgate *et al.* 1961). Tempest *et al.* (1967) were able to increase the viability under similar conditions by lowering the temperature and found that nitrogen-limited populations exhibited lower proportions of non-viable cells than carbon-limited populations.

Figure 5 demonstrates the range of possibilities of kinetic responses in the experiment described above. In case (1) the concentration of the growth-limiting substrate in the water permits the inoculated population to double within one retention time. A steady state is established, and the population will attain constancy. In case (2) the inoculated species does not grow and the cells are washed out at a rate equal to the dilution rate. Case (3) is the one described above and was used for the calculation of the actual growth rate. In case (4) the washout rate is higher than the dilution rate, possibly due to a loss of viability,

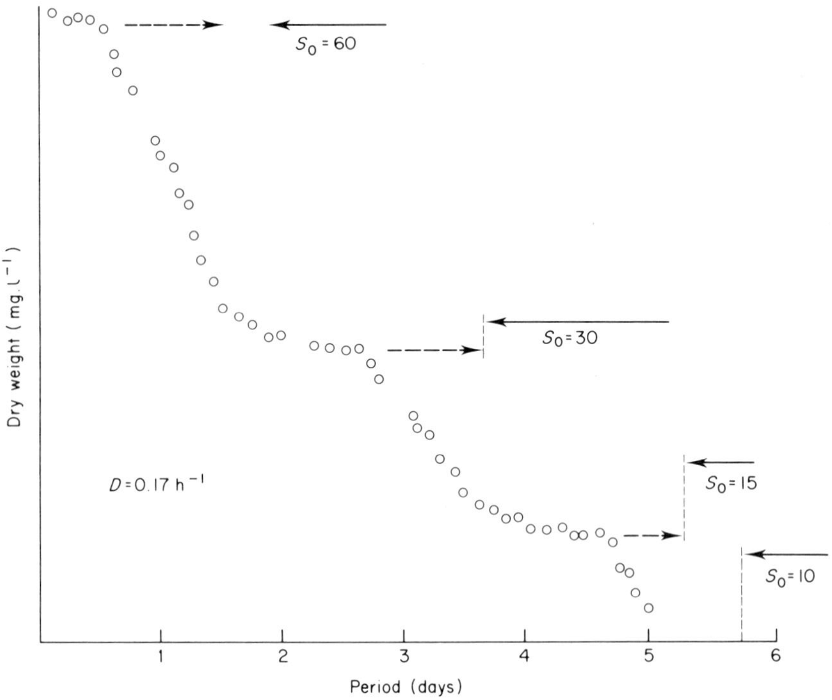

Fig. 4. Decrease of population density at a stepwise lowering of the limiting substrate concentration in the reservoir (S_0 in mg of lactate.l^{-1}). Expected and actually obtained steady state population densities are indicated by solid and broken arrows, respectively (from Jannasch & Mateles 1974).

presence of a phage, lysis for other reasons, bactericidal effects, or grazing. Examples for all four of these cases have actually been found.

If a bacterial isolate was chosen, which exhibited a colony type distinguishable at low magnification from other colonies on agar plates, it was possible to combine growth rate determinations with competition experiments (Jannasch 1969). This was done successfully with isolated marine strains of a *Spirillum* sp. and a pigment-forming mutant of *Serratia marcescens.* Table 1 shows their generation times in 'non-sterile' sea water run through the chemostat i.e. in the presence of the natural microbial population. These data are obtained at two dilution rates and are compared to the growth response of the test organisms and to filter sterilization and autoclaving of sea water. The relative data leave a wide margin for interpretations which are not the topic of the present discussion.

This kinetic approach has also been applied in describing the competition between the natural microbial population and one individual isolate for glycerol

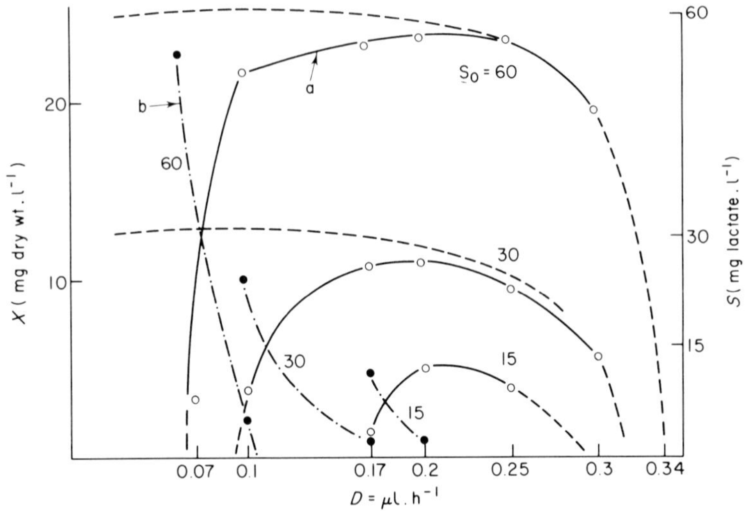

Fig. 5. Steady state population density (x, line a) and concentration of limiting substrate (S, line b) at steady state plotted versus dilution rate (D) for three concentrations of the limiting substrate in the reservoir (S_0); broken line, theoretical (from Jannasch & Mateles 1974).

Table 1

Generation times of test organisms in unsupplemented sea water

Organism	Generation times (h) in sea water at two retention times (R)					
	Unsterilized at R (h)		Filter sterilized at R (h)		Autoclaved at R (h)	
	6	12	6	12	6	12
Spirillum sp. (103)	75	78	52	58	40	48
	–	–	59	–	36	–
Serratia marcescens	110	130	140	170	72	60
	98	–	–	–	78	–
Achromobacter sp. (317)	0	0	53	71	24	26
	–	–	57	–	20	0

Both generation time and retention time are given in hours. Since growth measurements are based on viable counts, generation time ($G = 1/\mu$) is given rather than doubling time (t_d = n $2/\mu$). Experiments were run at two retention times, duplicates only at a six hour retention time. *Achromobacter* sp. was not grown in unsterilized sea water (from Jannasch 1969).

as a supplementary substrate (Table 2). Although the *Spirillum* strain was originally enriched and isolated on glycerol–sea water medium, the natural population of micro-organisms of the sea water sample used appeared to compete rather successfully. With increasing concentration of the supplementary substrate, the generation time of the test strain was increasingly affected in the presence of the competitors. This particular kinetic approach may be very useful in studies of the environmental effects and persistence of specific pollutants.

Table 2

Mean generation times of Spirillum *sp. in untreated (unsterile) and in autoclaved inshore sea water with and without an additional carbon source*

Sea water treatment	Added glycerol (mM)	Mean generation time (h)
Untreated	none	127
Autoclaved	none	98
Untreated	0.1	81
Autoclaved	0.1	17
Untreated	3.0	45
Autoclaved	3.0	3.5

Retention time in all experiments was 6 h (from Jannasch 1969).

5. Threshold Concentrations of Growth-limiting Substrates

Farthest removed from natural conditions of microbial growth are pure culture studies, and yet, questions of special ecological importance may only be studied in such systems. Most aquatic micro-organisms live in the presence of highly dilute substrates, and a cardinal problem has been the possible existence of a minimum concentration of a dissolved substrate below which a suspended bacterial population will be unable to maintain itself. The experimental approach was to lower the concentration of the growth-limiting substrate in the reservoir of a chemostat stepwise, by exchanging reservoir vessels, at a variety of dilution rates.

Figure 4 shows the population changes resulting from lowering the lactate concentration in the inflowing medium (S_0). The feed-line from the reservoir to the chemostat was small in volume so that the response to the next lower substrate concentration was delayed for only little more than two hours. The broken vertical lines indicate the observed discrepancy between the theoretical population density ($x = y(S_0 - S)$; y, yield coefficient; S, concentration of growth limiting substrate in the chemostat) and the actual level of population measured. Washout occurred in the presence of a relatively high substrate concentration (10 mg of lactate.l^{-1}).

This inability of the organism (*Spirillum serpens*) to utilize the available substrate below a certain concentration is assessable quantitatively and could be described mathematically (Jannasch 1965). It is dependent on the growth rate and showed a characteristic deviation from the theoretical population kinetics in an ideally optimal medium (Fig. 5). The decrease of population density at low dilution rates occurs at much higher substrate levels than in earlier experiments reported by Herbert (1958). In the latter, the phenomenon was interpreted as being caused by endogeneous respiration, and S approached zero at low dilution rates. In our studies, S increased at low dilution rates indicating a growth inhibitory effect. The same results were obtained with an energy or nitrogen source as the limiting substrates.

The ecological interest in this phenomenon is based on the fact that the medium, heavily aerated natural water, did not provide optimal growth conditions. The organism was found to be micro-aerophilic but was able to overcome the unfavourable effect of a high redox potential at high population densities i.e. high concentrations of the limiting substrate. At low substrate levels, growth was affected by an increasing rH. In turn, the population density became established at a lower steady state level than that calculated. Further lowering of S_0 led to washout of the culture in the presence of a reproducible threshold concentration of the substrate. If the rH was lowered artificially by an addition of ascorbic acid, the threshold concentration of lactate was also lowered. Thus, the threshold phenomenon can be explained by the need of a finite amount of substrate by the population to overcome suboptimal growth conditions and to maintain itself in the chemostat under the given conditions.

In batch culture, i.e. in the absence of a constant removal of cells, the existence of threshold concentrations cannot be detected. In a natural system, cells persist in the absence of growth, and a population would be able to recover from a periodical shortage of an essential substrate. The threshold phenomenon is an example showing that the chemostat, as a well defined open culture system and a tool for measuring kinetic parameters, is certainly not reproducing natural growth conditions.

The principle of threshold concentrations is important in species competition. A less micro-aerophilic organism would be able to utilize lactate to lower concentrations than the strain used in the above experiment. Indeed, when the culture at the washout point was inoculated with a sample of raw sea water, heavy turbidity occurred. The fact that certain amounts of organic substrates readily available to microbial utilization persist in natural waters, is commonly explained by the constant production during decomposition processes. It remains to be studied whether the threshold phenomenon provides an alternative explanation.

6. Conclusions

From this brief discussion of selected kinetic approaches in aquatic microbiology, some points in its favour and some limitations can be discussed.

Although natural populations in their undisturbed environment have been described as 'open systems', they are far removed from resembling a continuous culture. The latter cannot be envisaged as a reproduction or approximation of natural conditions, as sometimes suggested. This is best demonstrated by the strong selective pressure for that species growing at the highest rate under the given conditions, and the simultaneous displacement of all other species. The experimentally imposed indiscriminate removal of cells by dilution lowers the initial diversity of a natural population. There is no true equivalent of the dilution rate of a natural population in its environment where most factors, which eliminate cells or affect their viability, are selective.

In addition to certain characteristics of an open system, a natural microbial population has also those of a closed system, namely the partial accumulation of growth products and the resulting inconstancy of growth conditions. As a corollary, true steady states, as approached in chemostat cultures, are unlikely to occur in natural populations. Their situation is more closely described by 'transient states' i.e. increasing or decreasing populations of individual species at any given moment.

In short, the continuous culture technique produces artificial growth conditions with the aim of providing a mathematically managable system. Therefore, it has been used most profitably in pure culture work where true steady states can be achieved. The relatively simple relationships between experimental and cultural parameters apply in the steady state only. Transient states pose a completely different situation which has hardly been tackled yet for ecological purposes (Schaezler *et al.* 1971; Sinclair *et al.* 1971).

There are a few examples where the actual data obtained with natural populations appear to justify the use of simple steady state kinetics. Gaudy & Gaudy (1966) treated heterogeneous populations from whole sewage plants in this manner. As stated earlier "Wright and Hobbie's suggestion (1965) that Michaelis-Menten kinetics be used for calculating potential rates of substrate uptake by natural populations also relies on an apparent fit of data whenever obtained. Although the theoretical basis of determining enzyme activities in the required well-defined systems does not apply to this approach, it has been used with a certain justification as long as the heterogeneous natural population appears to behave like a pure culture. On the other hand, we must keep in mind that the relation between growth or uptake and the concentration of any essential nutrient has to result in a pseudohyperbolic curve, even in mixed populations and at variable rate limitation. But it is difficult or impossible to prove statistically from the scatter of data points whether they truly relate to the assumed single-reaction kinetics that permit extrapolation of valid constants as obtained, for instance, graphically from Lineweaver-Burk plots (Fig. 4). The significance of 'average' growth or uptake constants for natural populations appears doubtful, since prolonged exposure to any nutrient concentration will result in unpredictable changes of the population and its uptake characteristics" (Jannasch 1974).

The particular selection of kinetic studies discussed in this brief presentation does not include more complex prey–predator systems and some current work on the decomposition of organic materials in multistage continuous culture systems, on the effect of suspended particles on bacterial growth and on population genetics. Few published data exist on these topics, but from the experience gained over the last two decades it is safe to state that progress in aquatic microbiology has been substantially advanced by experimental studies in growth kinetics.

7. Acknowledgement

The preparation of this contribution was supported by the National Science Foundation, grants DES 75-15017 and OEC 75-21278. Contribution No. 3836 of the Woods Hole Oceanographic Institution.

8. References

CONTOIS, D. E. & YANGO, L. D. 1964 Abstracts of the American Chemical Society, 148th Meeting, Chicago.

GAUDY, C. P. L. & GAUDY, A. F. 1969 Control mechanisms operative in a natural microbial population selected for its ability to degrade L-lysine. III Effects of carbohydrates in continuous flow systems under shock load conditions. *Applied Microbiology* **18**, 790–797.

HARDER, W. & VELDKAMP, H. 1971 Competition of marine psychrophilic bacteria at low temperatures. *Antonie van Leeuwenhoek* **37**, 51–63.

HERBERT, D. 1958 Some principles of continuous culture. In *Recent Progress in Microbiology*, Proceedings of Symposia, 7th International Congress of Microbiology, Stockholm, 1958.

HERBERT, D., ELSWORTH, R. & TELLING, R. C. 1956 The continuous culture of bacteria: a theoretical and experimental study. *Journal of General Microbiology* **14**, 601–622.

HORNE, M. T. 1970 Coevolution of *Escherichia coli* and bacteriophages in chemostat culture. *Science, New York* **168**, 992–993.

ISHIDA, Y. & KADOTA, H. 1974 Ecological studies on bacteria in the sea and lake waters polluted with organic substances. I. Responses of bacteria to different concentrations of organic substances. *Bulletin of the Japanese Society of Scientific Fisheries* **40**, 999–1005.

JANNASCH, H. W. 1965 Starter populations as determined under steady state conditions. *Biotechnology and Bioengineering* **7**, 279–283.

JANNASCH, H. W. 1967 Enrichments of aquatic bacteria in continuous culture. *Archiv für Mikrobiologie* **59**, 165–173.

JANNASCH, H. W. 1968 Competitive elimination of Enterobacteriaceae from sea water. *Applied Microbiology* **16**, 1616–1618.

JANNASCH, H. W. 1969 Estimation of bacterial growth rates in natural waters. *Journal of Bacteriology* **99**, 156–160.

JANNASCH, H. W. 1974 Steady state and the chemostat in ecology. *Limnology and Oceanography* **19**, 716–720.

JANNASCH, H. W. & MATELES, R. I. 1974 Experimental bacterial ecology studied in continuous culture. *Advances in Microbial Physiology* **11**, 165–212.

KUENEN, J. G., BOONSTRA, J., SCHRODER, H. G. J. & VELDKAMP, H. 1977 Competition for inorganic substrates among chemoorganotrophic bacteria. *Microbial Ecology* **3**, 119–130.

MEERS, J. L. 1973 Growth of bacteria in mixed culture. *Critical Reviews in Microbiology* **2**, 139–184.

MEERS, J. L. & TEMPEST, D. W. 1968 The influence of extracellular products on the behaviour of mixed microbial populations in magnesium limited chemostat cultures. *Journal of General Microbiology* **52**, 309–317.

PAYNTER, M. J. B. & BUNGAY, H. R. 1971 The effect of microbial interactions on performance of waste treatment processes. In *Biological Waste Treatment,* ed. Canale, R. P. New York: Interscience.

POSTGATE, J. R. & HUNTER, J. R. 1962 The survival of starved bacteria. *Journal of General Microbiology* **29**, 233–263.

POSTGATE, J. R., CRUMPTON, J. E. & HUNTER, J. R. 1961 The measurement of bacterial viabilities by slide culture. *Journal of General Microbiology* **24**, 15–24.

POWELL, E. O. 1958 Criteria for growth of contaminants and mutants in continuous culture. *Journal of General Microbiology* **18**, 259–268.

SCHAEZLER, D. J., MCHARG, W. H. & BUSCH, A. W. 1971 Effect of the growth rate on the transient responses of batch and continuous microbial cultures. In *Biological Waste Treatment,* ed. Canale, R. P. New York: Interscience.

SINCLAIR, C. G., KING, W. R., RYDER, D. H. & TOPIWALA, H. H. 1971 Some difficulties in fitting dynamic models to experimental transient data in continuous culture. *Biotechnology and Bioengineering* **13**, 451–452.

TEMPEST, D. W., HERBERT, D. & PHIPPS, P. J. 1967 Studies on the growth of *Aerobacter aerogenes* at low dilution rates in a chemostat. In *Microbial Physiology and Continuous Culture,* eds. Powell, E. O. *et al.* London: H.M.S.O.

VAN GEMERDEN, H. & JANNASCH, H. W. 1971 Continuous culture of Thiorhodaceae. Sulfide and sulfur limited growth of *Chromatium vinosum. Archiv für Mikrobiologie* **79**, 345–353.

VELDKAMP, H. & JANNASCH, H. W. 1972 Mixed culture studies with the chemostat. *Journal of Applied Chemistry and Biotechnology* **22**, 105–123.

WRIGHT, R. T. & HOBBIE, J. E. 1965 The uptake of organic solutes in lake water. *Limnology and Oceanography* **10**, 22–28.

YEOH, H. T., BUNGAY, H. R. & KRIEG, N. R. 1968 A microbial interaction involving combined mutualism and inhibition. *Canadian Journal of Microbiology* **14**, 491–492.

Microbial Interactions Involving Protozoa

C. R. CURDS

Department of Zoology,
British Museum (Natural History),
Cromwell Road,
London, England

CONTENTS

1. Introduction

WHILE OUR KNOWLEDGE of micro-organisms has been largely gained from the study of isolated microbes, pure cultures of single organisms are rarely, if ever, found in the aquatic environment. This situation is naturally to be expected since experimental method demands the use of strict controls which are more easily arranged for populations of single species. In any event one must be able to understand the fundamental biology of the single organism before one can begin to decipher the relationships between several species. These practical constraints have resulted in a real lack of precise data concerning mixed populations although it is quite evident from the microbiological literature over the past decade that this is an area that fascinates microbiologists and has become a central theme of importance and discussion common to all microbiological disciplines.

Protozoa are rather more easily observed and identified in their natural habitat than are some other micro-organisms and this has resulted in many thousands of observations on associations between these and other microbes.

However in many cases the exact relationship between the organisms has not been determined. For the purpose of this chapter the relationship used by each original author will be used where the interaction has not been subsequently determined experimentally.

Terminology is of great importance in any discussion of interactions between organisms. A list of possible microbial interactions taken from Bungay & Bungay (1968) is given in Table 1. Although I do not personally favour some of their definitions, it is nevertheless a concise resumé. It will be evident throughout this chapter that it is often difficult to make a precise decision about what any particular interaction should be called. For example, in some cases the difference between predation and parasitism is one of semantics since the two organisms concerned may be of nearly equal size and this illustrates how interactions tend to merge with one another. It will be seen (Table 1) that the term 'symbiosis' has been omitted. Kirby (1941) pointed out that symbiosis should be used collectively to include commensalism, mutualism and parasitism although this has not been universally accepted. Here the term symbiosis will be used as originally intended (de Bary 1879) in its general form and not as a synonym of mutualism.

Table 1
Microbial interactions

Interaction	Definition
Neutralism	No interaction
Commensalism	One member benefits, other unaffected
Mutualism	Each member benefits from other
Competition	A race for nutrients and space
Amensalism	One adversely changes the environment for other
Parasitism	One organism steals from another
Predation	One organism ingests the other.

From Bungay & Bungay (1968).

Protozoa are known to interact in various ways with bacteria, fungi, viruses and other protozoa. Each interaction will be considered in sequence and numerous examples of each will be given. Finally the possible effects of some protozoan interactions on the overall aquatic microbial community will be discussed briefly.

2. Neutralism

It is difficult to envisage a situation where organisms living together do not affect each other in some way. Even if the organisms neither compete for a common nutrient nor promote environmental changes, they must as least compete for the same physical space. Neutralism is rarely reported in the literature which could be due either to the interaction being uncommon or to the fact that as a 'non-event' it tends not to be reported.

Burbanck & Williams (1963) seem to be the only authors to have described a possible neutralistic interaction among protozoa, where no significant differences in the rates of growth of the two ciliates *Colpidium colpidium* and *Paramecium aurelia* were found when grown alone or together in a bacteria-free medium. However, since the *Colpidium* population tended to form central aggregates and the *Paramecium* population tended to swim around the edges of the culture dish it is possible that lack of mixing was responsible for the neutralistic effect. No data concerning the concentrations of the populations were given and one might have expected some differences in the individual ciliate populations when grown singly or together.

3. Symbiosis

As stated in the introduction, the term symbiosis means rather different interactions to different people. For the present purpose symbiosis will be used in the general sense as originally intended by de Bary (1879), that is the simple "living together of dissimilarly named organisms". This author made no implications about either the spatial nature of the associates, or the entailment of benefit or harm to either or both organisms. Nevertheless later writers, particularly Europeans, have often inferred a closely apposed and mutually beneficial association of organisms. Here symbiosis will be used to include the three interactions mutualism, commensalism and parasitism. In many cases while the nature of a relationship is known to be symbiotic, in the general sense, it has often not been more precisely defined. In these cases the retention of the term symbiosis has a great practical advantage in that one can loosely refer to a relationship as being symbiotic until the association can be defined more precisely. A detailed discussion of the use of this term is given by Starr (1975).

(a) *Commensalism*

While protozoa appear not to be involved in the type of commensal relationship in which one species converts a macronutrient for the use of another, there are several protozoological examples of one organism excreting a growth factor essential or stimulatory for the growth of the second species. Many heterotrophic bacteria and blue-green algae are known to excrete growth factors such as vitamins and organisms having this ability appear to be widespread in nature. For example, Burkholder (1963) stated that bacteria capable of synthesizing and excreting biotin, thiamine, nicotinic acid and vitamin B_{12} comprise a significant portion of the marine microbial community. Similarly Erikson (1941) showed actinomycetes to be common in pond muds and soils which have also been reported (Stokstad *et al.* 1949) to produce vitamin B_{12} in abundance. Furthermore, some *Flavobacterium* species, common in the microflora of water, also form appreciable amounts of B_{12} (Petty & Matrishin 1949). Hutner & Provasoli (1951) have shown that many species of phytoflagellates belonging to genera such as *Euglena* and *Chlamydomonas* require vitamin B_{12} while others such as *Polytoma, Chilomonas* and *Polytomella* require vitamins such as thiamine, and they point out that the vitamin requirements of these flagellates were most likely to be furnished by bacteria and other micro-organisms in the aquatic environment.

Many reports have been published on the stimulation of one protozoon by the products of another and of protozoa producing self-stimulatory substances. For example, Mast & Pace (1938) found that the flagellate *Chilomonas paramecium* produced a heat-labile dialysable substance which stimulated the growth rate of members of the same species in axenic cultures. Similarly Kidder (1941) reported that the ciliate *Tetrahymena pyriformis* produced two heat-labile factors, one self-stimulatory and one self-inhibitory to its growth. More recently Lilly & Stillwell (1965) observed that the ciliate *Colpidium campylum* could produce a substance that raised the growth rate of other ciliates such as *P. caudatum* and various species of *Tetrahymena* by as much as 50%. Later Stillwell (1967) purified the stimulatory material and showed it to consist principally of RNA although lipids and proteins or peptides were also present in the complex which seemed to protect the RNA from the action of physico-chemical agents. Finally, there is also some published evidence which suggests that substances excreted by protozoa stimulate the activities of bacteria. Nikoljuk (1969) showed that amoebae and ciliates were capable of increasing the nitrogen capabilities of *Azotobacter chroococcum* due, it appeared, to the excretion of a heteroauxine, 3-indolylacetic acid. This work has been concentrated on the activities of soil organisms and there is no evidence of its importance in aquatic communities.

The growth of one organism upon the surface of another is a further example of commensalism, often called 'phoresis'. The adhesion of micro-organisms, usually bacteria, to protozoa is well documented but in most cases the association is incompletely defined. Most work has been carried out on the flagellated protozoa that inhabit the alimentary tract of certain insects, such as termites, where in some cases spirochaetes and bacteria attached to the surface of the host are actually responsible for the locomotion of the flagellate (see reviews by Kirby 1941; Ball 1969). With free-living protozoa, adhesion by bacteria seems to afford neither advantage nor disadvantage to the host. However, Kahl (1933) maintained that the adherent bacteria on some ciliates are advantageous symbionts and suggested that they contribute to the nutrition of their hosts; this was pure speculation and no modern data to substantiate his theory have been published.

Bacteria are reported to be attached to the host cells in various configurations and on all parts of the body; adhering by one end giving the appearance of a row of cilia, or more closely to the host's body along their full length. Fauré-Fremiet (1950*a*, *b*, 1951) for example, observed a closely-fitting dorsal layer of bacterial rods (Thiobacteriaceae?) on certain sand-dwelling ciliates of the genus *Centrophorella.* These bacteria were restricted to the ciliates and could not be detected on the surrounding sand particles. Similarly bacteria may be found sticking out at right angles to the stalks of peritrichous ciliates and covering the surfaces of holotrichous ciliates such as *Cristigera* (Kahl 1928, 1932). Phoretic bacteria have also been observed on chrysomonads (Geitler 1948), euglenoids (Tschermak-Woess 1950), *Volvox* (Hamburger 1958), a polytomid (Skuja 1958), ciliates (Collin 1912; Kahl 1928, 1932) and testate amoebae (Hedley & Ogden, pers. comm.). Although the adherent bacteria are commonly found in a wide variety of habitats they are more frequently reported from situations rich in organic materials where large populations of bacteria might normally be expected to occur.

Some protozoa are themselves phoretic organisms and recently the settlement of certain epiphytic ciliates has been shown to be more complex than most of the bacterial examples given above. Langlois (1975) demonstrated that the exudates of some marine filamentous green and red algae affect the settlement of the peritrich *Vorticella marina*, and this effect was enhanced by exudates of *Cladophora gracilis, Polysiphon harveyi* and *P. lanosa*, but was reduced by exudates of *Fucus spiralis* and *Scytosiphon lomentaria*. Langlois (1975) also suggested that peritrich telotrochs settle more readily on surfaces covered by bacteria, in a way similar to barnacle spat (Knight-Jones & Stevenson 1950).

Alexander (1971) describes another form of commensal relationship in which one species destroys a toxin or removes inhibitory factors from the local environment, which allows the growth of an adjacent species. The ability to destroy naturally occurring organic inhibitors or man-made pollutants by microbes is well known; for example, the highly toxic hydrogen sulphide is detoxified by being oxidized by the photosynthetic sulphur bacteria; thus allowing other sensitive organisms to survive. An experimental example of this type of commensal association is that of the ciliate *T. pyriformis* which grew in a chemostat vessel receiving a flow of medium containing 500 $mg.l^{-1}$ of phenol (G. L. Jones, pers. comm.). In this case the ciliate survived because the bacteria reduced the phenol concentration to below levels of detection (<1 $mg.l^{-1}$). Microbiological activities are well known to affect the local environment by lowering or raising the pH or E_h, by the removal or addition of oxygen and so on. The removal of oxygen and lowering of E_h by the metabolic activities of facultative anaerobic bacteria benefit oxygen-sensitive and strictly anaerobic protozoa and is the only way oxygen-sensitive protozoa such as *Trepomonas*, *Metopus*, *Saprodinium* and *Trimyema* can survive in the natural fresh water habitat.

Commensal relationships between other micro-organisms can also affect protozoa indirectly. Any microbial association that results in the growth of a population of organisms that are suitable as a food source for protozoa will have a beneficial effect upon the protozoan population. For example, exudates of fresh water algae such as *Nostoc*, *Calothrix*, *Chlorella*, *Anabaena*, *Tolypothrix* and *Anaenopsis* have all been shown to contain organic complexes that can be utilized by heterotrophic bacteria (Stewart 1963; Vela & Guerra 1966) and marine algae also produce complex polysaccharides which are used as carbon sources by marine bacteria. This is one area where field data exist to demonstrate the importance of algal populations in the input of carbon (Hellebust 1967) and nitrogen (Stewart 1963) in natural waters but no-one has yet monitored the response of the protozoa.

(b) *Mutualism*

Mutualism may be defined as "an association of organisms where each partner benefits from the presence of the other." Although a wide range of eukaryotic cells harbour intracellular, self-reproducing organisms that are clearly of foreign origin, in most cases the exact relationship between them has not been clearly established. For practical purposes all endosymbionts (other than those thought to be parasites) have been included in this section even though in the future, some may be shown not to be mutualists. The endosymbionts range from particles of unknown nature, viruses and bacteria to eukaryotic cells. Although there are many reports (Ball 1969) on the presence of virus-like particles in the nucleus and cytoplasm of many protozoa, most have not been identified adequately and the relationship between virus and host has certainly not been established.

(i) *Bacterial endosymbionts*

Rather more is known about the bacteria that are seen within the cytoplasm or nuclei of protozoa, and referred to either as symbionts or parasites, usually without specific evidence. In the 6th edition of Bergey's Manual of Determinative Bacteriology (Breed *et al.* 1948) the organisms found inside protozoa were all placed in an appendix to the Rickettsiales, but because of their greater similarity to bacteria were described in the 7th edition (Breed *et al.* 1957) in an addendum to the class Schizomycetes. In view of the paucity of exact data about these organism the editors of the 8th edition (Buchanan & Gibbons 1974) declined to position them in any taxonomic group but placed them in part 8 as they are all Gram negative organisms. Furthermore Buchanan & Gibbons (1974) state that "It is important that all bacteriologists recognize their existence for they represent an interesting group worthy of more detailed study." The subject has been reviewed by Beale *et al.* (1969), Preer (1975) and Preer *et al.* (1974). While bacteria have been noted in several protozoa the best studied are those in certain amoebae and ciliates.

There is still considerable discussion about the bacteria found in the giant amoeba *Pelomyxa palustris* and the reader should consult the review of Kirby (1941) for the earlier work. Hollande (1945) called the large rods found around the nucleus and glycogen masses of *P. palustris*, *Cladothrix*, but Keller (1949) thought they were the fruiting bodies of *Myxococcus pelomyxae* (Veley) Keller. The latter worker also recovered a second organism *Bacterium parapelomyxae* Keller from the amoeba. Leiner and his coworkers also studied the bacteria in *P. palustris* and although they did not name the organism, stated that it was probably essential for the existence of the amoeba. Leiner *et al.* (1951) claim that the bacterium plays a role in the carbohydrate metabolism and is important in nitrogen metabolism (Leiner & Wolhfeil 1954) but it is not known (Leiner *et al.* 1954) how the amoeba regulates growth of the bacteria. The presence of these rods has been suggested as a generic feature. However although close relatives, such as *Pelomyxa illinoisensis* do not contain this symbiont they do possess small membrane-covered, electron-dense bodies in the cytoplasm. These bodies may be infective (Daniels & Roth 1961) and perhaps correspond to the alpha granules of Andresen (1956) and to similar cytoplasmic bodies that have been noted in *Amoeba proteus* (Cohen 1957). Daniels (1964) and Rabinovitch & Plant (1962) have found that some of these particles are indeed infective organisms and suggest they are bacterial, viral or rickettsial in nature. The reported DNA synthesis and presence of RNA and DNA precursors in the cytoplasm of *A. proteus* (Plant & Sagan 1958) should be regarded with caution until the amoebae are freed from contaminants (Roth & Daniels 1961).

Several bacterial symbionts have been noted in ciliated protozoa. In *Paramecium* they may be in the macronucleus, micronucleus or cytoplasm (Wichterman 1940, 1945, 1953, 1954; Diller & Diller 1965; Wagtendonk *et al.* 1963) but whether these are mutualists or parasites is unclear. Fauré-Fremiet (1952) noted two species of *Euplotes* containing bacteria. Elimination of the latter by penicillin treatment resulted in the death of the ciliates. Although this evidence was not perhaps thoroughly convincing, Heckmann (1975) was able to save *Euplotes aediculatus*, after similar penicillin treatment to remove the bacterial symbiont omicron, by reinfecting the ciliate with the bacteria; this suggests that some bacteria are essential while others are not. Other endosymbiotic bacteria will be mentioned in Section 3(c).

(ii) *Algal endosymbionts*

Perhaps the best known symbionts of protozoa are the algae found in these animals from arctic, temperate and tropical habitats (McLaughlin & Zahl 1966).

The associations between algae and protozoa have long been recognized and *Paramecium bursaria* with its zoochlorellae and the radiolaria and foraminifera with their zooxanthellae are examples of a more or less continuous symbiotic relationship. The interaction between the two groups is believed to be mutualistic although in some cases the algae may only be commensals.

The algae, mainly Cyanophyceae, Chlorophyceae or Dinophyceae, are usually referred to as cyanellae, zoochlorellae or zooxanthellae, respectively, depending on whether they are blue-green, green or yellow in colour. No taxonomic precision is intended by these terms but the cyanellae are probably all blue-green algae, the zoochlorellae mainly *Chlorella* or *Pleurococcus* and the zooxanthellae mainly, but not exclusively, dinoflagellates. Table 2 indicates some protozoan genera known to harbour algal symbionts; although by no means exhaustive it indicates the wide range of protozoa involved. Infection appears to be by phagocytosis (Karakashian 1975) and the numbers of algae present can be very large. Sud (1968), for example, reported that the zoochlorellae in ciliates can account for between 10 and 56% of the volume of the host cells. The mechanism by which the numbers of symbionts are controlled is not usually understood and the actual degree of control varies considerably (Droop 1963).

By far the best known example of a mutualistic protozoan and alga is that of *P. bursaria* and *Chlorella* (see the review by Karakashian 1975). In contrast with the impressive detail known for certain other mutualistic systems (Smith *et al.* 1969), relatively little is known about the benefits of symbiosis to the ciliate host *P. bursaria*, and except for some recent data, the evidence relies heavily on observations on the growth of infected and aposymbiotic animals under various lighting and feeding schedules (Siegel 1960; Karakashian 1963; Pado 1965, 1967). Even less is known about the value of symbiosis to the algal partners in any association (Taylor 1973). Apparently the symbiotic association provides no significant benefit to infected paramecia when grown under optimal conditions (Siegel 1960; Karakashian 1963) but the presence of algae enhance the growth of infected cells once the bacterial food supply is depleted. The exact physiological basis of the algal contribution to the ciliate remains unclear but recent data (Brown & Nielsen 1974) have shown that *P. bursaria* containing algae can incorporate $^{14}CO_2$ to a greater extent than aposymbiotic animals. In these studies large amounts of radioactive carbohydrate were identified in the host fraction indicating that the algae release photosynthate *in situ.* In these experiments where the medium was maintained low in bacterial food, the algae were able to make a large contribution to the host's growth.

(c) *Parasitism*

Parasitism is an association that is difficult to define concisely. Alexander (1971) for example, stated that "A parasite is an organism that feeds on the cells, tissues or body fluids of another and usually larger organism, the host, which is

Table 2
Algae in protozoa

Host	Alga	Habitat	Authority
Mastigophorea			
Cryptella	Cy	FW	Pascher (1929*a*)
Cyanophora	Cy	FW	Korschikoff (1924) Hall & Claus (1963)
Noctiluca	–	M	Buchner (1953)
Peliaina	Cy	FW	Pascher (1929*a*)
Uroglena	Zch	FW	Lund (1953)
Leptodiscus	–	M	Nicol (1960)
Oikomonas	Cy	FW	Geitler (1959)
Sarcodina			
Acanthometra	Zx	M	Buchner (1953)
Acanthocystis	Zx	M	Buchner (1953)
Actinosphaerium	–	–	Buchner (1953)
Amoeba	Zch	FW	Pascher (1930)
Collozoum	Zx	M	Brandt (1885)
Difflugia	Zch	FW	Buchner (1953)
Globigerina	Zx	M	Buchner (1953)
Globigerinoides	Zx	M	Freudenthal *et al.* (1964)
Heliophrys	–	–	Buchner (1953)
Orbitolites	Zx	M	Doyle & Doyle (1940)
Paulinella	Cy	FW	Pascher (1929*a, b*)
Peneroplis	Zx	M	Winter (1907)
Sphaerozoum	Zx	M	Brandt (1885)
Trichosphaerium	Zx	M	Buchner (1953)
Acanthaires	Zx	M	Cachon-Enjumet (1961)
Heliosphaera	–	M	Nicol (1960)
Lithocerus	–	M	Nicol (1960)
Polycyttaires	Zx	M	Cachon-Enjumet (1961)
Sphaerellaires	Zx	M	Cachon-Enjumet (1961)
Sphaerocollides	Zx	M	Cachon-Enjumet (1961)
Thallassicola	–	M	Nicol (1960)
Ciliatea			
Climacostomum	Zch	FW	Sud (1968)
Condylostoma	*Chlamydomonas*	FW	Fauré-Fremiet (1958)
Cothurnia	–	–	Nicol (1960)
Euplotes	Zch	FW	Sud (1968), Curds (in press)
Frontonia	Zch	FW	Buchner (1953), Sud (1968)
Holosticha	Zch	FW	Sud (1968)
Mesodinium	–	–	Nicol (1960)
Ophrydium	Zch	FW	Buchner (1953)
Paraeuplotes	Zx	M	Wichterman (1942)
Paramecium	Zch	FW	Karakashian (1975)
Prorodon	Zch	FW	Sud (1968)
Scyphidia	–	–	Nicol (1960)
Stentor	Zch	FW	Tartar (1961)
Strombidium	Zch	M	Fauré-Fremiet (1948)
Tintinnus	–	M	Buchner (1953)
Trichodina	–	M	Buchner (1953)
Vorticella	–	–	Nicol (1960)

Cy, cyanellae; Zch, zoochlorellae; Zx, zooxanthellae; M, marine; FW, fresh water.

commonly injured in the process. The parasite is dependent, to a greater or lesser extent, on the host at whose expense it is maintained". Despite this lengthy statement it is still often difficult to decide what is a parasite and what is a predator. For example, are fungi growing at the expense of protozoa, parasites or predators? Often one term is applied more commonly than another as a matter of tradition. Ball (1969) included a wide range of organisms as parasites in his review but the term parasite has been limited here to organisms known to have a deleterious effect upon the host, and to those that are in intimate contact for considerable periods of time.

(i) *Bacterial parasites*

Several bacteria have been shown to affect protozoa adversely. In *Paramecium* bacterial parasites have been observed in the cytoplasm, macronucleus and micronucleus. For example, Wichterman (1940, 1953) reported a parasite, apparently of the genus *Holospora*, living in the macronucleus of *P. caudatum*; infected macronuclei were greatly enlarged and up to 80% of the culture was infected. Wichterman (1954) described another macronuclear parasite of *P. caudatum* where infected individuals showed obvious signs of cytoplasmic breakdown and nuclear abnormalities. Wichterman (1954) also described a zoochlorella-free *P. bursaria* containing bacteria-like spheres which prevented the division and conjugation of the host. Similarly Kirby (1942) described a bacterial parasite of *Vorticella* which caused macronuclear abnormalities while Davis (1942) found that the bacterial rods inside the suctorian *Trichophrya micropteri* were lethal in heavy infections. Drozanski (1956, 1963*a*, *b*, 1965) reported the behaviour of certain bacterial parasites of some soil amoebae such as *Acanthamoeba castellanii*; these Gram negative, flagellated rods could not be cultured outside the host cell nor in ciliates although they would reproduce within several soil amoebae and in two species of Mycetozoa. The amoebae engulfed the bacteria which then lysed the food vacuoles, filled the host, finally lysing the cell wall to infect other amoebae.

Since the discovery of the killer phenomenon in *P. auralia* by Sonneborn (1943) much has been done on the particles concerned which are now considered to be endosymbiotic bacteria. The subject has been reviewed by Beale 1954; Sonneborn 1959, 1961, 1965; Trager 1960; Hageman 1964; Ball 1969; Preer *et al.* 1974; Preer 1975, so only a brief account is given here. Sonneborn (1943) found that when cells of certain stocks of *P. auralia* were mixed the cells of one stock were killed; those which died he called 'sensitives' and those which survived, 'killers'. The ability to kill was found (Sonneborn 1945*a*) to be the result of particles in the cytoplasm which he called kappa particles. Although kappa was subsequently proved to be self-duplicating its ability to kill was shown to depend upon the presence of a dominant gene *K* in the macronucleus of the ciliate and the killing was due to the release of a toxin known as paramecin (Sonneborn 1945*b*; Beale 1954). Kappa particles were originally called viruses (Altenburg 1946*a*, *b*) and plasmagenes (Sonneborn 1946, 1947) but Preer (1948, 1950) noted that they were larger than viruses, were Feulgen positive with *Rickettsia*-like nuclei and were obligate symbionts. This evidence convinced Sonneborn (1961) that kappa were infectious intracellular organisms with a complex level of organization close to that of bacteria, and presented evidence concerning the nature of the interrelationship between kappa and host. He (Sonneborn 1961) found that kappa-bearing paramecia were less viable than sensitive strains when cultured under unfavourable conditions. Similarly, Soldo (1961) found that although killer strains could be grown axenically they had extra nutritional requirements and needed greater concentrations of other essentials than did sensitive stocks. Furthermore Burbanck & Martin (1965) were not able to grow killer strains axenically although animals artificially freed from kappa by antibiotic treatments grew immediately in the same media. This evidence suggests that kappa particles are parasitic in nature.

Since the discovery of kappa, other particles have been found in the cytoplasm of various stocks of *Paramecium.* Although most of these, like kappa, have the ability to kill sensitives they differ either in their structure or mode of action. These particles include mu, the mate-killing particle (Levine 1953; Siegel 1953), pi (Hanson 1954) and lambda (Schneller 1958; Soldo 1960). Very little is known about the nature of pi particles but it is thought they are mutants of kappa (Hanson 1954). The mate killers, mu, differ from kappa in morphology and mode of action, being Feulgen positive encapsulated rods which kill only at the time of conjugation; their morphology has led Beale & Jurand (1960) to suggest that they are cytoplasmic bacteria. Lambda particles are Gram negative rods containing both RNA and DNA, and release a substance into the medium which lyses sensitive paramecia (Schneller 1958; Butzel & Wagtendonk 1963). Antibiotics, particularly penicillin (Wagtendonk & Soldo 1965), effectively remove lambda particles and when this is complete the *Paramecium* has its folic acid requirement restored. *Paramecium* containing lambda do not require folic acid and in this respect it is beneficial to its host and should be regarded as a mutualist (Wagtendonk *et al.* 1963; Wagtendonk & Soldo 1965).

More modern work (see reviews by Preer *et al.* 1974 and Preer 1975) confirms that these endosymbionts are indeed bacteria. In addition to their general morphological similarities there is now a lot of other convincing evidence; for example, kappa has been shown to respire, to have the enzymes of the glycolytic and oxidative pathways, and to contain cytochromes unique to bacteria (Kung 1970, 1971). Mu has diaminopimelic acid, a substance only found in the cell walls of bacteria (Stevenson 1967). Lambda has bacterial flagella and the ribosomal RNA of mu is like that of bacteria and quite unlike that of *Paramecium.* Finally, the endosymbiotic bacteria have been described (Preer *et al.* 1974) and given binomial names based upon their DNA base ratio.

(ii) *Fungal parasites*

The great majority of fungi that attack protozoa are members of the Chytridiales. For some chytrids such as *Sphaerita* and *Nucleophaga* which infect parasitic protozoa, there is little doubt that the relationship is truly parasitic while for others, particularly those which attack free-living flagellates and amoebae, it is difficult to say whether the association is parasitic or predatory. With the possible exception of *Sphaerita* these fungi are ultimately lethal to their protozoan hosts. Sparrow (1960) gives a long list of fungal parasites with their hosts.

In the Phycomycetes, the fungus attacks from the outside by means of a tube from the encysted zoospore. Sometimes the complete fungus enters and develops within the host while in others the host-cell contents pass out into the fungus which remains outside. The zoopagales are soil-inhabiting fungi which attack naked and testate amoebae either by invasion and reproduction inside the host or by the production of a mycelium which traps the protozoa by means of an adhesive secretion. When caught, the root-like hyphae invade the trapped organism and ultimately digest it. The subject has been extensively reviewed by Duddington (1957) and Ball (1969).

(iii) *Protozoan parasites*

Protozoa are commonly parasitic in or on other protozoa and Table 3 lists some examples of this interaction. As in the fungi it is sometimes difficult to distinguish between a parasite and a predator since the parasite–predator is often as large as the host–prey. This difficulty is most commonly encountered in the Suctoria where the difference in size is often marginal. Most free-living suctoria feed by sucking the cell contents of the prey into the body via hollow tentacles. When a suctorian such as *Tokophrya*, attached to an inanimate substratum, feeds upon a ciliate prey such as *Tetrahymena*, the suctorian is a predator. However when *Tokophrya* is attached to the peritrich colony on which it feeds it is then called a parasite. The suctorian *Podophrya parasitica* is perhaps a clearer example of ectoparasitism since here numerous parasites may be attached to the larger host, the ciliate *Nassula ornata*, and chains of parasite and host may be formed (Fauré-Fremiet 1945). Several suctoria are endoparasitic and Canella (1957) has included a discussion of host choice by suctoria in his monograph.

Most major groups of protozoa have parasitic members that are found in free-living protozoa. However there seems to be only one parasitic foraminiferan (Le Calvez 1947) and surprisingly for an exclusively parasitic group, only one coccidian (Le Calvez 1939) both being parasites of foraminifera. Many parasites are specific to certain protozoan hosts but there are examples where the parasitism seems to be more of a matter of chance than design (López-Ochoterena 1962; Pérez Reyes & López-Ochoterena 1963). Parasitic dinoflagel-

Table 3

Protozoan parasites of protozoa

Parasite	Host	Habit	Authority
Astasia sp.	*Stentor coeruleus*	End, Cyt	Schönfeld (1959)
Astasia sp.	*Amoeba proteus*	End, Cyt	Fox (1946)
Leptomonas karyophilus	*Paramecium trichium*	End, Nuc	Gillies & Hanson (1962, 1963)
Parasitic dinoflagellates	Marine dinoglagellates, radiolaria & tintinnids	End, Cyt or Nuc	Grassé (1952)
Bodo sp.	*Collozoum* sp.	End, Cyt	Hollande & Enjumet (1953)
An amoeba	*Volvox aureus*	End, Cyt	Nieboer (1959)
Vahlkampfia pelomyxae	*Pelomyxa palustris*	End, Cyt	Hollande (1945), Hollande & Guilcher (1945)
An endamoeba	*Pelomyxa palustris*	End, Cyt & Nuc	Leiner *et al.* (1954)
Vahlkampfia discorbini	*Discorbis mediterraneanensis*	End, Cyt	Le Calvez (1940)
Entosolenia marginata	*Discorbis vilardeboanus*	Ect	Le Calvez (1947)
Amoebae	*Amoeba proteus*	End, Nuc	Wolska (1949)
Hypocoma parasitica	*Zoothamnium*	Ect	Chatton & Lwoff (1939)
Heterocoma hyperparasitica	*Trichophrya salparum*	Ect	Chatton & Lwoff (1939)
Blepharisma?	*Raphidiophrys* sp.	End, Cyt	Penard (1940)
Pottsia infusiorum	*Folliculina* sp.	End?	Andrews & Reinhard (1943)
Endosphaera engelmanni	*Trichodina spheroides*	End, Cyt	Laird (1953), Padnos & Nigrelli (1947)
Erastophrya chattoni	*Glossatella piscicola*	Ect	Fauré-Fremiet (1943)
Podophrya parasitica	*Nassula ornata*	Ect	Fauré-Fremiet (1945)
Sphaerophrya sol	*Epistylis*	–	Pérez Reyes & López-Ochoterena (1963)
Sphaerophrya sol	*Stylonychia*	–	López-Ochoterena (1962)
Sphaerophrya sol	*Paramecium*	–	Pérez Reyes & López-Ochoterena (1963)
Suctoria	*Urostyla* & *Paramecium*	End	Jankowski (1963)
Trophosphaera planorbulinae	*Planorbulina mediterranensis*	End	Le Calvez (1939)

End, endoparasite; Ect, ectoparasite; Cyt, cytoplasm; Nuc, nucleus.

lates are commonly found in the cytoplasm and nucleus of other dinoflagellates, in the cytoplasm of radiolaria and tintinnid ciliates. Here the parasites are clearly modified for a parasitic life style and are only recognizable as dinoflagellates when outside the host. These organisms are fully discussed by Chatton (in Grassé 1952). Parasitic dinoflagellates in radiolarians have until relatively recently been regarded as stages in the life cycle of the hosts (Hollande & Enjumet 1953; Hovasse & Brown 1953; Cachon-Enjumet 1961).

4. Amensalism

There are now innumerable reports that many microbes are able to synthesize and excrete organic or inorganic compounds *in vitro* that are harmful to the development of nearby microbial populations; considered by many to be a means of protection or offence. Several investigators have suggested that these methods are important in determining microbial distribution, population size and in governing the succession of micro-organisms. Although toxins and inhibitors are well documented for many microbial groups, this is not so in the Protozoa. Of course, micro-organisms that change the pH of their environment by the production of acids or alkalis, or that produce inorganic inhibitors such as carbon dioxide, oxygen or hydrogen sulphide will affect protozoan populations in the near vicinity. Amensalism may thus contribute to the success of one of a pair of competitors but otherwise it is a phenomenon distinct from competition.

Organic inhibitors known to be synthesized by micro-organisms may be of low potency (simple fatty acids, ethanol etc) and need to be present in high concentration to be effective, or highly potent complex organic inhibitors known collectively as antibiotics.

Bacteria, fungi and algae are all known to excrete substances toxic to protozoa *in vitro*. Heal & Felton (1970) demonstrated that small soil amoebae often rapidly encyst or develop large vacuoles and disintegrate in the presence of *Streptomyces* and *Streptomyces*-like *Nocardia*. Furthermore the microflora that support active reproduction of the amoeba *Acanthamoeba* tended to produce exudates which stimulated activity while those which were inedible or supported little amoeboid growth tended to produce exudates which inhibited or did not affect amoebae. Some *Nocardia* and *Mycobacterium* species caused the amoebae to encyst more rapidly than controls, and while *Penicillium claviforme* and *Aspergillus flavipes* were strongly inhibitory, other similar species were only weakly antagonistic. While all 14 of the *Streptomyces* spp. screened by Heal & Felton (1970) showed some antagonism to amoebae, exudates of *Streptomyces rimosus* caused disintegration within hours. Although these were laboratory experiments they tested the capacity of streptomycetes to produce toxic exudates when grown upon more natural substrates such as oak-leaf litter (radiation sterilized). In this case, even though the growth of *Streptomyces* spp.

was sparse, enough exudates were produced to kill amoebae. Other investigators have reported (Singh 1945; Knorr 1960; Groscop & Brent 1964; Curds & Vandyke 1966) that substances produced by bacteria are toxic to amoebae and ciliates.

Moikeha & Chu (1971) reported that blue-green algae such as *Lyngbya majuscula* produce exudates that could lyse the ciliate *Tetrahymena* and it has been suggested that this interaction could be used as a bioassay method for estimating the concentration of exudate which contains a dermatitis-producing principle. More recently, other algae have been reported (Langlois 1975) to exude substances which inhibit the settlement of telotrochs of marine peritrichs. The settlement of the vorticellids decreased when exposed to exudates of the alga *Ascophyllum nodosum* and was prevented by exudates of *Fucus spiralis* and *Scytosiphon lomentaria.* In nature, vorticellids do not settle on these algae but do so on other species whose exudates increase the percentage of telotroch settlement and survival. It is not known if any single-celled algae produce similar substances.

5. Competition

Competition may be defined as the struggle between organisms for an essential resource that is limited in the environment. While the resource may be a nutrient, space or light, most work has been concentrated on the competition for nutrients. In effect organisms are only in competition when the supply of a common resource is not sufficiently great to satisfy the needs of both. Two types of competition are commonly recognized – intraspecific competition which refers to rivalry between organisms of the same species and interspecific competition where the competition are of different species. Competition has been recognized as a potent force both in the survival of species in the natural habitat and in natural selection and there is a wealth of data available about micro-organisms (Clark 1965; De Bach 1967). However, information concerning protozoa is sparse and is mainly restricted to interspecific competition involving two protozoa competing for a common food source.

In this interaction the mathematical approach has been of value and preceeded experimental work. Volterra (1926), Lotka (1925, 1932) and Gause (1934*a*) all developed similar mathematical arguments which suggested that an organism with the highest rate of growth would succeed and displace the second organism. Analyses by Powell (1958) and Moser (1958) concerning the fate of contaminants or of mutants in continuous culture dealt with competition for a single limiting substrate and are equally applicable. Their theory predicts that provided the contaminant is not washed out before its first few divisions, the more rapidly growing organism should survive to displace the other. When the maximum specific growth rate of one organism is greater and its saturation constant is lower than its competitor, the first organism will always survive irrespective of dilution rate. However, should the maximum growth rate and the saturation constant of the first organism be greater than those of the second, then the outcome will depend upon the dilution rate at which the chemostat is operated. This theory has been shown to apply to bacteria (Meers 1971).

Gause (1934*b*, 1935) showed experimentally that when a fixed concentration of food supply (*Bacillus pyocyaneus*) was present in a batch culture of equal numbers of the two ciliates *P. aurelia* and *P. caudatum* then the latter species always survived and *P. aurelia* was totally eliminated, which agreed well with theory. However, Gause *et al.* (1934) later found that the situation was not as simple as originally stated there being, in fact, two phases of competition in their batch cultures; the first during active growth when there was competition for the utilization of a certain limited nutrient and the second after active growth had ceased when the populations began to decline. They found that if the decisive factor for competition was rapid utilization of food and growth then *P. caudatum* had the advantage but if resistance to waste products was the essential factor then *P. aurelia* could replace *P. caudatum.*

A similar study was made by Mučibabić (1957) who investigated competition for a common dissolved organic substrate between the flagellate *Chilomonas paramecium* and the ciliate *T. pyriformis*. It was found that the peak population numbers, the sizes and the shapes of both organisms were affected by the presence of the other. In mixed cultures the peak population size was significantly greater than when cultured alone but after stationary phase died out more rapidly. The numbers of *Tetrahymena* were also bigger in mixed cultures in the early stages of growth but by the onset of stationary phase this was reversed. Only after the death of *Chilomonas* did the ciliate population again reach the same numbers that it did in isolation. However when the results are expressed in terms of total biovolume, (rather than numbers which is highly misleading), the growth of the ciliate in mixed culture was not as good as in single cultures but later when *Chilomonas* had been sufficiently depleted the ciliate reached control population levels. During these experiments the growth rate of *Tetrahymena* was always greater than *Chilomonas* on the first day but later this situation was reversed. It should be remembered that a complex undefined growth medium was used for these studies and it could well be that the organisms were not competing for the same nutrient or that the values of the kinetic constants μ_m and K_s are such that *Tetrahymena* grows faster under conditions of high substrate concentration (high or equal μ_m, high K_s while *Chilomonas* grows faster under limited substrate conditions (low or equal μ_m, low K_s).

Some field observations concerning the survival of varieties of *P. aurelia* (Hairston 1958) have been explained by laboratory experiments (Hairston & Kellerman 1965). Hairston (1958) postulated from field data that competition takes place between var. 2 and 3 of the ciliate *P. aurelia* such that var. 2 was favoured in colder, less productive waters at all times and in richer waters only during cold weather whilst var. 3 was favoured in richer waters, particularly during warm weather. Later, Hairston & Kellerman (1965) showed in laboratory experiments that var. 2 displaced var. 3 at temperatures between 5 and 25°C even when var. 3 was given a 6 : 1 inoculation advantage. This end result was obtained most rapidly at 5°C, and var. 3 persisted longest at 15 and 20°C. In these experiments a four-fold increase in food supply (bacteria) had no effect upon the rate at which var. 2 predominated, suggesting that at both levels food was severely limiting to both populations. This is one of the few examples where field observations and laboratory experiments on protozoan interactions have been carried out and is particularly encouraging in view of the good agreement between the two approaches.

6. Predation

Protozoa can be predator or prey, and their latter role has already been discussed in Section 3(c)(ii). As predators, protozoa feed on a whole host of microbial prey such as viruses, bacteria, algae, yeasts, blue-green algae and other protozoa (Sandon 1932). Most observations have been of a qualitative nature but the last decade has seen a sudden surge of interest with an emphasis on the quantitative and dynamic aspects of the predatory activities of protozoa, particularly of ciliates. Presumably this work has been stimulated by general renewal of interest in the environment and the role that particular organisms might play in the ecological structure of various habitats.

(a) *Food preferences of protozoa*

Many authors have presented micro-organisms, either in suspension or as streaks on agar plates, to protozoa as the sole food supply. This approach has been used with ciliates (Hetherington 1933; Kidder & Stuart 1939; Burbanck 1942; Curds & Vandyke 1966; Barna & Weis 1973; Dive 1973*a*; Dive *et al.* 1974; Taylor & Berger in press) and with amoebae (Singh 1942, 1945, 1946; Heal 1963; Heal & Felton 1970). In most of these experiments bacteria were used as the prey organisms. This type of work clearly demonstrated that not all bacteria in isolation are suitable for the prolonged survival of all protozoa. Furthermore, Curds & Vandyke (1966) showed that while one bacterium may satisfy the dietary requirements of one ciliate it does not necessarily satisfy those of another, and this is borne out when the results of different workers are compared. Curds & Vandyke (1966) divided their bacterial prey into these broad categories – favourable, unfavourable and toxic – according to their effect upon the predators.

Different species of favourable prey have been found to be of different nutritive value to a predator, but nutritive value is measured differently by authors. For example, Curds & Vandyke (1966) used maximum specific growth rate as a measure of the value of a prey while Barna & Weis (1973) and Dive (1973*a*) estimated nutritive value in terms of the number of predator cells after four days–number of predators in the inoculum. The latter method is really a combination of growth rate and yield coefficient but in either case results showed that different bacteria support different growth rates, or are of varying nutritive value, to any given predator species. In general it is true to say that fresh water ciliates grow better on a diet of Gram negative rods while Gram positive organisms are either unfavourable or are of lower nutritive value (Barna & Weis 1973; Dive 1973*a*). This makes sense since most fresh water bacteria are Gram negative rods. However, Taylor & Berger (in press) have suggested that bacteria isolated from a pond in which four species of ciliate commonly occurred, were poorer food for these protozoa than were laboratory strains of bacteria, possibly because wild bacteria possess mechanisms which discourage predation. An alternative view is that the dominant bacteria in the pond, and therefore those most likely to be isolated, would be those least preferred by predators; those prey most preferred would be in lower concentrations because of predation.

Although many have studied protozoa feeding upon single prey species few workers have presented them with mixtures of microbial prey. Dive (1973*a*) showed that when the ciliate *Colpidium campylum* was growing at a low rate on Gram positive bacteria such as *Bacillus subtilis, B. cereus* or *B. licheniformis*, there was an immediate increase in growth rate when *Enterobacter cloacae* was added. A better approach was used by Barna & Weis (1973) who showed that a mixed bacterial flora supported a growth rate of *P. bursaria* that was intermediate to the growth rate as measured on cultures containing single bacterial species. These findings agree with those of my own laboratory (Shore, pers. comm.) where the growth rate of *T. pyriformis* when fed upon two different bacteria singly and in mixtures has been measured. More work of this nature is required rather than repetitive work of presenting single predator species with large numbers of prey species in isolation. Furthermore to find a different growth response on different prey organisms in isolation is not proof of selection as several authors have claimed. Selection can only be demonstrated when a choice of food is made available.

Micro-organisms other than bacteria have been presented to protozoa as food; Heal (1963) for example, presented four species of amoebae with 35 species of fungi and found that while all of the 19 yeasts supplied were eaten and supported amoeboid growth to varying degrees, the sporangiospores of the remaining fungal species, although ingested, did not. Only those of *Paecilomyces elegans* and *Polyscytalium fecundissimum* were favourable in this respect. Later,

Heal & Felton (1970) on presenting three amoebae *Acanthamoeba* sp., *Hartmanella astronyxis* and *Mayorella palestinensis* with 109 isolates of bacteria, actinomyces, fungi, yeasts and algae, found that the bacteria, yeasts and some *Nocardia* and *Mycobacterium* strains were eaten and supported amoeboid growth while *Streptomyces*, other *Nocardia* and *Mycobacterium* strains, mycelial fungi and algae were either inedible or supported little growth of the amoebae.

It has also emerged from this type of work that certain bacteria are positively toxic to certain protozoa; this has been reported on by Chatton & Chatton (1927), Kidder & Stuart (1939), Brent (1948), Groscop (1963), Groscop & Brent (1964), Curds & Vandyke (1966), Heal & Felton (1970), Barna & Weis (1973), Dive (1973*a*, *b*), Dive *et al.* (1974). In general it appears that bacterial pigment is the toxic agent; for example, the pigments of *Chromobacterium violaceum, Pseudomonas aeruginosa, Serratia marcescens* and others are very toxic. Dive *et al.* (1974) divided bacterial toxins into pigments, endotoxins and exotoxins and demonstrated that not all pigments are toxic to protozoa. The carotenoids of *Cytophaga, Flavobacterium*, pigmented Enterobacteria, *Pseudomonas putrifaciens* and of *Micrococcus* are not toxic to *Colpidium campylum.* It appears from the literature that the toxicity of a bacterium depends to a large extent upon the two organisms concerned. Finally it should be remembered that all these data have been obtained from laboratory studies and the significance of these observations when applied to the natural aquatic environment remains to be tested.

(b) *Food selection by protozoa*

There are far fewer data in the literature concerning positive food selection or avoidance than on food preference. Schaeffer (1910) appears to be the first to record selection when he observed that the ciliate *Stentor coeruleus* ingested 12 of the 15 *Phacus* cells provided but rejected all 13 of the sulphur particles presented at the same time. Further, *S. coeruleus* apparently was also able to descriminate between two species of *Phacus*, accepting *P. triqueter* but rejecting *P. longicaudus.* Other workers (Losina-Losinsky 1931; Gritner 1951; Bahr 1954) showed that coloured inanimate particles such as carmine would be accepted by *Paramecium* for a few days but rejected later. Bahr (1954) demonstrated that if the carmine was mixed with dead bacteria (*Bacillus subtilis*) both would first be accepted but later the carmine and then the bacteria were rejected resulting in the formation of empty food vacuoles and subsequent death. Empty food vacuole formation has also been noted in *P. caudatum* and *Histriculus vorax* (Curds & Vandyke 1966) when these ciliates were fed upon unfavourable bacteria. At first, *Paramecium* ingests the food but then rejects it by a reversal in ciliary beat prior to the closure of the food vacuole. Bahr (1954) also fed *Paramecium* on mixtures of living bacteria and noted that while *Bacillus subtilis, Staphylococcus aureus* and

Salmonella enteritidis were ingested others such as *Salmonella typhimurium* were ingested with difficulty, and *Escherichia coli* was rejected.

Some recent observations by Ricketts (1971) on the selection of particulate matter by *T. pyriformis* are rather more convincing than some of these earlier reports, and he showed that acid phosphatase activity increased after the uptake of useful food particles (yeasts or bacteria) but not after the ingestion of inert and useless latex particles. Furthermore, while fed *T. pyriformis* would rapidly ingest either useless or useful particles, starved cells would only ingest useful ones. However, the ciliate could be persuaded to accept latex if supplied in the presence of dissolved organic substances.

The recent study by Rapport *et al.* (1972) is based on a new definition of food preference (Rapport & Turner 1970) which enables food preference (selection) to be determined without any confusion with possible differences in 'prey catchability'. The method involves a comparison of the mean number of prey consumed when each prey species is present alone and in mixtures with other species. Using this method Rapport *et al.* (1972) showed that *Stentor coeruleus* preferred to eat *Chilomonas paramecium* and *T. pyriformis* when in the presence of either *Chlamydomonas reinhardti* or *Euglena gracilis*. These ideas are consistent with those of Hetherington (1932) who suggested that *S. coeruleus* had definite preferences for non-algal prey.

Further evidence on food selection has been furnished by Lee *et al.* (1966) who presented more than fifty ^{32}P- or ^{14}C-labelled species of protists to 10 species of foraminifera and found that although all were common inhabitants of the same ecological niche only certain ones were selected for ingestion. However, Muller & Lee (1969) later found that the identification of potential food organisms by these methods did not necessarily indicate whether or not they would support the growth of foraminifers.

Many carnivorous protozoa will feed upon certain prey protozoa to the exclusion of others, for example *Didinium nasutum* feeds exclusively upon *Paramecium* although Burbanck & Eisen (1960) found that it would not survive if the prey was fed upon single bacterial species. Other examples of specific feeding by carnivorous protozoa are given by Sandon (1932).

Recent work on the selective grazing of ciliates upon algal populations is of particular interest since the data were gathered from the field rather than artificial laboratory cultures. Goulder (1973) counted the numbers of the alga *Scenedesmus* inside natural populations of the ciliates *Loxodes magnus* and *L. striatus*. The larger *L. magnus* contained more algae than the other. While the concentration of the alga *S. quadricauda* was greater in the phytoplankton than in either *Loxodes* spp., the ratio of *S. denticulatus* to *S. quadricauda* was greater in the ciliates than in the plankton, data in agreement with the suggestion of Goulder (1972) that *L. magnus* selects *S. denticulata* in preference to *S. quadricauda*.

(c) *The stimulus for ingestion and food vacuole formation*

Bozler (1924) seems to be the first to suggest that the stimulus for the ingestion of particulate matter in ciliates is the presence of particles. Mast (1947) disagreed and stated that vacuoles are formed in the absence of particles. Food vacuole formation has subsequently been the subject of several papers, and most authors agree that in *T. pyriformis* the stimulus for formation is the presence of particles although some vacuoles are formed even in their absence. Uptake has been demonstrated by the use of many types of particles (Mueller *et al.* 1965; Elliott & Clemmons 1966; Chapman-Andresen & Nilsson 1968; Rasmussen & Kludt 1970; Ricketts 1971; Nilsson 1972) and uptake rates vary little with particle type. Rasmussen & Kludt (1970) demonstrated that the presence of inert particles was a prerequisite for rapid growth in axenic *Tetrahymena* cultures. The work of Ricketts (1972*a*, *b*) supported their observations since he showed that the volume of dissolved organic nutrients ingested per unit time increased by 300% in the presence of inert particles of 1–2 μm in diameter.

(d) *Feeding and growth kinetics*

Although microbiologists have long used batch and continuous-culture methods to estimate maximum specific growth rates, saturation constants and yield coefficients, protozoologists have only just begun to apply these classical methods to a study of the predatory activities of protozoa on other organisms. Proper & Garver (1966) were the first to employ microbiological methods to study protozoan predation. They supplied a few predators, including the ciliate *Colpoda steinii*, with various known concentrations of the prey *E. coli.* Ciliates were counted at frequent intervals and their peak populations obtained during batch culture were estimated. Provided that all of the prey is consumed (which is approximately true), then the slope of the line obtained when peak ciliate population was plotted against initial prey concentration gives an estimate of yield coefficient. Curds & Cockburn (1968) carried out a similar exercise in their batch-culture feeding study of *T. pyriformis* preying upon *Klebsiella aerogenes.* The yield coefficient can also be determined using continuous-culture methods if the steady state predator concentration is plotted against prey removal. Table 4 illustrates some values of yield coefficients of protozoa obtained by the use of these methods. The range of 0.2 to 0.78 is rather wide but it should be remembered that the actual value determined relies heavily upon the accuracy of the estimated weights of predator and prey populations which are particularly difficult to measure since both organisms are intimately mixed with each other. However even if the outside values are dismissed, the yield coefficient of protozoa is much greater than those generally quoted for invertebrate animals but certainly within the range normally given for aerobic bacteria.

In batch cultures the relationship between the specific growth rate of the predator and concentration of its prey can be investigated at the same time as yield coefficient. When the reciprocal of the growth rate is plotted against the reciprocal of the prey concentration (a Lineweaver-Burk plot) a linear relationship may be established. When the agreement is good the negative intercept in the y-axis is numerically equal to the saturation constant K_s, and the slope equals the maximum specific growth rate μ_m. The data required for such a plot can also be obtained from continuous cultures where some means of providing the predators with a constant flow of prey organisms is used. Here the prey concentration is measured at steady state when the predator's growth rate equals the dilution rate of the vessel.

Table 4

Numerical values of yield values of certain protozoa feeding on other microbes*

Predator	Prey	Culture type	Yield	Authority
Acanthamoeba sp.	*Saccharomyces cerevisiae*	B	0.37	Heal (1967)
Colpoda steinii	*Escherichia coli*	B	0.78	Proper & Garver (1966)
Colpoda steinii	*Anacystis nidulans*	C	2×10^{-9}†	Bader *et al.* (1976)
Paramecium caudatum	a 'yellow bacterium'	B	0.55	Tezuka (1974)
Tetrahymena pyriformis	*Klebsiella aerogenes*	B	0.50	Curds & Cockburn (1968)
Tetrahymena pyriformis	*Klebsiella aerogenes*	C	0.54	Curds & Cockburn (1971)
Tetrahymena pyriformis	*Aerobacter aerogenes*	B	0.73	Canale *et al.* (1973)
Tetrahymena vorax	*Escherichia coli*	B	0.20‡	Seto & Tazak (1971)

* On basis of weight to weight.
† On basis of *Anacystis* per *Colpoda.*
‡ On basis of carbon to carbon.

Whereas most authors (Proper & Garver 1966; Hamilton & Preslan 1969, 1970; Curds & Cockburn 1968, 1971; Canale *et al.* 1973) agree that a reciprical plot as described above gives an acceptable linear relationship there is some evidence to suggest that other population factors may also be involved. Cutler & Crump (1924) who were, perhaps, the first to report that the growth rate of a ciliate (*Colpidium colpoda*) is related to both bacterial and ciliate concentrations, found that the relationship between the growth rate of the ciliate and the number of bacteria available per ciliate took the form of a quadratic hyperbola. While Curds & Cockburn (1968) found that a similar relationship existed in batch cultures of *Tetrahymena* and bacteria the mean cell volume of the ciliate was more important in continuous-culture studies (Curds & Cockburn 1971). Curds & Bazin (in press) presented some data concerning the oscillatory predator–prey behaviour of amoebae of the slime mould *Dictyostelium discoideum* and its prey *E. coli*. Subsequent mathematical analysis of this transient system has shown (Bazin, pers. comm.) that the numbers of bacteria available per amoeba was the most significant factor in controlling the growth rate of the predator. There is little doubt that this is an area where more experimental work is needed preferably by using continuous-culture systems.

(e) *Dynamic behaviour of predator and prey*

The most widely used theoretical framework for the study of predator–prey dynamics is the Lotka-Volterra equations derived independently by Lotka (1925) and Volterra (1926) which are non-linear and cannot be solved directly. However it is possible to simulate solutions for these equations by numerical methods, on an analogue computer or by mathematical analysis. Since a review of these methods and theory is being published shortly (Curds & Bazin, in press) they will not be repeated here. One of the basic assumptions made in the Lotka-Volterra equations is clearly erroneous, viz. the assumption that the relationship between predator growth rate and prey concentration is linear. Recent workers, however, commonly assume Monod (1942) kinetics although several others (Contois 1959; Kono & Asai 1969; Jost *et al.* 1973*a,b*) have suggested various modifications. The Lotka-Volterra equations were originally applied to batch cultures but are now more commonly used to model continuous-culture systems. It is clear from these models that there are three stable solutions to the equations (Gause 1934*a*; Canale 1969, 1970; Curds 1971) a stable node, a stable focal point and an unstable node, depending upon the relative values of the rates and kinetic constants used. In a stable node situation the predator and its prey asymptote towards a steady state while the populations achieve stable levels after a series of damped oscillations in the case of a stable focal point. The third solution, which is the one most commonly predicted, is the unstable focal point where both populations pass through a series of complementary limit-cycle oscillations (constant wavelength and amplitude).

In testing the Lotka-Volterra models, Gause (1934*a*) used the ciliate predator *Didinium* and its prey *Paramecium*, and without exception found that the prey was totally consumed in his simple batch system. Co-existence was however finally obtained if a few *Paramecium* cells were added just prior to the extinction of the *Didinium*. Later Gause *et al.* (1936) and Flanders (1948) demonstrated that co-existence could also be obtained when the prey are afforded a refuge. More recently Luckinbill (1973) found that if the frequency of encounters between predator and prey was reduced by the addition of methyl cellulose to slow down movement, then co-existence could take place so that at least several oscillations could be obtained with *Didinium* dying before *Paramecium*. Furthermore, the addition of *Paramecium* at regular intervals produced sustained oscillations. This final observation illustrates the value of continuous-culture systems for investigating predator–prey dynamics. Batch cultures have a single fixed initial quantity of substrate available and are time dependent whereas substrate continually enters a chemostat and dynamic behaviour can be studied over long periods of time. Only a single reactor is necessary for dynamic studies of this type and normally a medium containing a fixed concentration of limiting nutrient for the prey species is pumped into the reactor where the predator feeds upon the actively growing prey.

Two groups of protozoan predators, slime mould amoebae and ciliates, have been studied using continuous-culture methods. One of the most obvious properties of these biological systems is that although strictly controlled, they exhibit highly non-linear responses which do not correspond to the simple cyclical changes predicted by theory. In some cases (*Anon.* 1969; Tsuchiya *et al.* 1972) this behaviour may be partly attributable to the measurement of cell numbers, rather than biomass, and Williams (1971) has shown the former to be much more variable than the latter. Furthermore, two-organism systems tend to be particularly sensitive to minor perturbations. In general however, the biological populations oscillate and often with damped amplitude (Tsuchiya *et al.* 1972; Van der Ende 1973; Bazin's data in Curds & Bazin, in press).

Several independent studies have been carried out on the dynamic behaviour of *T. pyriformis* and bacteria such as *Klebsiella aerogenes* or *E. coli.* Curds & Cockburn (*Anon.* 1969) were the first to publish some data which showed complementary oscillations but these were not sufficiently comprehensive to evaluate the possibilities of damping. Although Van der Ende (1973) observed damped oscillations Jost *et al.* (1973*a*) obtained sustained oscillations with a similar system. The data of Canale *et al.* (1973) are rather more difficult to interpret since their populations varied in an erratic manner and do not show the expected periodicity in sucrose concentration.

While Jost *et al.* (1973*a*) found that the Monod (1942) model correctly predicted the occurrence of sustained oscillations their prediction of the minimum bacterial population density was smaller than observed. Curds (1971) made a similar observation when attempting to model earlier results (*Anon.* 1969). Recently, Bonomi & Fredrickson (1976) have proposed a model which takes account of bacterial growth on the walls of the culture vessel which commonly occurs in chemostats (Van der Ende 1973) and found that they could thereby explain some of the differences noted between predictions and experimental data. Furthermore they state that bacteria on the walls of the vessel do not even need to be visible to the naked eye to have a significant effect upon the suspended populations. Nevertheless, Tsuchiya *et al.* (1972) found that a simple Monod (1942) model was sufficient to explain their slime mould amoebae data but Bonomi & Fredrickson (1976) suggest that this is because amoebae would be able to ingest bacteria growing on surfaces while ciliates would not.

(f) Effects of predation

The obvious effect of predation is the reduction of the prey species. For example, Curds *et al.* (1968) were able to show that effluents, obtained from activated-sludge plants artificially kept free from protozoa, contained much larger concentrations of freely suspended bacteria than did those with ciliate predators present. Later Curds & Fey (1969) demonstrated that ciliated protozoa are partly responsible for the disappearance of the faecal bacterium *E. coli* during the activated-sludge process; and similar work by Javornický and Prokešová (1963) showed that the development of protozoa in isolated reservoir water samples markedly reduced the numbers of bacteria present.

In addition to the reduction in numbers of prey species, several investigators have noted that the purification of sewage, as measured by the rate of oxygen consumption accompanying its degradation, occurs more rapidly and proceeds further in the presence of protozoa and bacteria than when bacteria are present alone (Butterfield *et al.* 1931; Javornický & Prokešová 1963). In addition, Nasir (1923), Cutler & Bal (1926) and Hervey & Greaves (1941) have reported that protozoa are capable of stimulating nitrogen fixation by *Azotobacter chroococcum.* Darbyshire (1972*a, b*) has confirmed these early reports and has shown that nitrogen fixation by that bacterium is increased in the presence of the ciliate *Colpoda steinii* at temperatures between 15 and 25°C, but less nitrogen is fixed at 28°C. In view of the known oxygen sensitivity of *Azotobacter* during nitrogen fixation (Parker 1954; Postgate 1971) it was suggested that the ciliates may help to decrease the dissolved oxygen concentration in mixed cultures to levels more suitable for nitrogen fixation at medium temperatures. At 28°C the respiratory activity of *Azotobacter* was thought to be sufficient by itself to lower the dissolved oxygen levels, particularly since the solubility of oxygen at this temperature is lower.

Unfortunately the system studied by Darbyshire (1972*a, b*) is a rather specialized example and his results cannot be used to explain the rather general phenomenon of predation appearing to stimulate bacterial activity. For example, Johannes (1965) commented on the effects of predation thus: "grazing ciliates prevent bacteria from reaching self-limiting numbers; the bacterial populations are thus kept in a prolonged state of physiological youth, and their rate of assimilation of organic materials is greatly increased". It is evident that this and other authors believe that simple predation stimulates bacterial activity. It is true that predation increases the growth rate of the prey (Curds 1971) but this is because substrate concentration rises as the result of the fall in prey concentration. This would not explain the data and statement of Johannes (1965) since consumption of substrate is related to the product of growth rate and biomass; this is an area where carefully controlled experimental work is required before any conclusions can be reached.

From the theoretical point of view, Curds (1974) has indicated the importance of predation by protozoa on the structure and behaviour of a microbial community, and he modelled several complex microbial food chains and webs to consider the fates of soluble substrates, heterotrophic bacteria and predatory protozoa. Computer simulations were then used to study the dynamic behaviour of these organisms using a single-stage chemostat as the model aquatic system. The food chains included a series dealing with the effect of predation by ciliates upon bacteria competiting for the same limiting nutrient and the concurrent utilization of nutrient by competiting organisms in the presence and absence of predators. Computer simulation suggests that predatory activities by protozoa could promote the survival of two competing bacteria whereas theory and practice would normally suggest (Moser 1958; Powell 1958; Meers 1971) that one would be washed out of the system in the absence of a predator. Concurrent utilization of food supply by bateria and protozoa, modelled according to the experimental data of Stumm-Zollinger (1966), in some cases predicted complex non-repetitive dynamic behaviour in the populations. In the light of these simulations it was concluded that predation by protozoa could help to create a diverse population of heterotrophic organisms and furthermore suggested that since non-repetitive variations could be obtained using simple models perhaps the microbial ecologist should expect, rather than be surprised at, highly complex dynamic behaviour in natural aquatic systems.

7. References

ALEXANDER, M. 1971 *Microbial Ecology*. New York & Chichester: John Wiley & Sons.

ALTENBURG, E. 1946*a* The "viroid" theory in relation to plasmagenes, viruses, cancer and plastids. *American Naturalist* **80**, 559–567.

ALTENBURG, E. 1946*b* The symbiont theory in explanation of the apparent cytoplasmic inheritance in *Paramecium. American Naturalist* **80**, 661–662.

ANDRESEN, N. 1956 Cytological investigations on the giant amoeba *Chaos chaos. Compte rendu des travaux du Laboratoire Carlsberg, Copenhague Sér. Chim.* **29**, 435–555.

ANDREWS, E. A. & REINHARD, E. G. 1943 A folliculinid associated with a hermit crab. *Journal of the Washington Academy of Science* **33**, 216–223.

ANON. 1969 *Water Pollution Research*. Ministry of Technology, 1968. p. 153. London: H.M.S.O.

BADER, F. G., TSUCHIYA, H. M. & FREDRICKSON, A. G. 1976 Grazing of ciliates on blue-green algae: Effects of light shock on the grazing relation and on the algal population. *Biotechnology and Bioengineering* **18**, 333–348.

BAHR, H. 1954 Untersuchungen über die Rolle der Ciliaten als Bakterienvernichter im Rahmen der biologischen Reinigung des Abwassers. *Zeitschrift für Hygiene und Infektionskrankheiten* **139**, 160–181.

BALL, G. H. 1969 Organisms living on and in Protozoa. In *Research in Protozoology*, Vol. 3, ed. Tze-Tuan Chen. Oxford: Pergamon Press.

BARNA, I. & WEIS, D. S. 1973 The utilization of bacteria as food for *Paramecium bursaria. Transactions of the American Microscopical Society* **92**, 434–440.

DE BARY, A. 1879 *Die Erscheinung der Symbiose*. Strassburg: Verlag von Karl J. Trübner.

BEALE, G. H. 1954 *The genetics of* Paramecium aurelia. Cambridge Monographs in Experimental Biology No. 2. Cambridge: Cambridge University Press.

BEALE, G. H. & JURAND, A. 1960 Structure of the mate-killer (mu) particles in *Paramecium aurelia*, stock 540. *Journal of General Microbiology* **23**, 243–252.

BEALE, G. H., JURAND, H. A. & PREER, J. R. 1969 The classes of endosymbiont of *Paramecium aurelia. Journal of Cell Science* **5**, 65–91.

BONOMI, A. & FREDRICKSON, A. G. 1976 Protozoan feeding and bacterial wall growth. *Biotechnology and Bioengineering* **18**, 239–252.

BOZLER, E. 1924 Über die Morphologie der Ernährungsorganelle und der Physiologie der Nahrungsaufnahme bei *Paramecium caudatum* Ehrenberg. *Archiv für Protistenkunde* **49**, 163–215.

BRANDT, K. 1885 Die kolonienbilden Radiolarien des Golfes von Neapel. *Fauna und Flora des Golfes von Neapel, Berlin* **13**, 65–71.

BREED, R. S., MURRAY, E. G. D. & HITCHENS, A. P. 1948 *Bergey's Manual of Determinative Bacteriology*, 6th Edn. Baltimore: Williams & Wilkins Co.

BREED, R. S., MURRAY, E. G. D. & SMITH, N. R. 1957 *Bergey's Manual of Determinative Bacteriology*, 7th Edn. Baltimore: Williams & Wilkins Co.

BRENT, M. M. 1948 The antibiotic action of chromogenic bacteria on the culture of *Entamoeba histolytica.* M.S. Thesis. Northwestern University, Evanston, Illinois.

BROWN, J. A. & NIELSEN, P. J. 1974 Transfer of photosynthetically produced carbohydrate from endo-symbiotic chlorellae to *Paramecium bursaria. Journal of Protozoology* **21**, 569–570.

BUCHANAN, R. E. & GIBBONS, N. E. 1974 *Bergey's Manual of Determinative Bacteriology*, 8th Edn. Baltimore: Williams & Wilkins Co.

BUCHNER, P. 1953 *Endosymbiose der Tiere mit pflanzlichen Mikroorganismen.* Basle: Birkhäuser.

BUNGAY, H. R. & BUNGAY, M. L. 1968 Microbial interactions in continuous culture. *Advances in Applied Microbiology* **10**, 269–290.

BURBANCK, W. D. 1942 Physioloy of the ciliate *Colpidium colpoda.* I. The effect of various bacteria as food on the division rate of *Colpidium colpoda. Physiological Zoology* **15**, 342–362.

BURBANCK, W. D. 1942 Physiology of the ciliate *Colpidium colpoda.* I. The effect of *Paramecium aurelia* as food for *Didinium nasutum. Journal of Protozoology* **7**, 201–206.

BURBANCK, W. D. & MARTIN, V. L. 1965 The effect of the food of *Paramecium aurelia*, syngen 4, stock 51, mating type VII on its role in symbiosis. In *Progress in Protozoology*, Abstracts of the Second International Conference on Protozoology, London, 1965, pp. 114–115.

BURBANCK, W. D. & WILLIAMS, D. B. 1963. The coexistence of *Colpidium colpidium* (Schew.) and *Paramecium aurelia* Ehrenberg under monoxenic conditions. In *Progress in Protozoology*, Abstracts of the First International Conference on Protozoology, Prague, 1961, pp. 304–307.

BURKHOLDER, P. R. 1963 In *Symposium on Marine Microbiology*, ed. Oppenheimer, C. H. Illinois: C. C. Thomas.

BUTTERFIELD, C. T., PURDY, W. C. & THERIAULT, E. J. 1931 Experimental studies of natural purification in polluted water. IV. The influence of plankton on the biochemical oxidation of organic matter. *Public Health Reports, Washington* **46**, 393–426.

BUTZEL, H. M. & VAN WAGTENDONK, W. J. 1963 Some properties of the lethal agent found in cell-free fluids obtained from cultures of lambda-bearing *Paramecium aurelia*, syngen 8, stock 299. *Journal of Protozoology* **10**, 250–252.

CACHON-ENJUMENT, M. 1961 Contribution á l'étude des Radiolaires Phaeodariés. *Archives de zoologie expérimentale et générale* **100**, 151–237.

CANALE, R. P. 1969 Predator-prey relationships in a model for the activated process. *Biotechnology and Bioengineering* **11**, 887–907.

CANALE, R. P. 1970 An analysis of models describing predator-prey interaction. *Biotechnology and Bioengineering* **12**, 353–378.

CANALE, R. P., LUSTIG, T. D., KEHRBERGER, P. M. & SALO, J. E. 1973 Experimental and mathematical modelling studies of protozoan predation on bacteria. *Biotechnology and Bioengineering* **15**, 707–728.

CANELLA, M. F. 1957 Studi e richerche sui tentaculiferi nel quadro della biologia generale. *Annali dell' Università di Ferrara*. Sez. III. Biologia Animale, **1**, 259–716.

CHAPMAN-ANDRESEN, C. & NILSSON, J. R. 1968 On vacuole formation in *Tetrahymena pyriformis* GL. *Compte rendu des travaux du Laboratoire de Carlsberg, Copenhague* **36**, 405–432.

CHATTON, E. & CHATTON, M. 1927 Sur le pouvoir cytolytique immédiat des cultures de quelques bactéries chromogènes. *Compte rendu des séances de la Société de Biologie* **97**, 289–292.

CHATTON, E. & LWOFF, A. 1939 Sur la systématique de la tribu des Thigmotriches Rhynchoidés. Les deux familles des Hypocomidae Bütschli et des Ancistrocomidae n. fam., les deux genres nouveaux, *Heterocoma* et *Parhypocoma*. *Compte rendu des séances de la Société de Biologie* **209**, 429–431.

CLARK, F. E. 1965 The concept of competition in microbial ecology. In *Ecology of Soil-Borne Plant Pathogens*, eds Baker, K. F. & Snyder, W. C. Berkeley: University of California Press.

COHEN, A. I. 1957 Electron microscope observations of *Amoeba proteus* in growth and inanition. *Journal of Biophysical and Biochemical Cytology* **3**, 859–866.

COLLIN, B. 1912 Étude monographique sur les Acinétiens. II. Morphologie, Physiologie, Systematique. *Archives de zoologie expérimentale et générale* **51**, 1–457.

CONTOIS, D. E. 1959 Kinetics of bacterial growth: relationship between population density and specific growth rate of continuous cultures. *Journal of General Microbiology* **21**, 40–50.

CURDS, C. R. 1971 A computer-simulation study of predator-prey relationships in a single-stage continuous-culture system. *Water Research* **5**, 793–812.

CURDS, C. R. 1974 Computer simulations of some complex microbial food chains. *Water Research* **8**, 769–780.

CURDS, C. R. 1977 Morphological and nomenclatural notes on three members of the Euplotidae (Hypotrichida, Ciliatea). *Bulletin of the British Museum (Natural History), London* **31** (*6*), 267.

CURDS, C. R. & BAZIN, M. J. 1977 Protozoan predation in batch and continuous culture. In *Advances in Aquatic Microbiology*, Vol. 1, eds Droop, M. R. & Jannasch, H. W. London & New York; Academic Press.

CURDS, C. R. & COCKBURN, A. 1968 Studies on the growth and feeding of *Tetrahymena pyriformis* in axenic and monoxenic culture. *Journal of General Microbiology* **54**, 343–358.

CURDS, C. R. & COCKBURN, A. 1971 Continuous monoxenic culture of *Tetrahymena pyriformis*. *Journal of General Microbiology* **66**, 95–108.

CURDS, C. R. & FEY, G. 1969 The effect of ciliated protozoa on the fate of *Escherichia coli* in the activated-sludge process. *Water Research* **3**, 853–867.

CURDS, C. R. & VANDYKE, J. M. 1966 The feeding habits and growth rates of some freshwater ciliates found in activated-sludge plants. *Journal of Applied Ecology* **3**, 127–137.

CURDS, C. R., COCKBURN, A. & VANDYKE, J. M. 1968 An experimental study on the role of the ciliated protozoa in the activated-sludge process. *Water Pollution Control* **67**, 312–329.

CUTLER, D. W. & BAL, D. V. 1926 Influence of protozoa on the process of nitrogen fixation by *Azotobacter chroococcum*. *Annals of Applied Biology* **13**, 516–534.

CUTLER, D. W. & CRUMP, L. M. 1924 The rate of reproduction in artificial culture of *Colpidium colpoda* Part III. *Biochemical Journal* **18**, 905–912.

DANIELS, E. W. 1964 Electron microscopy of centrifuged *Amoeba proteus*. *Journal of Protozoology* **11**, 281–290.

DANIELS, E. W. & ROTH, L. E. 1961 X-irradiation of the giant amoeba, *Pelomyxa illinoisensis*. III. Electron microscopy of centrifuged organisms *Radiation Research* **14**, 66–82.

DARBYSHIRE, J. F. 1972*a* Nitrogen fixation by *Azotobacter chroococcum* in the presence of *Colpoda steini* – I. The influence of temperature. *Soil Biology and Biochemistry* **4**, 359–369.

DARBYSHIRE, J. F. 1972*b* Nitrogen fixation by *Azotobacter chroococcum* in the presence of *Colpoda steini* – II. The influence of agitation. *Soil Biology and Biochemistry* **4**, 371–376.

DAVIS, H. S. 1942 A suctorian parasite of the small mouth Black Bass, with remarks on other suctorian parasites of fishes. *Transactions of the American Microscopical Society* **61**, 309–327.

DE BACH, P. 1967 The competitive displacement and coexistence principles. *Annual Review of Entomology* **11**, 183–212.

DILLER, I. C. & DILLER, W. F. 1965 Isolation of an acid-fast organism from axenic cultures of *Paramecium aurelia* 139. *Transactions of the American Microscopical Society* **84**, 152–153.

DIVE, D. 1973*a* Nutrition holozoique de *Colpidium campylum* phenomenes de selection et d'antagonisme avec les bactéries. *Water Research* **7**, 695–706.

DIVE, D. 1973*b* Action de pigments exocellulaires sécrétés par *Pseudomonas aeruginosa* sur la croissance et la division du cilié *Colpidium campylum*. *Protistologica* **9**, 315–318.

DIVE, D., DUPONT, C. & LECLERC, H. 1974 Nutrition holozoique de *Colpidium campylum* aux de bactéries pigmentées on synthétisant des toxins. *Protistologica* **10**, 517–525.

DOYLE, W. L. & DOYLE, M. M. 1940 The structure of zooxanthellae. *Carnegie Institute of Washington Publications* **517**, 129–142.

DROOP, M. R. 1963 Algae and invertebrates in symbiosis. In *Symbiotic Associations*, eds Nutman, P. S. & Mosse, B. 13th Symposium of the Society for General Microbiology. Cambridge: Cambridge University Press.

DROŻAŃSKI, W. 1956 Fatal bacterial infection in soil amoebae. *Acta microbiologica polonica* **5**, 315–317.

DROŻAŃSKI, W. 1963*a* Studies of intracellular parasites of free-living amoebae. *Acta microbiologica polonica* **12**, 3–8.

DROŻAŃSKI, W. 1963*b* Observations on intracellular infection of amoebae by bacteria. *Acta microbiologica polonica* **12**, 9–24.

DROŻAŃSKI, W. 1965 Fatal bacterial infection of small free-living amoebae and related organisms. In *Progress in Protozoology*, Abstracts of the 2nd International Conference on Protozoology, London, 1965.

DUDDINGTON, C. L. 1957 The predaceous fungi and their place in microbial ecology. In *Microbial Ecology*, eds Williams, R. E. O. & Spicer, C. C. 7th Symposium of the Society for General Microbiology. Cambridge: Cambridge University Press.

ELLIOTT, A. M. & CLEMMONS, G. L. 1966 An ultra-structural study of ingestion and digestion in *Tetrahymena pyriformis*. *Journal of Protozoology* **13**, 311–323.

ERIKSON, D. 1941 Studies on some lake-mud strains of micromonospora. *Journal of Bacteriology* **41**, 277–300.

FAURÉ-FREMIET, E. 1943 Commensalisme et adaptation chez un Acinétien; *Erastophrya chattoni* n.g. n.sp. *Bulletin de la Société zoologique de France* **68**, 145–147.

FAURÉ-FREMIET, E. 1945 *Podophrya parasitica* nov. sp. *Bulletin biologique de la France et de la Belgique* **79**, 85–97.

FAURÉ-FREMIET, E. 1948 Le rhythme de marée du *Strombidium oculatum* Gruber. *Bulletin biologique de la France et de la Belgique* **82**, 3–23.

FAURÉ-FREMIET, E. 1950*a* Caulobacteriés epizoiques associées aux *Centrophorella* (Ciliés, holotriches). *Bulletin de la Société zoologique de France* **75**, 134–137.

FAURÉ-FREMIET, E. 1950*b* Ecologie des ciliés psammophiles littoraux. *Bulletin biologique de la France et de la Belgique* **84**, 35–75.

FAURÉ-FREMIET, E. 1951 The marine sand-dwelling ciliates of Cape Cod. *Biological Bulletin of the Marine Biological Laboratory, Woods Hole* **100**, 59–70.

FAURÉ-FREMIET, E. 1952 Symbiontes bactériens des ciliés du genre *Euplotes. Compte rendu hebdomadaire de séances de l'Académie des sciences* **235**, 402–403.

FAURÉ-FREMIET, E. 1958 Le cilié *Condylostoma tenuis* n.sp. et son algue symbiote. *Hydrobiologia* **10**, 43–48.

FLANDERS, S. E. 1948 A host parasite community to demonstrate balance. *Ecology* **29**, 123.

FOX, K. 1946 A possible endo-parasite of *Amoeba proteus. Micro Notes* **2**, 3–6.

FREUDENTHAL, H. D., LEE, J. J. & KOSSOY, V. 1964 Cytochemical studies of the zooxanthella from the planktonic foraminifera *Globigerinoides rubra. Journal of Protozoology* **11** Suppl., 12.

GAUSE, G. F. 1934*a The Struggle for Existence*. Baltimore: Williams & Wilkins Co.

GAUSE, G. F. 1934*b* Experimental analysis of Vito Volterra's mathematical theory of the struggle for existence *Science, New York* **79**, 16–17.

GAUSE, G. F. 1935 Experimentalle Untersuchungen über die Konkurrenz zwischen *Paramecium caudatum* und *Paramecium aurelia. Archiv für Protistenkunde* **84**, 207–224.

GAUSE, G. F., NASTUKOVA, O. K. & ALPATOV, W. W. 1934 The influence of biologically conditioned media on the growth of a mixed population of *Paramecium caudatum* and *P. aurelia. Journal of Animal Ecology* **34**, 222–230.

GAUSE, G. F., SMARAGDOVA, N. P. & WITT, A. A. 1936 Further studies of interaction between predator and prey. *Journal of Animal Ecology* **5**, 1–18.

GEITLER, L. 1948 Symbiosen zwischen Chrysomonaden und knospenden bakterienartigen Organismen sowie Beobachtungen über Organisationseigentüm-lichtkeiten der Chrysomonaden. *Osterreichische Botanische Zeitscrift, Wien* **95**, 300–324.

GEITLER, L. 1959 Syncyanosen. In *Encyclopedia of Plant Physiology*, Vol. 11, ed. Ruhland, W. Berlin: Springer Verlag.

GILLIES, C. & HANSON, E. D. 1962 A flagellate parasitizing the ciliate macronucleus. *Journal of Protozoology* **9** Suppl., 15.

GILLIES, C. & HANSON, E. D. 1963 A new species of *Leptomonas* parasitizing the macronucleus of *Paramecium trichium. Journal of Protozoology* **10**, 467–473.

GOULDER, R. 1972 Grazing by the ciliated protozoan *Loxodes magnus* on the alga *Scenedesmus* in a eutrophic pond. *Oikos* **23**, 109–115.

GOULDER, R. 1973 Observations over 24 hours on the quantity of algae inside grazing ciliated protozoa. *Oecologia* **13**, 177–182.

GRASSÉ, P. P. 1952 *Traité de Zoologie*, Vol. 1. Parts 1, 2. Paris: Masson et Cie.

GRITNER, I. 1951 Zur Nahrungs auswahl der Infusorien, untersucht an *Paramecium caudatum* und *Stentor coeruleus* Ehrenberg. *Biologisches Zentralblatt* **70**, 128–151.

GROSCOP, J. A. 1963 A comparative study of the effects of selected strains of pigmented microorganisms on five species of small soil amoebae. M.S. Thesis, Bowling Green State University, Ohio.

GROSCOP, J. A. & BRENT, M. M. 1964 The effects of selected strains of pigmented microorganisms on small free-living amoebae. *Canadian Journal of Microbiology* **10**, 579–584.

HAGEMAN, R. 1964 Plasmatische Vererbung. In *Genetik, Grundlagen, Ergebnisse und Probleme in Einzeldarstellungen*, No. 4. Jena: Gustav Fischer Verlag.

HAIRSTON, N. G. 1958 Observations on the ecology of *Paramecium*, with comments on the species problem. *Evolution, Lancaster, Pa.* **12**, 440–450.

HAIRSTON, N. G. & KELLERMAN, S. L. 1965 Competition between varieties 2 and 3 of *P. aurelia.* The influence of temperature in a food-limited system. *Ecology* **46**, 134–139.

HALL, W. T. & CLAUS, G. 1963 Ultrastructural studies on the blue-green algal symbiont in *Cyanophora paradoxa* Korschikoff. *Journal of Cell Biology* **19**, 551–563.

HAMBURGER, B. 1958 Bakteriensymbiose bei *Volvox aureus* Ehrenberg. *Archiv für Mikrobiologie* **29**, 291–300.

HAMILTON, R. D. & PRESLAN, J. E. 1969 Cultural characteristics of a pelagic marine hymenostome ciliate, *Uronema* sp. *Journal of Experimental Marine Biology and Ecology* **4**, 90–99.

HAMILTON, R. D. & PRESLAN, J. E. 1970 Observations on the continuous culture of a planktonic phagotrophic protozoan. *Journal of Experimental Marine Biology and Ecology* **5**, 94–104.

HANSON, E. D. 1954 Studies on kappa-like particles in sensitives of *Paramecium aurelia*, variety 4. *Genetics, Princeton, New Jersey* **39**, 229–239.

HEAL, O. W. 1963 Soil fungi as food for amoebae. In *Soil Organisms*, eds Doeksen, S. & van der Drift, J. Amsterdam: North-Holland Publishing Co.

HEAL, O. W. 1967 Quantitative feeding studies on soil amoebae. In *Progress in Soil Biology*, eds Graff, O. & Satchell, J. E. Amsterdam: North-Holland Publishing Co.

HEAL, O. W. & FELTON, M. J. 1970 Soil amoebae: their food and their reaction to microflora exudates. In *Animal Populations in Relation to their Food Resources*, ed. Watson, A. Symposium of the British Ecological Society. Oxford: Blackwell Scientific Publications.

HECKMANN, K. 1975 Omikron, ein essentieller Endosymbiont von *Euplotes aediculatus. Journal of Protozoology* **22**, 97–104.

HELLERBUST, J. A. 1967 Excretion of organic compounds by cultured and natural populations of marine phytoplankton. In *Estuaries*, ed. Lauff, G. H. Washington: American Association for the Advancement of Science.

HERVEY, R. J. & GREAVES, J. E. 1941 Nitrogen fixation by *Azotobacter chroococcum* in the presence of soil protozoa. *Soil Science* **51**, 85–100.

HETHERINGTON, A. 1932 The constant culture of *Stentor coeruleus. Archiv für Protistenkunde* **76**, 118–129.

HETHERINGTON, A. 1933 The culture of some holotrichous ciliates. *Archiv für Protistenkunde* **80**, 255–280.

HOLLANDE, A. 1945 Biologie et reproduction de Rhizopodes des genres *Pelomyxa* et *Amoeba* et cycle évolutif de l'*Amoebophilus destructor* nov. gen., nov. sp., Chrytidinée (Sic., Chytridinée) parasite de *Pelomyxa palustris* Greeff. *Bulletin biologique de la France et de la Belgique* 79, 31–66.

HOLLANDE, A. & ENJUMET, M. 1953 Contribution à l'étude biologique des Sphaerocollides (Radiolaires collodaires et Radiolaires polycyltaires) et de leures parasites. Partie I: Thalassicollidae, Physematidae, Thalassophysidae. *Annales des sciences naturelles Zoologie Sér II* **15**, 99–183.

HOLLANDE, A. & GUILCHER, Y 1945 Les amibes du genre *Pelomyxa*: Ethologie, structure cycle évolutif parasites. *Bulletin de la Société zoologique de France* **70**, 53–56.

HOVASSE, R. & BROWN, E. M. 1953 Contribution à la connaissance des Radiolaires et de leurs parasites syndinens. *Annales de sciences naturelles Zoologie Sér 11* **15**, 405–438.

HUTNER, S. H. & PROVASOLI, L. 1951 The phyto-flagellates. In *Biochemistry and Physiology of Protozoa,* Vol. 1, ed. Lwoff, A. New York & London: Academic Press.

JANKOWSKI, A. W. 1963 Pathology of Ciliophora. II. Life cycles of suctorian parasites in *Urostyla* and *Paramecium* (in Russian). *Tsitologiya i Genetika, Academiya Nauk Ukrainskoi S.S.R., Kiev.* 5, 428–439.

JAVORNICKÝ, P. & PROKEŠOVÁ, V. 1963 The influence of protozoa and bacteria upon the oxidation of organic substances in water. *Intenationale Revue der Gesamten Hydrobiologie* **48**, 335–350.

JOHANNES, R. E. 1965 Influence of marine protozoa on nutrient regeneration. *Limnology and Oceanography* **10**, 434–442.

JOST, J. L., DRAKE, J. F., FREDRICKSON, A. G. & TSUCHIYA, H. M. 1973*a* Interactions of *Tetrahymena pyriformis, Escherichia coli, Azotobacter vinelandii* and glucose in a minimal medium. *Journal of Bacteriology* **113**, 834–840.

JOST, J. L., DRAKE, J. F., TSUCHIYA, H. M. & FREDRICKSON, A. G. 1973*b* Microbial food chains and food webs. *Journal of Theoretical Biology* **41**, 461–484.

KAHL, A. 1928 Die Infusorien (Ciliata) der Oldesloer Salzwasser stellen. *Archiv für Hydrobiologie* **19**, 50–123.

KAHL, A. 1932 Urtiere oder Protozoa. I: Wimpertiere oder Ciliata (Infusoria), eine Bearbietung der freilebenden und ectocommensalen Infusorien der Erde, unter Ausschluss der marinen Tintinnidae. In *Die Tierwelt Deutschlands*, Teil 18. Jena: G. Fischer.

KAHL, A. 1933 Ciliata libera et ectocommensalia. In *Die Tierweld der Nord und Ostsee,* Leif 23, Teil II, C3, eds Grimpe, G. & Wagler, E.

KARAKASHIAN, S. J. 1963 Growth of *Paramecium bursaria* as influenced by the presence of algal symbionts. *Physiological Zoology* **36**, 52–68.

KARAKASHIAN, S. J. 1975 Symbiosis in *Paramecium bursaria*. In *Symbiosis*, eds Jennings, D. H. & Lee, D. L. 29th Symposium of the Society for Experimental Biology. Cambridge: Cambridge University Press.

KELLER, H. 1949 Untersuchungen über die intrazellulären Bakterien von *Pelomyxa palustris* Greeff. *Zeitschrift für Naturforschung, Wiesbaden* **4B**, 293–297.

KIDDER, G. W. 1941 Growth studies on ciliates. The acceleration and inhibition of ciliates grown in biochemically conditioned medium. *Physiological Zoology* **14**, 209–226.

KIDDER, G. W. & STUART, C. A. 1939 Growth studies on ciliates. I. The role of bacteria in the growth and reproduction of *Colpoda. Physiological Zoology* **12**, 329–340.

KIRBY, H. 1941 Organisms living on and in Protozoa. In *Protozoa in Biological Research*, eds Calkins, G. N. & Summers, F. M. New York: Columbia University Press.

KIRBY, H. 1942 A parasite of the macronucleus of *Vorticella. Journal of Parasitology* **28**, 311–314.

KNIGHT-JONES, E. W. & STEVENSON, J. P. 1950 Gregariousness during settlement in the barnacle *Elminius modestus* Darwin. *Journal of the Marine Biological Association of the United Kingdom* **29**, 281–297.

KNORR, M. 1960 Verusche über die biologische Sperre gegen Bakterien und Viren bei vertikaler Bodeninfiltration. *Schweizerische Zeitschrift für Hydrologie, Basel* **22**, 493–502.

KONO, T. & ASAI, T. 1969 Kinetics of fermentation processes. *Biotechnology and Bioengineering* **11**, 293–321.

KORSCHIKOFF, A. A. 1924 Protistologische Beobachtung I. *Cyanophora paradoxa. Russkiĭ archiv protistologii, Moskva* **3**, 57–74.

KUNG, C. 1970 The electron transport system of kappa particles from *Paramecium aurelia* stock 51. *Journal of General Microbiology* **61**, 371–378.

KUNG, C. 1971 Aerobic respiration of kappa particles from *Paramecium aurelia. Journal of Protozoology* **18**, 328–332.

LAIRD, M. 1953 The Protozoa of New Zealand intertidal zone fishes. *Transactions of the Royal Society of New Zealand* **81**, 79–143.

LANGLOIS, G. A. 1975 Effect of algal exudates on substratum selection by motile telotrochs of the marine peritrich ciliate *Vorticella marina. Journal of Protozoology* **22**, 115–123.

LE CALVEZ, J. 1939 *Trophosphaera planorbulinae* n.gen., n.sp. protiste parasite due Foraminifére *Planorbulina mediterranensis* d' Orb. *Archives de zoologie expérimentale et générale* **80**, 425–443.

LE CALVEZ, J. 1940 Une amibe, *Vahlkampfia discorbini*, n.sp. parasite du Foraminifère *Discorbis mediterranensis* (d' Orbigny). *Archives de zoologie expérimentale et générale* **81**, 123–129.

LE CALVEZ, J. 1947 *Entosolenia marginata*, Foraminifère *Discorbis vilardeboanus. Compte rendu hebdomadaire de séances de l'Academie des sciences* **224**, 1448–1450.

LEE, J. J., MCENERY, M. PIERCE, S., FREUDENTHAL, H. D. & MULLER, W. A. 1966 Tracer experiments in feeding littoral foraminifera. *Journal of Protozoology* **13**, 659–670.

LEINER, M. & WOHLFEIL, M. 1954 Das symbiontische Bakterium in *Pelomyxa palustris* Greeff. III. *Zeitschrift für Morphologie und Okologie der Tiere, Berlin* **42**, 529–549.

LEINER, M., WOHLFEIL, M. & SCHMIDT, D. 1951 Das symbiontische Bakterium in *Pelomyxa palustris* Greeff. I. *Zeitschrift für Naturforschung, Wiesbaden* **6**B, 158–170.

LEINER, M., WOHLFEIL, M. & SCHMIDT, D. 1954 *Pelomyxa palustris* Greeff. *Annales des sciences naturelles* **16**, 537–594.

LEVINE, M. 1953 The diverse mate-killers of *Paramecium aurelia*, variety 8; their interrelations and genetic basis. *Genetics, Princeton* **38**, 561–578.

LILLY, D. M. & STILLWELL, R. 1965 Probiotics: growth-promoting factors produced by microorganisms. *Science, New York* **147**, 747–748.

LÓPEZ-OCHOTERENA, E. 1962 Protozoairios ciliados de México. I. *Stylonychia mytilus* Ehrenberg, 1838 y *Sphaerophrya sol* Metchinikoff, 1864. Un caso de parasitismo entre protozoairos. *Acta zoologica mexicana* **6**, 1–6.

LOSINA-LOSINSKY, L. K. 1931 Zur Ernährungsphysiologie der Infusorien; Untersuchungen über Nahrungsauswahl und Vermehrung bei *Paramecium caudatum*. *Archiv für Protistenkunde* **74**, 18–120.

LOTKA, A. J. 1925 *Elements of Physical Biology*. Baltimore: Williams & Wilkins Co.

LOTKA, A. J. 1932 The growth of mixed populations: two species competing for a common food supply. *Journal of the Washington Academy of Science* **22**, 461–469.

LUCKINBILL, L. S. 1973 Coexistence in laboratory populations of *Paramecium aurelia* and its predator *Didinium nasutum*. *Ecology* **54**, 1320–1327.

LUND, J. W. G. 1953 New or rare British Chrysophyceae II. *Hyalobryon polymorphum* n.sp. and *Chrysonebula holmesii* sp. *New Phytologist* **52**, 114–123.

MCLAUGHLIN, J. A. & ZAHL, P. 1966 Endozoic algae. In *Symbiosis*, Vol. 1, ed. Henry, S. New York & London: Academic Press.

MAST, S. O. 1947 The food vacuole in *Paramecium*. *Biological Bulletin of the Marine Biological Laboratory, Woods Hole* **92**, 31–72.

MAST, E. C. & PACE, D. M. 1938 The effect of substances produced by *Chilomonas paramecium* on the rate of reproduction. *Physiological Zoology* **11**, 359–382.

MEERS, J. L. 1971 Effect of dilution rate on the outcome of chemostat mixed culture experiments. *Journal of General Microbiology* **67**, 359–361.

MOIKEHA, S. N. & CHU, G. W. 1971 Dermatitis – producing alga *Lyngbya majuscula* Gomont in Hawaii. H. Biological properties of the toxic factor. *Journal of Phycology* **7**, 8–13.

MONOD, J. 1942 *Recherches sur la Croissance des Cultures Bactériennes*. Paris: Hermann & Cie.

MOSER, H. 1958 The dynamics of bacterial populations maintained in the chemostat. *Carnegie Institute of Washington Publications* **614**, 1–136.

MUČIBABIĆ, S. 1957 The growth of mixed populations of *Chilomonas paramecium* and *Tetrahymena pyriformis*. *Journal of General Microbiology* **16**, 561–571.

MUELLER, M., RÖHLICH, P. & TÖRÖ, I. 1965 Studies on feeding and digestion in protozoa VII. Ingestion of polystyrene latex particles and its early effect on acid phosphatase in *Paramecium multi micronucleatum* and *Tetrahymena pyriformis*. *Journal of Protozoology* **12**, 27–34.

MULLER, W. A. & LEE, J. J. 1969 Apparent indispensability of bacteria in foraminiferan nutrition. *Journal of Protozoology* **16**, 471–495.

NASIR, S. M. 1923 Some preliminary investigations on the relationship of protozoa to soil fertility with special reference to nitrogen fixation. *Annals of Applied Biology* **10**, 122–133.

NICOL, J. A. C. 1960 *The Biology of Marine Animals*. New York & Chichester: John Wiley & Sons, pp. 610–623.

NEIBOER, H. J. 1959 *Volvox aureus*, historisch overzicht, kweken, zwemmen, orientatie. *Levende Natur* **62**, 65–71, 88–94.

NIKOLJUK, V. F. 1969 Some aspects of the study of soil protozoa. *Acta protozoologica* **7**, 99–109.

NILSSON, J. R. 1972 Further studies on vacuole formation in *Tetrahymena pyriformis* G.L. *Compte rendu des travaux du Laboratoire de Carlsberg, Copenhague* **39**, 83–110.

PADNOS, M. & NIGRELLI, R. F. 1947 *Endosphaera engelmanni* endoparasitic in *Trichodina spheroidesi* infecting the puffer, *Sphaeroides maculatus. Zoologica, New York* **32**, 169–172.

PADO, R. 1965 Mutual relation of protozoans and symbiotic algae in *Paramecium bursaria*. I. The influence of light on the growth of symbionts. *Folia biologica, Praha* **13**, 173–182.

PADO, R. 1967 Mutual relation of protozoans and symbiotic algae in *Paramecium bursaria*. II. Photosynthesis. *Acta Societatis botanicorum Poloniae* **36**, 97–108.

PARKER, C. A. 1954 Effect of oxygen on the fixation of nitrogen by *Azotobacter. Nature, London* **173**, 780–781.

PASCHER, A. 1929*a* Über einige Endosymbiosen von Blaualgen in Einzellern. *Jahrbuch für wissenschaftliche Botanik, Berlin* **71**, 386–403.

PASCHER, A. 1929*b* Über die Natur der Blaugrünen Chromatophoren des Rhizopoden *Paulinella chromatophora. Zoologischer Anzeiger, Leipzig* **81**, 189–194.

PASCHER, A. 1930 Eine neue stigmatisierte und phototaktische Amoebe. *Biologisches Zentralblatt* **50**, 1–7.

PENARD, E. 1940 Protozooaires et psychologie. *Archives des sciences physiques et naturelles, Genève* **22**, 160–175, 179–200, 203–226, 265–289.

PÉREZ REYES, R. & LÓPEZ-OCHOTERENA, E. 1963 *Sphaerophrya sol* (Ciliata: Suctoria) parasitic in some Mexican ciliates. *Journal of Parasitology* **49**, 697.

PETTY, M. A. & MATRISHIN, M. 1949 Society of American Bacteriologists 49th Gen Meeting, p. 470.

PLANT, W. & SAGAN, L. A. 1958 Incorporation of thymidine in the cytoplasm of *Amoeba proteus. Journal of Biophysical and Biochemical Cytology* **4**, 843–844.

POSTGATE, J. 1971 Relevant aspects of the physiological chemistry of nitrogen fixation. In *Microbes and Biological Productivity*, eds Hughes, D. E. & Rose, A. H. Cambridge: Cambridge University Press.

POWELL, E. O. 1958 Criteria for the growth of contaminants and mutants in continuous culture. *Journal of General Microbiology* **18**, 259–268.

PREER, J. R. 1948 A study of some properties of the cytoplasmic factor "kappa" in *Paramecium aurelia*, variety 2. *Genetics, Princeton* **33**, 349–404.

PREER, J. R. 1950 Microscopically visible bodies in the cytoplasm of the "killer" strains of *Paramecium aurelia. Genetics, Princeton* **35**, 344–362.

PREER, J. R. 1975 The hereditary symbionts of *Paramecium aurelia*. In *Symbiosis*, eds Jennings, D. H. & Lee, D. L. 29th Symposium of the Society for Experimental Biology. Cambridge: Cambridge University Press.

PREER, J. R., PREER, L. B. & JURAND, A. 1974 Kappa and other endosymbionts in *Paramecium aurelia. Bacteriological Reviews* **38**, 113–163.

PROPER, G. & GARVER, J. C. 1966 Mass culture of the protozoa *Colpoda steinii. Biotechnology and Bioengineering* 7, 287–296.

RABINOVITCH, M. & PLANT, W. 1962 Cytoplasmic DNA synthesis in *Amoeba proteus.* II. On the behaviour and possible nature of the DNA – containing elements. *Journal of Cell Biology* **15**, 535–540.

RAPPORT, D. J. & TURNER, J. E. 1970 Determination of predator food preferences. *Journal of Theoretical Biology* **26**, 365–372.

RAPPORT, D. J., BERGER, J. & REID, D. B. W. 1972 Determination of food preference of *Stentor coeruleus. Biological Bulletin of the Marine Biology Laboratory, Woods Hole* **142**, 103–109.

RASMUSSEN, L. & KLUDT, T. A. 1970 Particulate material as a prerequisite for rapid cell multiplication in *Tetrahymena pyriformis. Experimental Cell Research* **59**, 457–463.

RICKETTS, T. R. 1971 Endocytosis in *Tetrahymena pyriformis.* The selectivity of uptake of particles and the adaptive increase in cellular acid phosphatase activity. *Experimental Cell Research* **66**, 49–58.

RICKETTS, T. R. 1972*a* The induction of endocytosis in starved *Tetrahymena pyriformis. Journal of Protozoology* **19**, 373–375.

RICKETTS, T. R. 1972*b* The interaction of particulate material and dissolved foodstuffs in food uptake by *Tetrahymena pyriformis*. *Archiv für Mikrobiologie* **81**, 344–349.
ROTH, L. E. & DANIELS, E. W. 1961 Infective organisms in the cytoplasm of *Amoeba proteus*. *Journal of Biophysical and Biochemical Cytology* **9**, 317–323.
SANDON, H. 1932 *The food of protozoa*. Egyptian University Faculty of Science Publ. 1. Cairo: Misr-Sokkar Press.
SCHAEFFER, A. A. 1910 Selection of food in *Stentor coeruleus* (Ehr.). *Journal of Experimental Zoology* **8**, 75–132.
SCHNELLER, M. V. 1958 A new type of killing action in a stock of *Paramecium aurelia* from Panama. *Proceedings of the Indiana Academy of Science* **67**, 302–303.
SCHÖNFELD, C. 1959 Über das parasitische Verhalten einer *Astasia* – Art in *Stentor coeruleus*. *Archiv für Protistenkunde* **104**, 261–264.
SETO, M. & TAZAKI, T. 1971 Carbon dynamics in the food chain system of glucose – *Escherichia coli* – *Tetrahymena vorax*. *Japanese Journal of Ecology* **21**, 179–188.
SIEGEL, R. W. 1953 A genetic analysis of the mate-killer trait in *Paramecium aurelia*, variety 8. *Genetics, Princeton* **38**, 550–560.
SIEGEL, R. W. 1960 Hereditary endosymbiosis in *Paramecium bursaria*. *Experimental Cell Research* **19**, 239–252.
SINGH, B. N. 1942 Selectivity of bacterial food by soil flagellates and amoebae. *Annals of Applied Biology* **29**, 18–22.
SINGH, B. N. 1945 The selection of bacterial food by soil amoebae, and the toxic effects of bacterial pigments and other products on soil protozoa. *British Journal of Experimental Pathology* **26**, 316–325.
SINGH, B. N. 1946 A method of estimating the number of soil protozoa especially amoebae, based on their differential feeding on bacteria. *Annals of Applied Biology* **33**, 112–119.
SKUJA, H. 1958 Eine neue vorwiegend sessil oder rhizopodial auftrende synbacteriotische Polytomee aus einem Schwefelgewässer. *Svensk botanisk tidskrift* **52**, 379–390.
SMITH, D., MUSCATINE, L. & LEWIS, D. 1969 Carbohydrate movement from autotrophs to heterotrophs in parasitic and mutualistic symbiosis. *Biological Reviews* **44**, 17–90.
SOLDO, A. T. 1960 Cultivation of two strains of *Paramecium aurelia* in axenic medium. *Proceedings of the Society for Experimental Biology and Medicine* **105**, 612–615.
SOLDO, A. T. 1961 The use of particle-bearing *Paramecium* in screening for potential antitumour agents. *Transactions of the New York Academy of Science* **23**, 653–661.
SONNEBORN, T. M. 1943 Gene and cytoplasm I. The determination and inheritance of the killer character in variety 4 of *Paramecium aurelia*. *Proceedings of the National Academy of Science of the United States of America* **29**, 329–338.
SONNEBORN, T. M. 1945*a* The dependence of the physiological action of a gene on a primer and the relation of primer to gene. *American Naturalist* **79**, 318–319.
SONNEBORN, T. M. 1945*b* Gene action in *Paramecium*. *Annals of Missouri Botanical Garden, St. Louis* **32**, 213–221.
SONNEBORN, T. M. 1946 Experimental control of the concentration of cytoplasmic genetic factors in *Paramecium*. *Cold Spring Harbor Symposium on Quantitative Biology* **11**, 236–255.
SONNEBORN, T. M. 1947 Recent advances in the genetics of *Paramecium* and *Euplotes*. *Advances in Genetics* **1**, 263–358.
SONNEBORN, T. M. 1959 Kappa and related particles in *Paramecium*. *Advances in Virus Research* **6**, 229–356.
SONNEBORN, T. M. 1961 Kappa particles and their bearing on host-parasite relations. *Perspective in Virology* **2**, 5–12.
SONNEBORN, T. M. 1965 The metagon: RNA and cytoplasmic inheritance. *American Naturalist* **99**, 279–307.
SPARROW, F. K. 1960 *Aquatic Phycomycetes*. Ann Arbour: University of Michigan Press.
STARR, M. P. 1975 A generalized scheme for classifying organismic associations. In *Symbiosis*, eds Jennings, D. H. & Lee, D. L. 29th Symposium of the Society for Experimental Biology. Cambridge: Cambridge University Press.

STEWART, W. D. P. 1963 Liberation of extracellular nitrogen by two nitrogen-fixing blue-green algae. *Nature, London* **200**, 1020–1021.
STEVENSON, I. 1967 Diaminopimelic acid in the mu particles of *Paramecium aurelia. Nature, London* **215**, 434–435.
STILLWELL, R. 1967 *Colpidium* – produced RNA as a growth stimulant for *Tetrahymena. Journal of Protozoology* **14**, 19–22.
STOKSTAD, E. L. R., JUKES, T. H., PIERCE, J., PAGE, A. C. & FRANKLIN, A. L. 1949 The multiple nature of the animal protein factor. *Journal of Biological Chemistry* **180**, 647–654.
STUMM-ZOLLINGER, E. 1966 Effects of inhibition and repression on the utilization of substrates by heterogeneous bacterial communities. *Applied Microbiology* **14**, 654–664.
SUD, G. C. 1968 Volumetric relationships of symbiotic zoochlorellae to their hosts. *Journal of Protozoology* **15**, 605–607.
TARTAR, V. 1961 The biology of *Stentor.* New York & London: Pergamon Press.
TAYLOR, D. L. 1973 Algal symbionts of invertebrates. *Annual Review of Microbiology* **27**, 171–187.
TAYLOR, W. D. & BERGER, J. 1977 Growth responses of cohabiting ciliate protozoa to various prey bacteria. *Canadian Journal of Zoology* (in press).
TEZUKA, Y. 1974 An experimental study on the food chain among bacteria, *Paramecium* and *Daphnia. Internationale Revue der gesamten Hydrobiologie und Hydrographie, Leipzig* **59**, 31–37.
TRAGER, W. 1960 Intracellular parasitism and symbiosis. In *The Cell*, Vol. 4, eds Brachet, J. & Mirsky, A. E. New York & London: Academic Press.
TSCHERMAK-WOESS, E. 1950 Über eine Synbacteriose und ähnliche Symbiosen. *Österreiche Botanische Zeitschrift, Wien* **97**, 188–206.
TSUCHIYA, H. M., DRAKE, J. F., JOST, J. L. & FREDRICKSON, A. G. 1972. Predator-prey interactions of *Dictyostelium discoideum* and *Escherichia coli* in continuous culture. *Journal of Bacteriology* **110**, 1147–1153.
VAN DEN ENDE, P. 1973 Predator-prey interactions in continuous culture. *Science, New York* **181**, 562–564.
VELA, G. R. & GUERRA, C. N. 1966 On the nature of mixed cultures of *Chlorella pyrenoidosa* TX71105 and various bacteria. *Journal of General Microbiology* **42**, 123–131.
VOLTERRA, V. 1926 Variazoni e fluttuazioni dei numeri d'individui in specie animali conviventi. *Atti dell' Academia nazionale dei Lincei, Memorie, Roma* **2**, 31.
WAGTENDONK, W. J. VAN, CLARK, J. D. & GODOY, G. A. 1963 The biological status of lambda and related particles in *Paramecium aurelia. Proceedings of the National Academy of Sciences of the United States of America* **50**, 835–838.
WAGTENDONK, W. J. VAN, & SOLDO, A. T. 1965 Endosymbionts of ciliated protozoa. In *Progress in Protozoology*, 2nd International Conference on Protozoology, London, 1965.
WICHTERMAN, R. 1940 Parasitism in *Paramecium caudatum. Journal of Parasitology* **26** Suppl., 29.
WICHTERMAN, R. 1942 A ciliate from a coral of Tortugas and its symbiotic algae. *Papers from the Tortugas Laboratory, Carnegie Institution, Washington, D.C.* **33**, 107–111.
WICHTERMAN, R. 1945 Schizomycetes parasitic in *Paramecium bursaria. Journal of Parsitology* **31** Suppl., 25.
WICHTERMAN, R. 1953 *The biology of* Paramecium. New York: Blakiston.
WICHTERMAN, R. 1954 The common occurrence of micronuclear variation during binary fission in an unusual race of *Paramecium caudatum. Journal of Protozoology* **1**, 54–59.
WILLIAMS, F. M. 1971 Dynamics of microbial populations. In *Systems Analysis and Simulation Ecology*, ed. Patten, B. C. New York & London: Academic Press.

WINTER, F. W. 1907 Zur Kentniss der Thalamophoren. *Archiv für Protistenkunde* **10,** 1–113.

WOLSKA, J. 1949 The small amoebas in the plasm of *Amoeba proteus* Pall. *Annales Universitatis Mariae Curie-Sklodowska, Lublin* **4C**, 137–147.

New Aquatic Budding and Prosthecate Bacteria and Their Taxonomic Position

P. HIRSCH, M. MÜLLER AND H. SCHLESNER

Institut für Allgemeine Mikrobiologie der Universität, Kiel, Germany

CONTENTS

1. Introduction

ALTHOUGH the microbial composition of the aquatic environment has been studied for a long time (Lauterborn 1916; Koppe 1923; Utermöhl 1924; Benecke 1933; Henrici & Johnson 1935; Huber-Pestalozzi 1938; Hirsch & Rheinheimer 1968; Staley 1968; Caldwell & Tiedje 1975*a, b*; Kuznetsov 1975; Rheinheimer 1975), the development of new observation techniques and special enrichment conditions (La Riviere 1965; Pfennig 1969; Gorlenko 1970; Hirsch & Pankratz 1970; Hirsch 1973) has yielded many new bacterial types of great physiological diversity. Morphologically distinct, 'unusual' bacteria (which were easily recognizable) have also been described, (Utermöhl 1924; Henrici & Johnson 1935; Hirsch & Rheinheimer 1968; Staley 1968; Gorlenko 1970; Hajdu 1974) but unfortunately, most of them have never been cultured in the laboratory.

An early exception was the budding bacteria of the genus *Hyphomicrobium.* These could be isolated and cultured quite easily (Hirsch & Conti 1964*a, b*; Sperl & Hoare 1971; Attwood & Harder 1972) and their cells were soon found to respond dramatically to alterations in the environmental conditions (Hirsch & Conti 1964*a, b*): their morphology and cell size varied greatly. Their natural occurrence in different shapes and sizes has led to the notion that such 'pleomorphic' micro-organisms might be potential 'bioindicators' in aquatic ecosystems (Hirsch 1968).

Morphologically distinct and highly differentiated bacteria were found to be more common among micro-organisms inhabiting the water surface, i.e. the neuston layer, as compared to those of the lower parts of the water column. They were also found to be numerous among those bacteria growing attached to solid, submerged surfaces, and they were frequent in extreme aquatic environments (Schulz & Hirsch 1973). A reason for this could be the fact that these conditions induce greater polarity of the bacterial cell itself. Such environments, therefore, seemed to be ideal sources for the isolation of new and morphologically distinct forms.

After the development of special techniques for the descriptive observation of 'unusual' bacteria, attempts were made to cultivate and isolate such new forms. Generally, the most successful enrichment method consisted of *not* adding any nutrients to the natural water samples. Thus, direct incubation of surface water samples for long periods of time yielded several new and morphologically interesting types, while others developed better in samples with only traces of peptone, yeast extract or vitamins, or other organic compounds. The present communication is a first account of such enrichment and isolation procedures and of some of the interesting new bacteria thus isolated.

2. Materials and Methods

(a) *Sampling*

Water samples were taken directly into sterilized jars or by using conventional van Dohrn-type water samplers, and were transported immediately to the laboratory, usually with cooling of the samples. In some cases inoculation of enrichment cultures was made directly at the sampling site.

(b) *Observations*

Living specimens were studied with agar-coated glass slides and phase-contrast light microscopy. Preparations were made as follows: clean glass slides were coated thinly with molten 2% Bacto agar (Difco) cooled to *ca.* 60°C. After draining off the excess agar, the slides were dried dust-free and in a horizontal position. For microscopy, three differently sized, small droplets of the bacterial suspension were placed on the dry agar surface and a cover glass placed gently on each of these. Excess sample water was avoided or blotted off. The swelling agar moved the bacteria into a narrow layer adjacent to the cover slip and additionally surrounded each cell, thus providing optimal focussing conditions (N. Pfennig, pers. comm.).

(c) *Electron microscopy*

Specimens dried on formvar-coated, 300 mesh copper grids were Pt-C shadowed at an angle of *ca.* 12°. For negative staining, predried cells were treated with 0.5 or 1% phosphotungstic acid, pH 7.2, to which 0.05% of bovine serum albumen had been added. In some cases the cells were prefixed with buffered glutaraldehyde to give a final concentration of 2%.

(d) *Enrichments*

The simplest enrichment condition consisted of incubating at 20–25°C the 50–100 ml samples in sterile 300–500 ml Erlenmeyer flasks, covered with loose cotton plugs. After various periods of time the microbial population was monitored, and streaks were made on agar as soon as desired organisms appeared in relatively high numbers. This occurred sometimes only after 1 year. Some special enrichments were performed in a similar way, except that organic compounds were added to give a final concentration of 0.005–0.1% (w/v). Sterilization was mostly by autoclaving (sugars, amino acids) or filtration (vitamins, other labile compounds). Some sugars, sugar alcohols and chitin were sterilized by adding ethyl ether which was then allowed to evaporate, as proposed for the study of streptomycetes (Shirling & Gottlieb 1966). Incubation was at 20, 25, 30 or 43°C.

One enrichment technique was especially successful for cultivating the attaching bacteria: the bottom of a 20 cm diam. glass Petri dish was covered with a 2 cm layer of sterilized water agar before adding the water sample (Fig. 1). Sterilized cover glasses ± water agar coat were placed vertically and partially into the agar layer and monitored from time to time. As soon as the

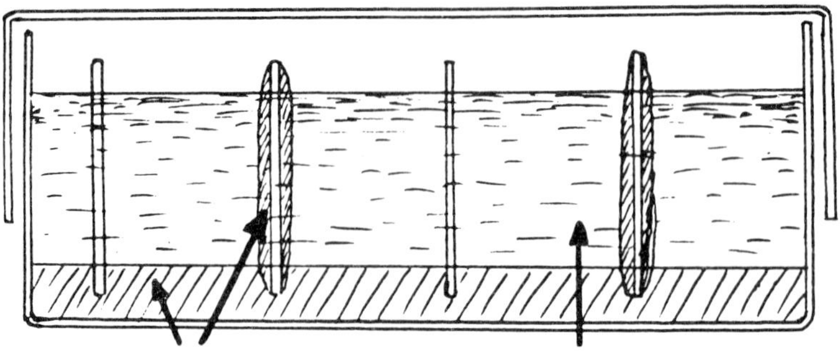

Fig. 1. Enrichment technique for attaching budding bacteria (the water agar – Petri dish method). Clean, agar-coated cover slips are placed vertically into the bottom layer of water agar and then the dish is filled with sample water to nearly cover the glass plates.

desired bacteria had appeared on such sample cover glasses, other parallel cover glasses were taken out for agar streak isolations. A high density of attached growth was observed on such submerged surfaces.

(e) *Isolations*

Water samples or enrichments were immediately streaked on agar media poor in organic nutrients, such as the sample water to which 1.5–1.8% of Bacto agar had been added, or on PYVG agar as suggested by Staley (1968) and Van Ert & Staley (1971) and modified by using 1.8% of Bacto agar. Moderate halophiles or halotolerant bacteria from the Kiel Bight of the Baltic Sea were grown in or on PYVG medium made up with 25% of artificial sea water. For isolations from a hyperthermal, hypersaline Solar Lake in Israel the same medium was employed, but it was prepared with aged, paper-filtered solar lake water (salinity 7.2%). Further cultures of organisms from this site were kept on media prepared with doubly concentrated artificial sea water. Plates were incubated at 20, 25, 30 or 43°C, as for the enrichment cultures. Light was usually excluded.

Colonies of morphologically interesting, 'unusual' bacteria were found by picking off, with sterile tooth picks, a large number of different colony types which were then inspected microscopically. With the same tooth pick inoculum, prior to microscopic observation, a 'master plate' was inoculated to detect possible further growth of a desired organism.

(f) *Identification procedures*

Generally, conventional techniques were employed (Skerman 1967; Norris & Robbins 1971).The carbon- or nitrogen-utilization spectrum was determined by culturing in medium 337 (Hirsch & Conti 1964*a*) or PYVG (van Ert & Staley, 1971; Staley 1968). Sugar compounds were added at a concentration of 0.2–1.0% (w/v). Requirements for Na^+, Cl^-, Ca^{2+}, Mg^{2+} or $HCO_3{}^{2-}$ were tested by substituting individual components of the artificial sea water in the medium by corresponding ions of equal valency. Thus, replacement pairs were Na^+ and K^+, or Ca^{2+} and Mg^{2+}. Chloride ions had to be replaced by $SO_4{}^{2-}$, and $HCO_3{}^{2-}$ was removed by allowing freshly heated medium to cool down in the complete absence of CO_2.

Growth and generation times were estimated from optical density measurements at 620–640 nm. CO tolerance was tested for by cultivating the bacteria on medium 337 or PYVG in a desiccator with an atmosphere of 30% CO, 15% O_2 and 0.5% CO_2, the remainder being nitrogen. pH optimum curves were obtained by buffering medium 337 with a 4% phosphate buffer (Umbreit *et al.* 1957). Temperature optimum curves were prepared by growing the test strains in triplicate glass tubes placed in an aluminium temperature gradient block (+13 to +50°C) kindly provided by D. Klein (Fort Collins, Colorado, USA).

(g) *DNA base ratio determinations*

DNA was extracted according to Marmur (1961) and the base ratio determinations (T_m) were made with an SP 1800 spectrophotometer (Pye Unicam Ltd, Cambridge) equipped with an SP 876 Temperature Programme Controller, according to the method of Moore & Hirsch (1973*b*). Organisms of group I.4 had to be treated with 25% SDS at 60°C for 30 min to release the DNA.

3. Description of Strains Isolated

Budding and prosthecate bacteria were seen in all samples from every site investigated. Successful enrichments and isolations were performed in several different ways (Table 1). The Petri dish method with a bottom layer of water agar was quite advantageous for the isolation of attaching bacteria. Generally, incubating water samples for extended periods of time, perhaps for a year or longer, resulted in growth of many unusual bacteria. A total 30 pure cultures was obtained by employing various methods.

These unusual bacteria could be grouped according to their formation of tapered prosthecae, of hyphae with branches, or of holdfast materials (Table 1). All strains discussed here are budding bacteria. Additionally, many truely stalked bacteria (*Caulobacter* spp. etc.) were isolated; these forms will be described elsewhere.

Table 1

Origins and methods of isolation of various types of newly recognized, budding and prosthecate bacteria

Group	Number of strains isolated	General properties: Budding	Prosthecae ± tapered	Hyphae ± branching	Excreted stalks are fibrous	Origin of inoculum	Isolation technique	Figure number	Probable genus or species name
I.1	3	+	–	–	±	1, 2, 4	A	2	*Blastobacter*
I.2	5	+	–	–	–[a]	5	B, C	3	*Pasteuria*
I.3	–	+	–	–	–	6	–	4	*Planctomyces guttaeformiss*
I.4	3	+	–	–	+	2, 5	B, C	5	–
I.5	7	+	–	–	–[a]	4–7	C	6a–c	–
I.6	–	+	–	–	+	2, 6	–	6d	*Planctomyces bekefii* Gim.
II.1	1	+[b]	+	–	–	5	C	7	*Stella* sp.
II.2	–	+[b]	+	–	+	3	–	8	–
II.3	3	+[b]	+	?	–	5	C, D	9	*Prosthecomicrobium*
III.1	2	+	–	+	–	5	B, D	11	*Hyphomicrobium*, *Hyphomonas*
III.2	1	+	–	+	–	5	D	12	*Hyphomicrobium*
III.3	4	+	–	+	–	8	A, D	13	
III.4	1	+	–	+	–	5, 6	C	14	–

1, Höftsee; 2, Plußee; 3, Blunkersee; 4, Pond near Westensee; 5, Kiel Bight of Baltic Sea (Germany); 6, Wintergreen Lake, Michigan; 7, Forest Pond, Augusta (Michigan, USA); 8, Solar Lake near Eilat (Israel).

A, surface water samples directly streaked out; B, Petri dish enrichment with water agar as shown in Fig. 1; C, samples fortified with 0.0025–0.1% of an organic compound; D, samples incubated without addition.

a, a 'holdfast' polymer is formed; b, formation of tapered prosthecae is a budding process (Hirsch 1974).

Many of the new pure cultures could not be identified with any bacteria known or described in *Bergey's Manual of Determinative Bacteriology* (Buchanan & Gibbons 1974). For these organisms new names will have to be proposed and published together with their complete descriptions.

In the following sections we present some preliminary descriptions of our pure cultures as well as of forms frequently encountered in natural water samples.

(a) *Group I.1* (Blastobacter; *strains Mü-161A, -222, -216*)

Members of this group are rod-shaped bacteria which form buds on one of their cell poles. This reproductive pole is sometimes slightly thinner than the other (Fig. 2). Two strains were isolated from surface waters of small Holstein lakes (Mü-161A, Höftsee; Mü-222, Plußsee). One strain (Mü-216) came from a small forest pond near Westensee (Kiel). The isolation technique was to streak samples directly on dilute agar media or to concentrate these by centrifugation before streaking them out.

The three strains differ only slightly from each other. One (Mü-216) is non-motile, while the other two form swarmer cells. Colonies on PYVG agar are colourless to whitish. The bacteria of this group are aerobic chemoheterotrophs; they show a broad pH optimum curve around pH 7.0. Good growth occurs on

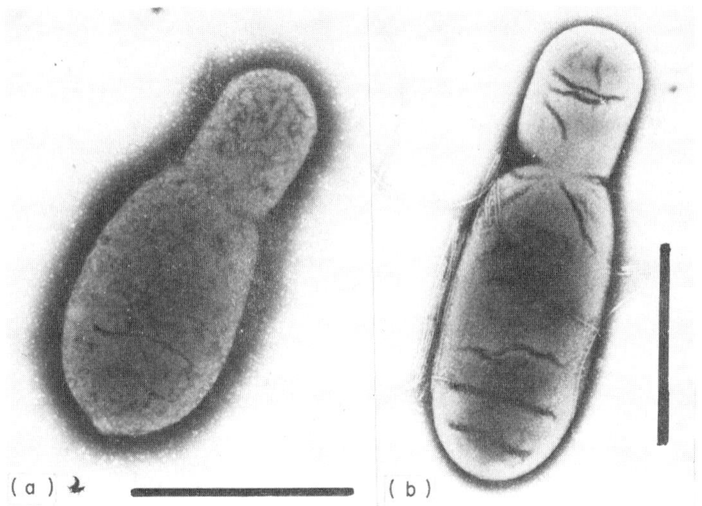

Fig. 2. *Blastobacter*-like, budding bacteria of group I.1 showing young, polar buds. Electron micrographs; cells negatively stained. Magnification bar represents 1 μm. (a) Strain Mü-161A; (b) strain Mü-222.

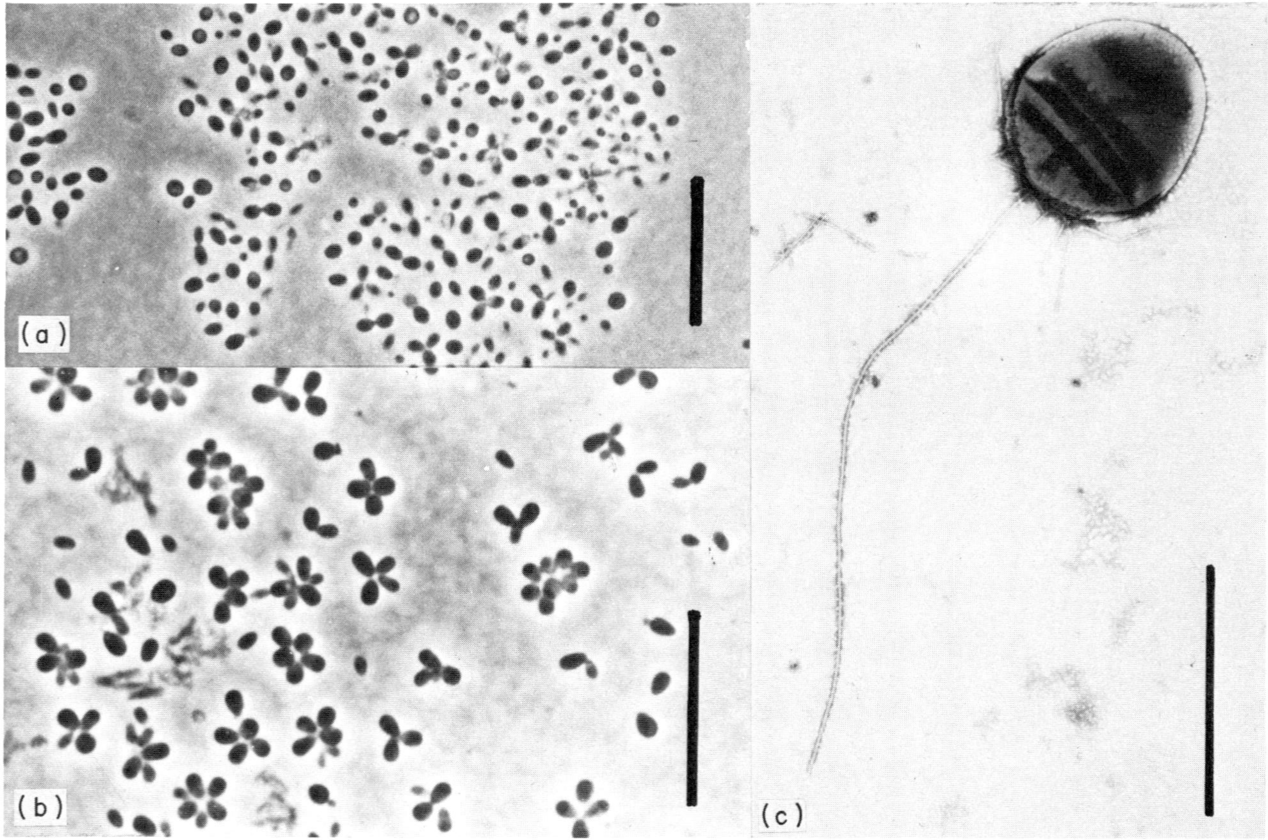
(a)
(b)
(c)

1% (w/v) of glucose, maltose, lactose, galactose, mannitol, fructose or ribose. They also grow well on 0.1% (w/v) of glutamic acid or histidine. Ammonium, Bacto peptone, or Bacto yeast extract are utilized as N sources. All strains are catalase positive, and peroxidase positive or oxidase positive; all show a distinct tolerance of 30% CO.

The DNA base ratios were found to be 60.4 ± 0.5 (Mü-161A); 58.9 ± 1.2 (Mü-216) and 64.5 ± 0.6 Mol% GC for Mü-222.

A certain similarity exists between these three strains and *Blastobacter* spp., bacteria which are still little known and which were rather incompletely described in *Bergey's Manual of Determinative Bacteriology* (Buchanan & Gibbons 1974).

(b) *Group I.2* (Pasteuria *spp.; strains Schl-1, -29, -31, -32, -116*)

Five strains isolated from surface brackish water of the Kiel Bight had drop-shaped cells with flagellated, polar buds. Rosette formation was frequently observed. Three of these strains were obtained by adding 0.1% (w/v) of chitin to the sample and simply incubating it at room temperature. One strain (Schl-31) came from an enrichment with 0.1% of mannitol, and strain Schl-116 was isolated from a water agar enrichment culture into which coverslips had been placed (Fig. 1).

The five strains show certain similarities with members of the genus *Pasteuria* (Staley 1973; Buchanan & Gibbons 1974). Their colonies are white to colourless (Schl-116) or pink. There appears to be a capsule and there is a polar holdfast polymer. One strain (Schl-1) was studied more closely. It required Na^+, Ca^{2+}, Cl^-, HCO_3^{2-} and possibly Mg^{2+} for growth. The temperature optimum was 28–30°C and the optimum was followed by a sharp decline, the maximum was found to be between 31.5 and 34°C. Further studies will have to be made to compare these isolates with *Pasteuria ramosa,* isolated recently by Staley (1973).

(c) *Group I.3* (Planctomyces guttaeformis)

Although pure cultures have not yet been obtained, organisms of this type occur in many samples and enrichment cultures. They show certain similarities to *Pasteuria* spp. except that their cells are more elongated and often surrounded by a voluminous capsule to form spherical coenobia (Fig. 4). There is a possibility that these bacteria may be identical to the *Marssoniella*, an enigmatic cyanobacterium or fungus (Komarek & Vavra 1968).

Fig. 3. *Pasteuria* spp. (group I.2) from brackish water of the Kiel Bight. (a) and (b) Phase-contrast light micrographs of rosettes and aggregates of strains Schl-1, and Schl-32, respectively. Magnification bar: 15 μm. (c) Electron micrograph of a strain Schl-1 swarmer cell with filamentous appendage (flagellum?). Negative stain; bar: 1 μm.

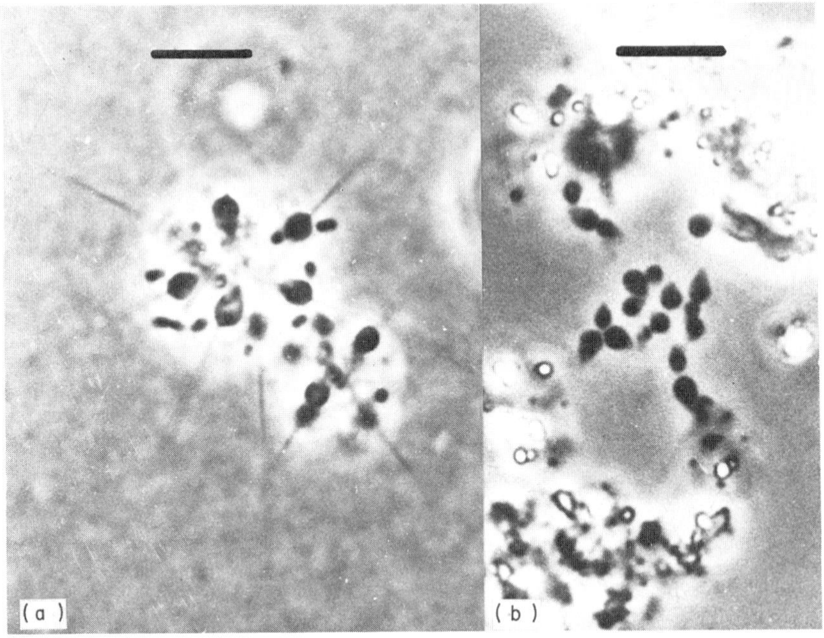

Fig. 4. Aggregating, budding fresh water bacteria of group I.3 resembling *Planctomyces guttaeformis.* The form on the left (a) with long, rigid, terminal spines and subpolar buds. (b) Resembles *Marssoniella.* Phase-contrast light micrographs; bar: 5 μm.

(d) *Group I.4 (strains Mü-279, Mü-290 and Schl-130)*

Three strains of attaching, budding bacteria have conical cells with short stalks consisting of an excreted bundle of fibres (Fig. 5). Morphologically similar forms were seen and depicted by A. T. Henrici in his classical study of attached water bacteria (Henrici & Johnson 1935). Two of these strains (Mü-279 and Mü-290) came from the Plußsee, and the third strain (Schl-130) was isolated from a brackish water sample of the Kiel Bight. In all cases the samples were stored for a long period of time; two strains were obtained by using the water agar–Petri dish method.

The colonies were pink or white (Schl-130). Their temperature optima were around 30°C and the pH optimum was 6.2–7.0. Strains Mü-279 and Mü-290 utilized glucose, maltose, lactose, galactose or saccharose. The following compounds or mixtures were utilized as N sources by the same two strains: NH_4^+, peptone, yeast extract or various amino acids. The GC ratio of two strains was 54.4 (Mü-279) and 51.0 mol% GC (Mü-290), respectively.

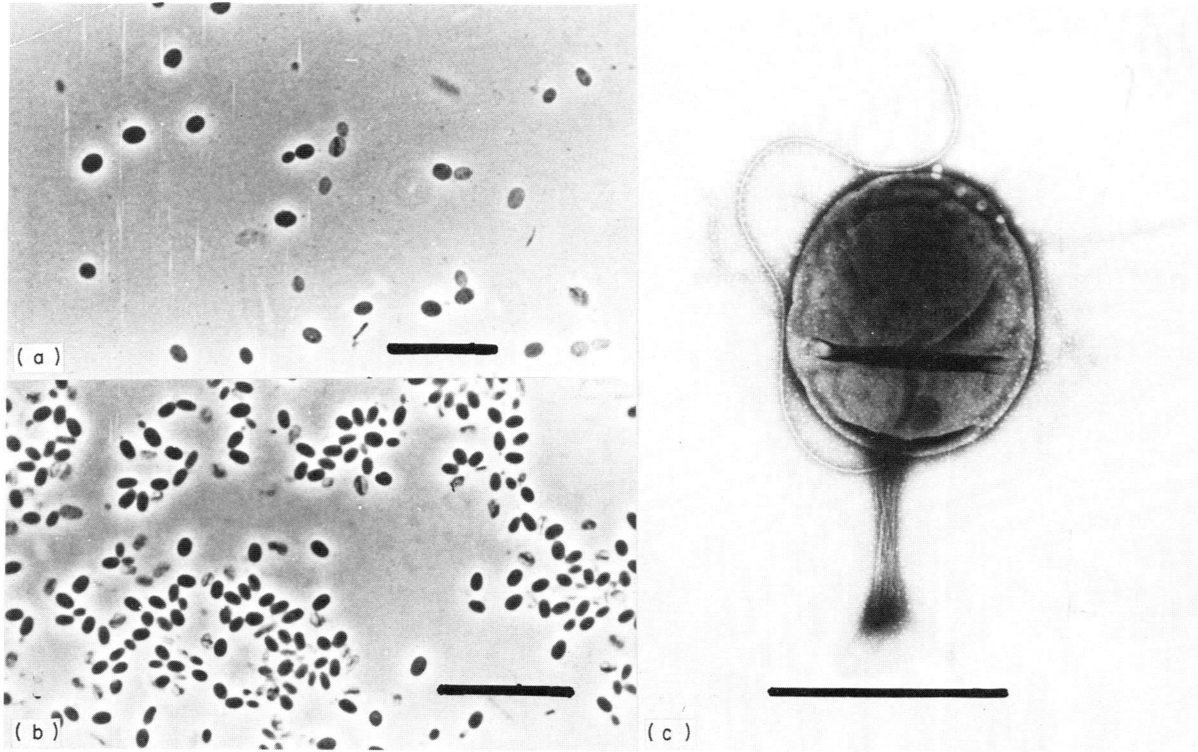

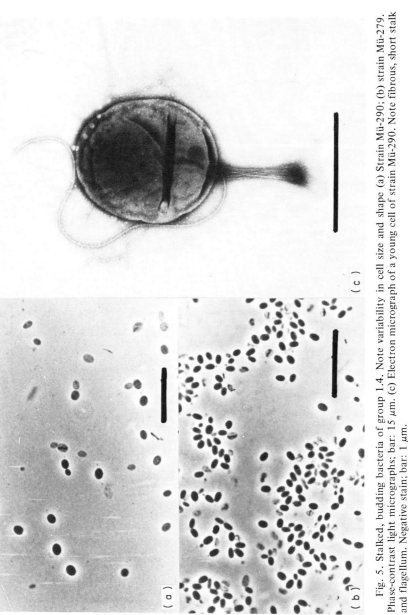

Fig. 5. Stalked, budding bacteria of group 1.4. Note variability in cell size and shape (a) Strain Mü-290; (b) strain Mü-279. Phase-contrast light micrographs; bar: 15 μm. (c) Electron micrograph of a young cell of strain Mü-290. Note fibrous, short stalk and flagellum. Negative stain; bar: 1 μm.

(e) *Group I.5* (*strains Hi-A to Hi-F and Schl-120*)

Bacteria of this group often form buds or at least cells which are considerably smaller than their parent cells. However, the possibility that they might also grow and divide like other regular bacteria cannot be excluded (Fig. 6). Short, rod-shaped cells attach with a holdfast polymer and form polar buds with a pointed distal cell pole. The buds sometimes carry flagella. Isolation was successful with enrichment cultures from a Michigan Forest Pond (near Augusta) to which 0.005% (w/v) of Bacto peptone had been added. One strain (Schl-120) came from the surface water of the Kiel Bight. Studies on the physiology of these organisms (7 strains) are in progress.

(f) *Group I.6* (Planctomyces bekefii)

The drumstick-shaped bacteria of this group have been observed in many aquatic sites – usually during a time when there was a decay of a water bloom (Kristiansen 1971). Rosettes with or without iron deposits on the stalks were frequently seen in samples and enrichments from Wintergreen Lake (Michigan), or the lakes Plußsee and Höftsee in Germany. The spherical cells produce their buds subpolarly, and the electron microscope reveals the complete coverage of the cell surface with thin, rigid, long spines (Fig. 6d). Unfortunately, this organism has not yet been obtained in pure culture.

(g) *Group II.1* (Stella *sp.; strain Schl-41*)

The bacteria of this group have pointed prosthecae. Strain Schl-41 grows as a 5- to 6-pronged star, not unlike a Star-of-David. After division by cross-septation, two 3-pronged crowns remain, each of which regenerates the missing half (Fig. 7). These bacteria seemed to be flat, but other shapes were also observed. Evidently the form and presence, as well as numbers of prosthecae, depend on the growth conditions. Morphologically, strain Schl-41 was clearly identical with bacteria of the genus *Stella*, described previously to grow only in the presence of humic substances (Vasil'eva *et al.* 1974).

Our *Stella* strain-41 was not dependent on humic substances. It was obtained from the water surface of brackish water of the Kiel Bight. It was isolated as a chalky-white colony and came from an enrichment to which 0.05% (w/v) of casein hydrolysate had been added and which had been incubated for about one year.

Stella Schl-41 was non-motile. The cells contained a storage granule. Growth was excellent on PYVG medium made up with 25% artificial sea water. The generation time was 6 h under these conditions and at 34°C. The temperature

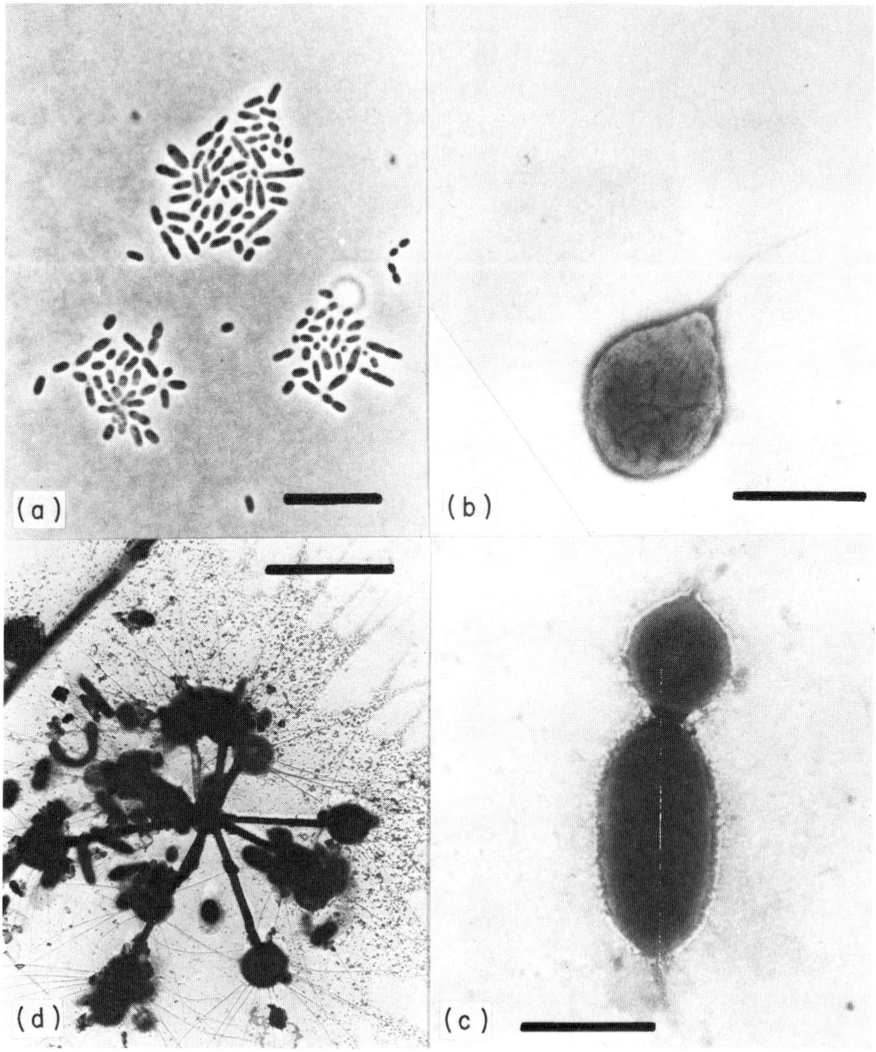

Fig. 6. (a-c) Budding bacteria of group I.5 as grown on PYVG medium. (a) Small aggregates with cells of various age, some of which appear to divide normally. Phase-contrast light micrograph; bar: 10 μm. (b) Swarmer cell (?) of strain Schl-120. Note long, terminal filament. Negative stain; bar: 1 μm. (c) Budding cell of strain Hi-KBS. Negative stain; bar: 1 μm.

(d) *Planctomyces bekefii* Gimesi, a planctonic bacterium originally described as a fungus. Coenobium of cells from Wintergreen Lake, Michigan. Partial encrustation with ferric oxide hydrate. Note subterminal buds and the long, rigid spines on the spherical cells. Pt-C shadow of 12°; bar: 3 μm.

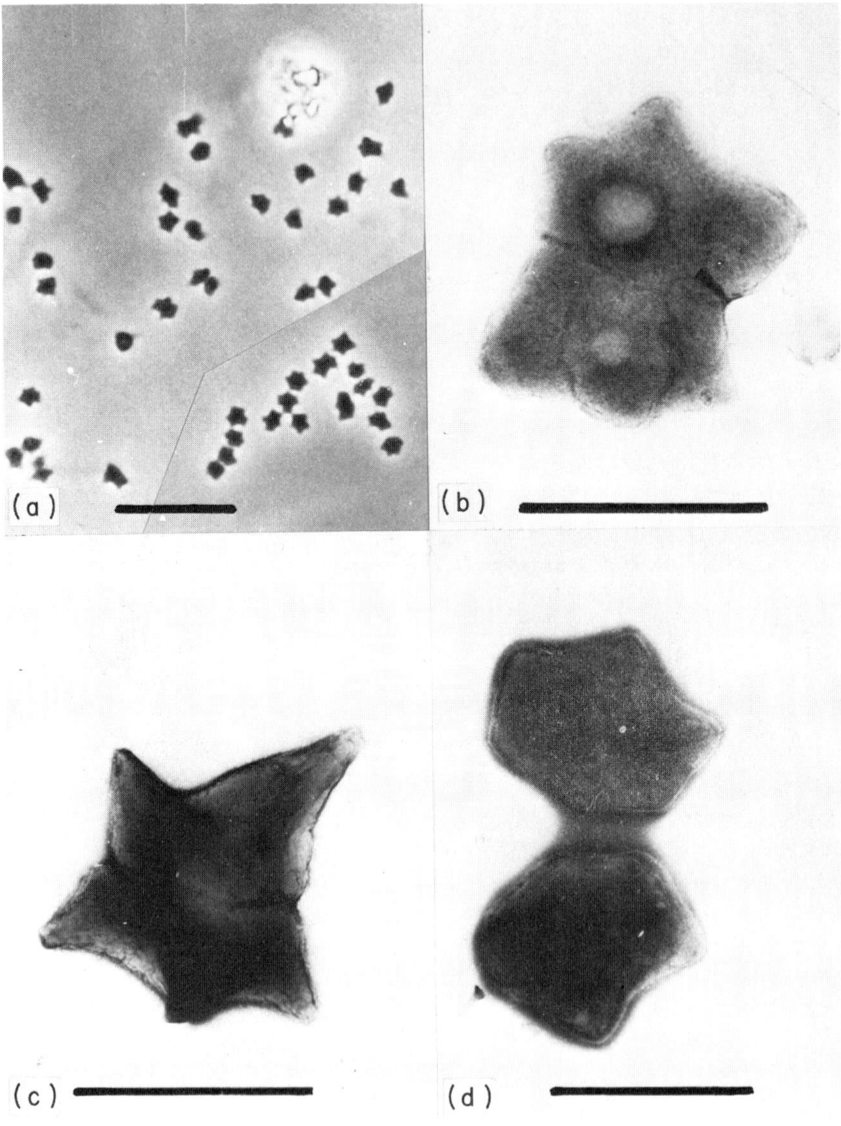

Fig. 7. *Stella* sp. (strain Schl-41; group II.1). (a) Star-shaped cells grown on organic medium PYVG. Phase-contrast light micrograph; bar: 8 μm. (b-d) Cells in various stages of division. Note storage granules. Negative stain; bar: 1 μm.

optimum was 31–36°C, the maximum 41–43.5°C, the minimum below 12°C. The pH optimum was between 7.0 and 7.8. *Stella* Schl-41 required Ca^{2+} and Na^{+} for growth.

(h) *Group II.2*

Occasionally we have found bacteria resembling *Stella* but attached to a surface by means of a long stalk (Fig. 8). A flagellum indicates motility. This type of organism occurred in enrichments from the Blunkersee, but pure cultures have not yet been obtained.

(i) *Group II.3 (strains Schl-16, -127 and -129)*

Bacteria with long and distinct prosthecae (cellular appendages, which usually taper at the end) have been described as *Prosthecomicrobium* spp. (Staley 1968). Three strains from the surface water of the Kiel Bight had such properties, but there were also prosthecae of even greater length, which would have to be called hyphae. These three strains therefore appear to be intermediate between the truly hyphal forms (*Hyphomicrobium, Pedomicrobium* etc.) and the 'normal' prosthecomicrobia. One strain was obtained from a water sample to which 0.0025% of Bacto peptone had been added (Schl-129); another strain grew in a sample enriched with 0.05% of glucose (Schl-16). The third strain (Schl-127) came from a sample stored for 2 years without any additions. All three strains differ in their shapes and lengths of prosthecae, in colony pigmentation and in their carbon utilization spectrum. The temperature optima were near 31°C. Strain Schl-16 grew well at 30°C with a generation time of 6–7 h.

(j) *Group III.1 (hyphomicrobia; strains Schl-89 and Schl-125)*

The budding and hyphal bacteria of the genus *Hyphomicrobium* have been studied intensively for the past 14 years (Hirsch & Conti 1964*a, b*; Conti & Hirsch 1965; Sperl & Hoare 1971; Attwood & Harder 1972; Mandel *et al.* 1972; Moore & Hirsch 1972; Hirsch 1974). Approximately 120 pure cultures now exist in our laboratory. New enrichments were made recently (Fig. 10), and these yielded hyphomicrobia with intensive rosette formation (strain Schl-125, Fig. 11). The enrichment of this strain was started with brackish surface water from the Kiel Bight which was incubated over water agar containing manganese chloride in a fashion similar to that shown in Fig. 1.

The cell form of this new strain is different from that of the other hyphomicrobia isolated previously with C-1 compounds: the cells are slender, elongated drops with thin, slender buds at the tips of rarely-branched hyphae. Swarmers are motile. The optimal temperature was found to be 24–28°C;

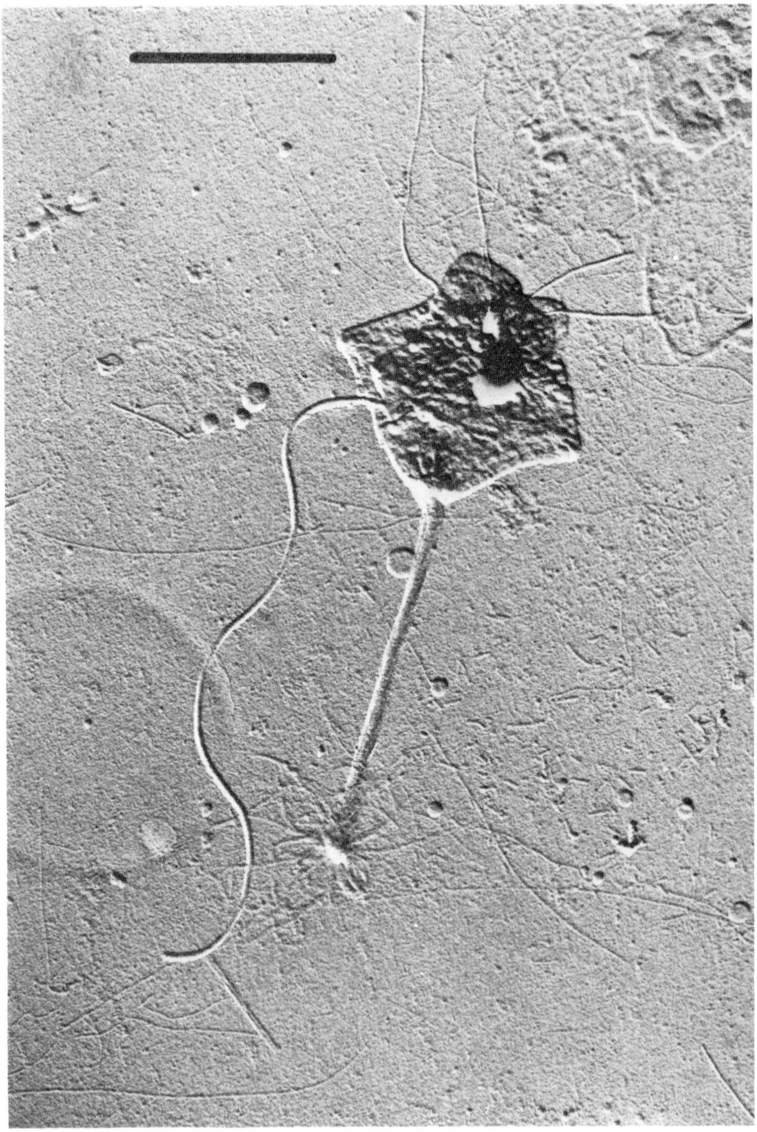

Fig. 8. A prosthecate bacterium of group II.2 with a fibrous stalk and a flagellum. Such forms were frequently seen in enrichments inoculated with water from the Blunkersee. Note terminal holdfast fibres on the stalk. Pt-C shadowed preparation; bar: 1 μm.

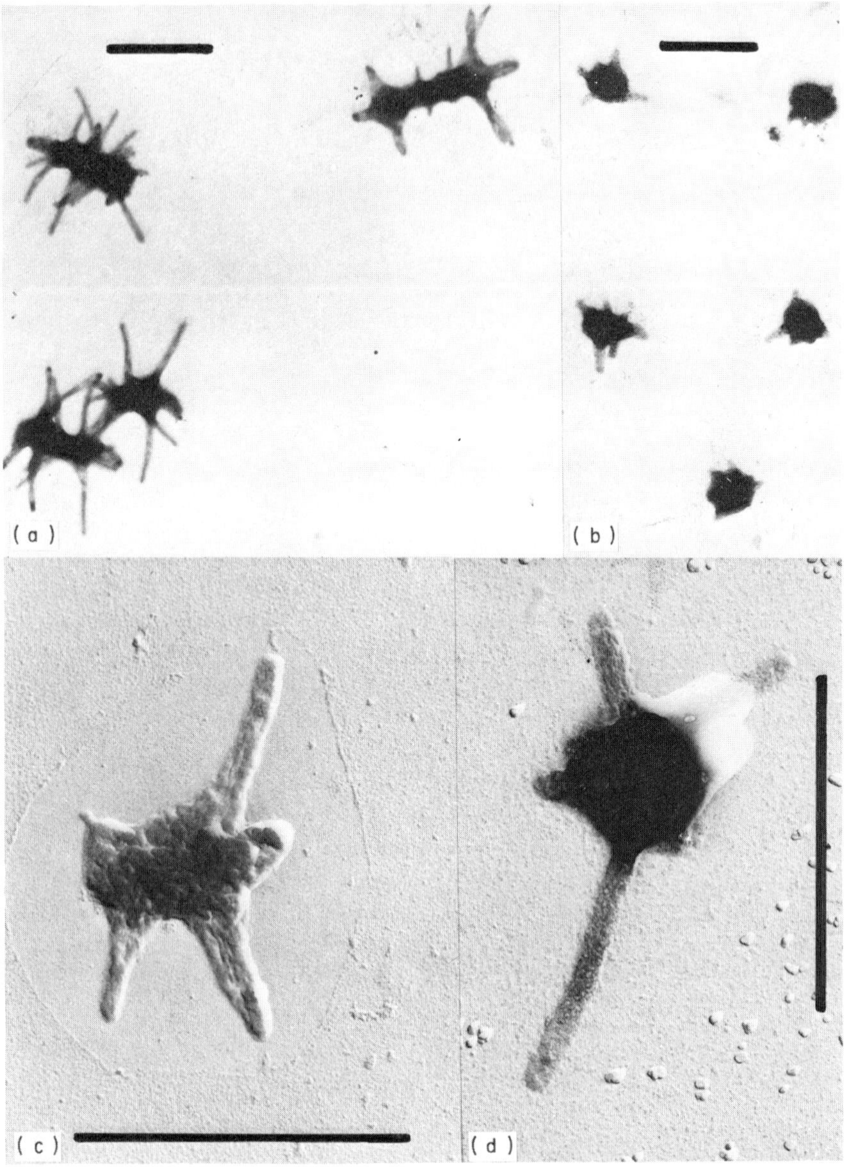

Fig. 9. Prosthecate bacteria of group II.3 from brackish water. (a and c) Strain Schl-127; (b and d) strain Schl-16. Negatively stained preparations (a and b) or Pt-C shadowed specimens (c and d). Bar: 2 μm.

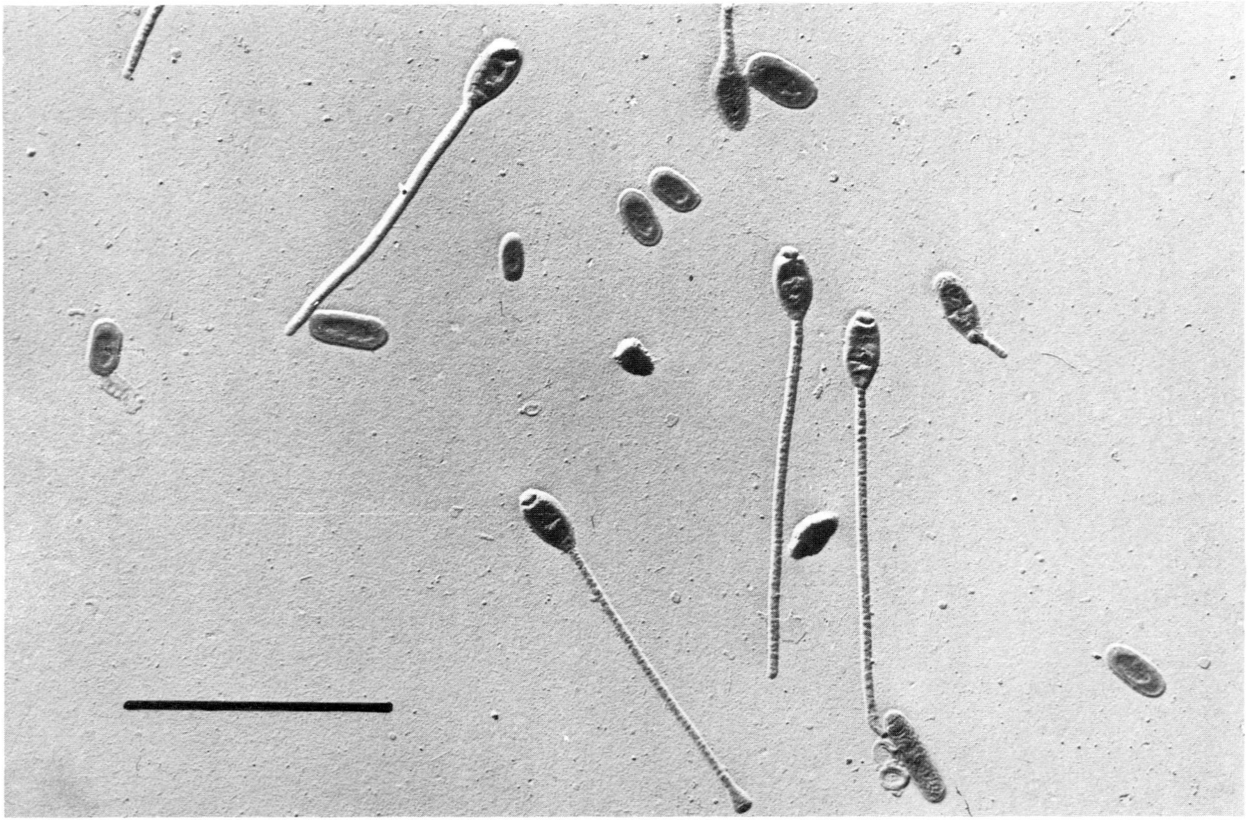

Fig. 10. *Hyphomicrobium* spp. from an enrichment with 0.005% Bacto peptone inoculated with surface water from the Plußsee. Shadowed preparation of the enrichment culture surface pellicle; bar: 5 μm.

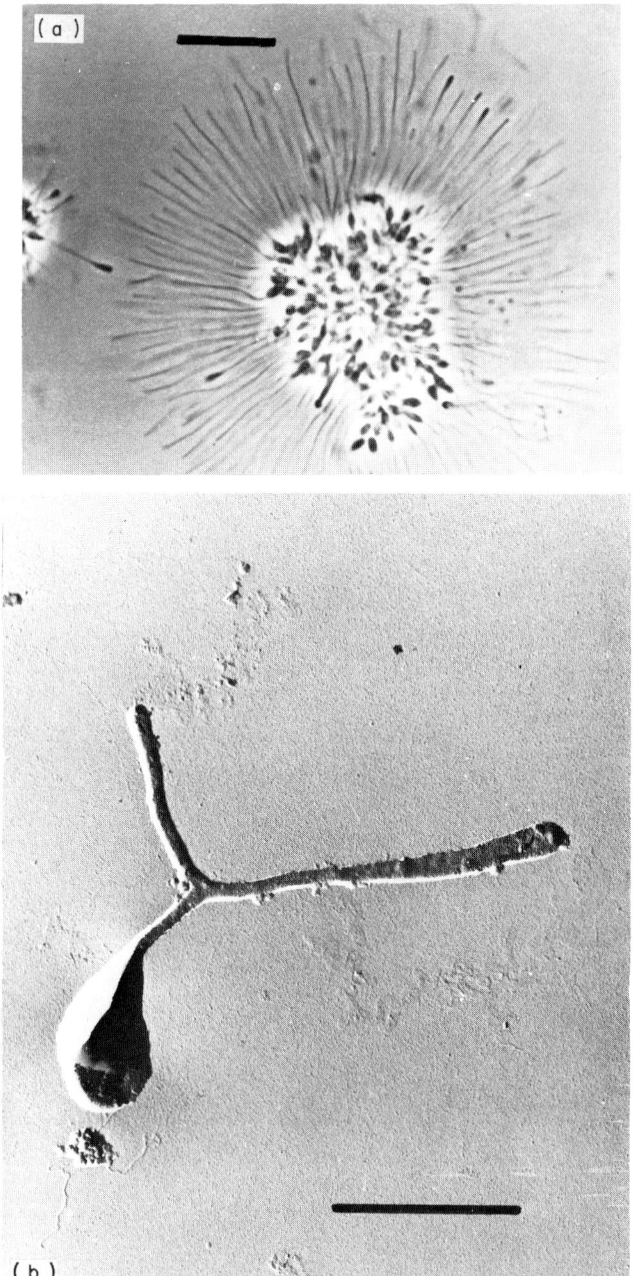

Fig. 11. *Hyphomicrobium* sp. strain Schl-125 (group III.1) from brackish water of the Kiel Bight. (a) Dense rosette with hyphae radiating outward are quite common. Phase-contrast light micrograph; bar: 10 μm. (b) Cell with branched hypha showing thickening of the hyphal tip (beginning of bud formation?). Shadowed preparation; bar: 2 μm.

growth was optimal on PYVG medium containing NH_4^+ and 25% artificial sea water.

From the same site, but from a depth of 2 m, came a short *Hyphomicrobium* which resembled *H. neptunium* in having stout, almost spherical cells with short and rarely-branched hyphae (strain Schl-89). This culture was isolated after incubating for 1 year a sample to which nothing had been added. Growth was excellent in PYVG medium with artificial sea water.

(k) *Group III.2* ("Hyphomicrobium"; *strain Schl-128*)

Another bacterium somewhat resembling hyphomicrobia came from the 2 m water sample of the Kiel Bight, after incubation for 1 year. This organism had very thin, almost cylindrical cells, usually with hyphae on both cell poles (Fig. 12). The hyphae were up to 30 μm long, straight and quite rigid; occasionally they were branched. The buds never had flagella and tended to form chains. Buds also occasionally produced side branches, thus creating forked cells. The colonies were red on PYVG medium. Growth was stimulated by adding 0.1% (w/v) of formate to this medium or by incubation in an atmosphere containing methanol vapour.

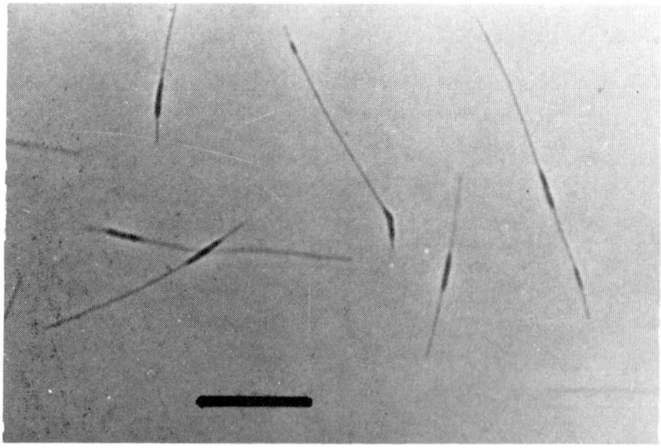

Fig. 12. Bacterium resembling *Hyphomicrobium* spp. but with rigid 'hyphae' and non-motile buds (strain Schl-128; group III.1). Phase-contrast light micrograph; bar: 10 μm.

(l) *Group III.3* (*strains Hi-40/7; Hi-40/8; Hi-41/6 and Hi-41/7*)

An investigation of the heterotrophic microflora of the Solar Lake, near Eilat, Israel, during March 1976, demonstrated, in several layers of this hypersaline, hyperthermal lake, the presence of strange, hyphal, budding bacteria. Frequently forked cells had three hyphae, only one of which budded at a given time. In their natural habitat, these bacteria tolerate temperatures from 20 to 43°C. They grew at a wide range of oxygen concentrations, and even with low concentrations of sulphide. Enrichments were set up by using various media, but in most cases these cells were rapidly overgrown by other bacteria. However, when samples from 1.5 or 3.5 m depth were streaked directly on PYVG medium (made up with doubly-concentrated artificial sea water) small reddish colonies appeared after prolonged incubation at 43°C in the dark, either aerobically or anaerobically. These colonies were easily re-streaked and purified, and growth in liquid media gave some information on the life cycle of these bacteria (Fig. 13). Tetraedral or drop-shaped cells produce a hypha from one cell 'corner'. After forming a terminal bud, this hypha shrinks and another 'corner' produces a new hypha for the next budding process (Fig. 13a, b). The formation of a third hypha from the third 'corner' is again preceeded by the shrinkage of the second one (Fig. 13c). It therefore appears as if each hypha can produce only one bud. Branching of the hyphae occurs, as well as forking of the mother cells. A detailed study of the life cycle of this organism is in preparation. These organisms showed similarities with strains of the genus *Pedomicrobium*, the budding and hyphal bacteria from soil or water which deposit Fe or Mn oxides on the cell surface (R. Gebers, pers. comm.).

(m) *Group III.4* (*strain Schl-107*)

Bacteria of the last group are also budding forms which grow hyphae. However, only one pure culture has been obtained so far, and thus a brief description will have to suffice.

The short, rod-shaped cells produce 1–8 hyphae subpolarly around one cell pole, or distributed over the whole cell surface. Some of these thus resemble a minute *Hydra* (Fig. 14). A bud is formed terminally at the same, hypha-producing pole. This site of budding is strikingly different from that of other hyphal, budding bacteria where the bud originates from hyphal tips only.

The *Hydra*-like hyphal and budding organisms were seen frequently in the neuston layers of Michigan marl lakes (Lawrence Lake, Cassidy Lake, Wintergreen Lake etc.). Strain Schl-107 came from the Kiel Bight and was enriched for by adding 0.1% of glucosamine to the sample. It produces turbidity in pure culture and forms red colonies on PYVG agar with 25% of artificial sea water. The bacterium is Gram negative, has a pH optimum between 6.6 and 8.2, an optimum temperature between 28 and 32°C, is aerobic, light independent and does not seem to form visible storage products.

4. Discussion

The morphologically 'unusual' bacteria described here are all easily recognizable by simple phase-contrast light microscopy – "if one knows what to look for"; and if one knows which structures are real and which ones are likely to be

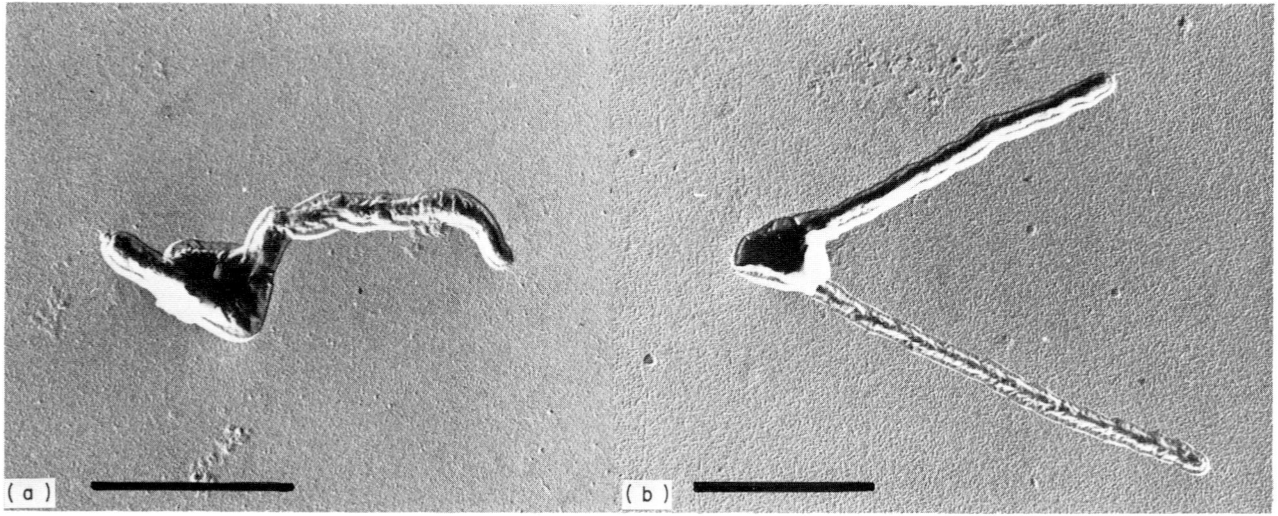
(a)
(b)

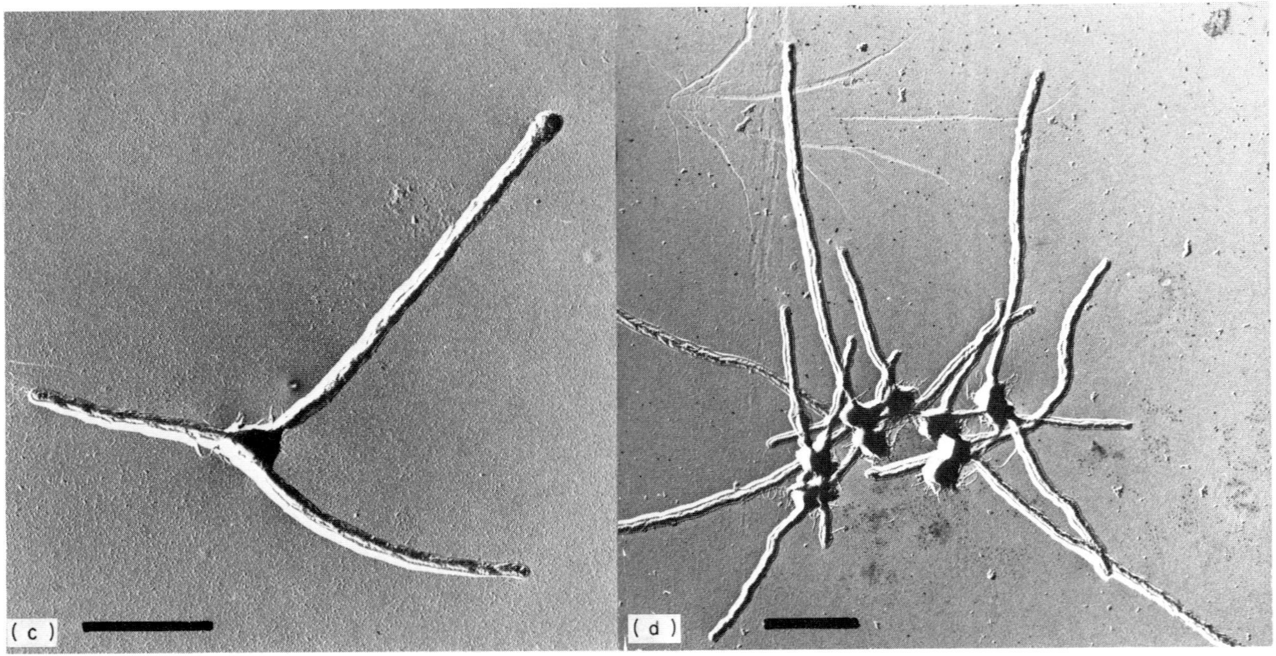

Fig. 13. Budding, hyphal bacteria from the hypersaline, hyperthermal Solar Lake near Eilat, Israel (strain Hi-41/7; group III.3). (a) Formation of second hypha. (b) First hypha shrunken. (c) Formation of bud on third hypha. (d) Microcolony of older cells. Electron micrographs of shadowed preparations; bar: 2 μm.

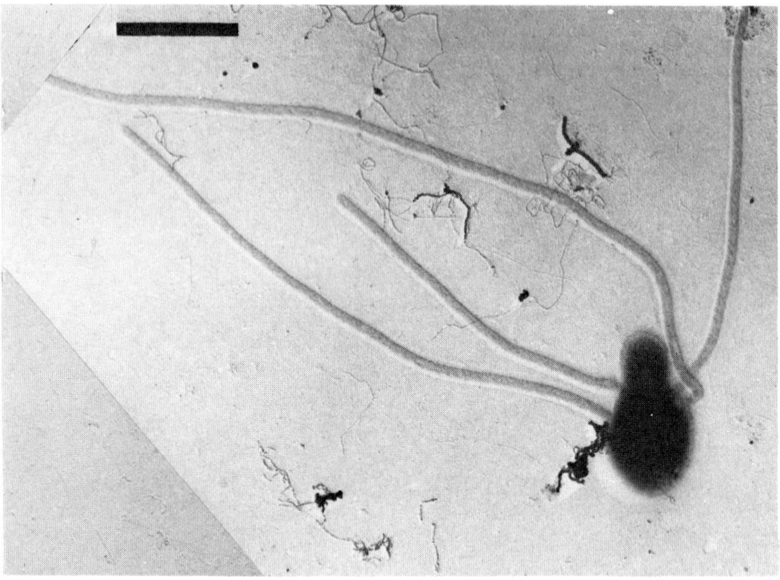

Fig. 14. Budding, hyphal bacterium from the neuston of a Michigan marl lake (group III.4). The mother cell has 4 subpolar hyphae (formed on one cell pole), and a polar bud is being formed, thus giving the cell an appearance of a minute *Hydra.* Shadowed preparation; bar: 1 μm.

artifacts. The problem of recognition begins with the uncertainty of their morphology in an altered environment. Our isolations and pure culture studies are the first steps toward a better understanding of the pleomorphism or life cycles of these bacteria. It must again be pointed out that our strains were usually isolated with only little addition of nutrients to the enrichment culture. Yet, in pure culture, growth was remarkably good on media containing up to 1% (w/v) of a carbon source. Hence the 'oligocarbophily' of these strains may be really an 'oligocarbotolerance' and their growth in the enrichment cultures may have resulted from lack of competition rather than from other specific enrichment conditions.

Our new, 'odd' bacterial strains show only little and superficial morphological similarity to each other. With still more information on hand, and with more strains studied, we will be able to link them to each other or to known bacterial genera of our present taxonomic system. More information on their physiology will also enable us to study their ecology quantitatively, which in turn may lead to more isolations.

The differences in G + C ratios also indicate only weak relationships among these budding forms. The process of budding may, therefore, have to be considered to be of polyphyletic origin.

Whatever their activities in the natural ecosystem may be, all of these bacteria were easily recognizable. This fact alone seems to make them ideal objects for studies on bioindication. Their characteristic and often beautiful shapes could also serve as a stimulant for young, enthusiastic microbiologists with an urge to discover and with an ability to become enchanted still with the extraordinary diversity nature has created (Fig. 15).

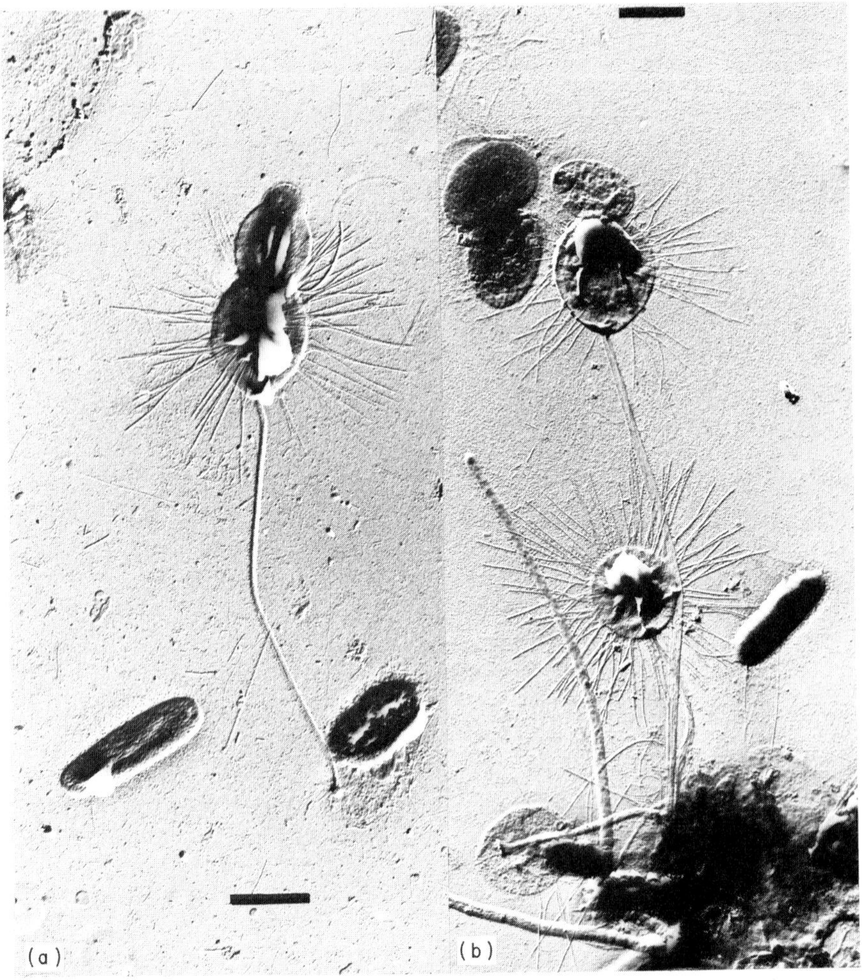

Fig. 15. Stalked, budding bacteria and other forms from the Plußsee. Water samples were fortified with 0.005% of Bacto peptone (a) or 0.005% of Bacto yeast extract (b). Note the rigid spines similar to those of *Planctomyces* spp. Shadowed preparations of the enrichments surface layers after several months of incubation at room temperature, bar: 1 μm.

5. Acknowledgements

We gratefully acknowledge the competent technical assistance of Mrs B. Doose, A. Gräter, E. Heiden, S. Klopp and H. Marsen. The present research was supported by a grant from the Deutsche Forschungsgemeinschaft to one of us (P.H.).

6. References

ATTWOOD, M. M. & HARDER, W. 1972 A rapid and specific enrichment procedure for *Hyphomicrobium* spp. *Antonie van Leeuwenhoek* **38**, 369–378.

BENECKE, W. 1933 Bakteriologie des Meeres. In *Handbuch der Biologischen Arbeitsmethoden* Abt. 9, Teil 5, ed. Abderhalden.

BUCHANAN, R. E. & GIBBONS, N. E. (eds) 1974 *Bergey's Manual of Determinative Bacteriology*, 8th Edn. Baltimore: Williams & Wilkins Co.

CALDWELL, D. E. & TIEDJE, J. M. 1975*a* A morphological study of anaerobic bacteria from the hypolimnia of two Michigan lakes. *Canadian Journal of Microbiology* **21**, 362–376.

CALDWELL, D. E. & TIEDJE, J. M. 1975*b* The structure of anaerobic bacterial communities in the hypolimnia of several Michigan lakes. *Canadian Journal of Microbiology* **21**, 377–385.

CONTI, S. F. & HIRSCH, P. 1965 Biology of budding bacteria. III. Fine structure of *Rhodomicrobium* and *Hyphomicrobium* spp. *Journal of Bacteriology* **89**, 503–512.

GORLENKO, V. M. 1970 A new phototrophic green sulphur bacterium *Prosthecochloris aestuarii*: nov. gen.; nov. spec. *Zeitschrift für allgemeine Microbiologie* **10**, 147–149.

HAJDU, L. 1974 Observations on the bacterium species *Planctomyces guttaeformis* Hortob. (Caulobacteriales). *Archiv für Hydrobiologie* **74**, 172–185.

HENRICI, A. T. & JOHNSON, D. E. 1935 Studies of fresh water bacteria. II. Stalked bacteria, a new order of Schizomycetes. *Journal of Bacteriology* **30**, 61–93.

HIRSCH, P. 1968 Gestielte und knospende Bakterien: Spezialisten für C-1 Stoffwechsel an nährstoffarmen Standorten. *Mitteilungen der internationalen Vereinigung für Limnologie* **14**, 52–63.

HIRSCH, P. 1973 The ecology of *Thiopedia* spp. *Abstracts of a Symposium on Prokaryotic, Photosynthetic Microorganisms, Freiburg*, p. 77.

HIRSCH, P. 1974 Budding bacteria. *Annual Review of Microbiology* **28**, 391–444.

HIRSCH, P. & CONTI, S. F. 1964*a* Biology of budding bacteria. I. Enrichment, isolation and morphology of *Hyphomicrobium* spp., *Archiv für Mikrobiologie* **48**, 339–357.

HIRSCH, P. & CONTI, S. F. 1964*b* Biology of budding bacteria. II. Growth and nutrition of *Hyphomicrobium* spp. *Archiv für Mikrobiologie* **48**, 359–367.

HIRSCH, P. & PANKRATZ, S. H. 1970 Study of bacterial populations in natural environments by use of submerged electron microscope grids. *Zeitschrift für allgemeine Microbiologie* **10**, 589–605.

HIRSCH, P. & RHEINHEIMER, G. 1968 Biology of budding bacteria. V. Budding bacteria in aquatic habitats: occurrence, enrichment and isolation. *Archiv für Mikrobiologie* **62**, 289–306.

HUBER-PESTALOZZI, F. G. 1938 Das Phytoplankton des Süszwassers. Systematik und Biologie 1. Teil: Allgemeiner Teil: Blaualgen Bakterien Pilze. In *Die Binnengewässer*, ed. Thienemann, **16**, 260–261.

KOMAREK, J. & VAVRA, J. 1968 In memoriam of *Marssoniella* Lemm. 1900. *Archiv für Protistenkunde* **111**, 12–17.

KOPPE, F. 1923 Die Schlammflora der ostholsteinischen Seen und des Bodensees. *Archiv für Hydrobiologie* **14**, 619–672.

KRISTIANSEN, J. 1971 On *Planctomyces bekefii* and its occurrence in Danish lakes and ponds. *Botanisk Tidsskrift* **66**, 293–302.

KUZNETSOV, S. J. 1975 *The Microflora of Lakes and its Geochemical Activity.* (Translation of Russian Original 1970). Texas: University of Texas Press.

LA RIVIÈRE, J. W. M. 1965 Enrichment of colorless sulfur bacteria. In *Anreicherungskultur und Mutantenauslese*, Proceedings of a Symposium, Göttingen, 1965. Stuttgart: G. Fischer Verlag, pp. 17–27.

LAUTERBORN, R. 1916 Die sapropelische Lebewelt. Ein Beitrag zur Biologie des Faulschlammes natürlicher Gewässer. *Verhandlungen des Naturhistorisch – Medizinischen Vereins zu Heidelberg* **13**, 395–481.

MANDEL, M., HIRSCH, P. & CONTI, S. F. 1972 Deoxyribonucleic acid base compositions of hyphomicrobia. *Archiv für Mikrobiologie* **81**, 289–294.

MARMUR, J. 1961 A procedure for the isolation of deoxyribonucleic acid from microorganisms. *Journal of Molecular Biology* **3**, 208–218.

MOORE, R. L. & HIRSCH, P. 1972 DNA base sequence homologies of some budding and prostheoate bacteria. *Journal of Bacteriology* **110**, 256–261.

MOORE, R. L. & HIRSCH, P. 1973*a* First generation synchrony of isolated *Hyphomicrobium* swarmer populations. *Journal of Bacteriology* **116**, 1447–1455.

MOORE, R. L. & HIRSCH, P. 1973*b* Nuclear apparatus of *Hyphomicrobium. Journal of Bacteriology* **116**, 1447–1455.

NORRIS, J. R. & RIBBONS, D. W. (eds) 1971 *Methods in Microbiology*, Vol. 5A. London & New York: Academic Press.

OLÁH, J., HAJDU, L. & ELEKES, K. 1972 Electron microscopic investigation of natural bacterial populations in the water and sediment of Lake Balaton and Belsö. *Annals of Biology Tihany (Hungary)* **39**, 123–129.

PFENNIG, N. 1969 *Rhodopseudomonas acidophila* sp.n., a new species of the budding purple nonsulfur bacteria. *Journal of Bacteriology* **99**, 597–602.

RHEINHEIMER, G. 1975 *Mikrobiologie der Gewässer* **2**, Jena: G. Fischer Verlag.

SCHULZ, E. & HIRSCH, P. 1973 Morphologically unusual bacteria in acid bog water habitats. *American Society for Microbiology* (Abstracts). G. 207.

SHIRLING, E. B. & GOTTLIEB, D. 1966 Methods for characterization of *Streptomyces* species. *International Journal of Systematic Bacteriology* **16** (3) 313–340.

SKERMAN, V. B. D. 1967 *A Guide to the Identification of the Genera of Bacteria*, 2nd edn. Baltimore: Williams & Wilkins Co.

SPERL, G. T. & HOARE, D. S. 1971 Denitrification with methanol: a selective enrichment for *Hyphomicrobium* species. *Journal of Bacteriology* **108**, 733–736.

STALEY, J. T. 1968 *Prosthecomicrobium* and *Ancalomicrobium*: new prosthecate freshwater bacteria. *Journal of Bacteriology* **95**, 1921–1942.

STALEY, J. T. 1973 Budding bacteria of the *Pasteuria-Blastobacter* group. *Canadian Journal of Microbiology* **19**, 609–614.

UMBREIT, W. W., BURRIS, R. H. & STAUFFER, I. F. 1957 *Manometric Techniques.* Minneapolis: Burgess Publishing Co.

UTERMÖHL, H. 1924 Phaeobakterien (Bakterien mit braunen Farbstoffen) *Biologisches Zentralblatt* **43**, 605–610.

VAN ERT, M. & STALEY, J. T. 1971 Gas-vacuolated strains of *Microcyclus aquaticus. Journal of Bacteriology* **108**, 236–240.

VASIL'EVA, L. V., LAFIZKAYA, T. N., ALEKSANDRUSCHKINA, N. J. & KRASSILNIKOVA, E. N. 1974 The physiological-biochemical peculiarities of the prosthecate bacteria *Stella humosa* and *Prosthecomicrobium* sp. (Translation, Russian). *Isvestiya Akademii Nauk, S.S.S.R. Biological Series* **5**, 699–714.

The Identification of Some Gram Negative Heterotrophic Aquatic Bacteria

D. M. GIBSON, MARGARET S. HENDRIE, N. C. HOUSTON AND G. HOBBS

Torry Research Station, 135 Abbey Road, Aberdeen, Scotland

CONTENTS

1. Introduction

IN 1960, Shewan *et al.* presented a scheme for the identification of certain genera of aerobic heterophic, Gram negative bacteria, especially the Pseudomonadaceae. This scheme, based on relatively few bacteriological tests, had been developed to identify isolates from the marine environment and from spoiling fish. It has since been used on isolates from other environments and has proved useful in obtaining preliminary identification to genus level. Shewan *et al.* (1960) recognized that it was an identification scheme not based on nor expressing any taxonomic relationships between the groups. They also recognized that further work was needed to expand the scheme so that the groups could be more accurately defined and further subdivided.

The scheme arranged the aerobic heterotrophic, Gram negative, rod-shaped bacteria into four broad groups on the basis of motility and flagella, oxidase reaction (Kovacs 1956) and pigment production: (a) motile with polar flagella, oxidase positive (*Pseudomonas, Aeromonas, Vibrio* etc.); (b) motile with peritrichous flagella, oxidase negative (Enterobacteriaceae); (c) non-motile, non-pigmented short stout rods (*Achromobacter, Alcaligenes*) and (d) non-motile with yellow pigmented colonies (*Flavobacterium, Cytophaga*). The organisms in (a) were further subdivided on the basis of diffusible pigment production in the medium of King *et al.* (1954), reaction in O-F medium (Hugh & Leifson 1953) and sensitivity to 0/129 (2:4-diamino-6:7-di-*iso*-propyl pteridine) (Shewan *et al.* 1954) into *Pseudomonas* (4 groups), *Aeromonas* and *Vibrio.* The scheme was modified later (Hendrie *et al.* 1964) to identify oxidase positive, non-fermentative rods, motile with peritrichous flagella (*Achromobacter* and *Alcaligenes*). The above tests still form the basis of the preliminary division of the Gram negative, heterotrophic bacteria (Table 1). In identifying the motile types, much emphasis is placed on the type of flagellation – whether polar or peritrichous. In general this is constant for a genus or a species, but there is a growing amount of evidence that, for some species, the type of flagella observed is dependent on the growth conditions (Allen & Baumann 1971; W. Hodgkiss, pers. comm.) or on the age of the culture when examined (Quigley & Colwell 1968). Much caution is needed when interpreting flagella stains or electron microscope preparations (cf. Rhodes 1965).

The developments in taxonomy and in identification schemes since 1960 are the result of the application of new approaches and techniques e.g. numerical analysis; molecular biology and genetic studies; investigation of intermediary matabolism. Much of the change has been due to Sneath's (1957*a*) theories on classification and the work on numerical taxonomy (Sneath 1957*b*; Sokal & Sneath 1963) which have been extensively applied. While some genera are differentiated particularly well, we may question its value when dealing with collections of strains possessing and sharing many negative characters. Standards for sample size, number of tests performed and analysis of the data have been described in detail (Lockhart & Liston 1970; Sneath 1972).

2. *Pseudomonas* and Allied Genera

The Shewan *et al.* (1960) scheme divides the pseudomonads into four groups. Subsequent studies have shown that strains which fall into groups I and II (i.e. those which are oxidative in glucose O-F medium) come within the description of the genus *Pseudomonas* as defined in the 8th edition of *Bergey's Manual of Determinative Bacteriology* (Buchanan & Gibbons 1974), having a DNA base composition in the range 58–70 mol % GC. Strains which fall into groups III/IV of the Shewan *et al.* (1960) scheme are not homogeneous with respect to their DNA base composition, some have values within the range for *Pseudomonas* (including *Ps. acidovorans, Ps. alcaligenes, Ps. maltophilia, Ps. pseudoalcaligenes* and *Ps. testosteroni*), whilst others have a much lower value – in the range 40–50 mol % GC. It has been evident for some time that some named *Pseudomonas* species such as *Ps. atlantica* and *Ps. putrefaciens* fall within the latter range (De Ley & Van Muylem 1963; Mandel 1966; Hill 1966; Rosypal & Rosypalová 1966; Herbert *et al.* 1971) and can no longer be regarded as belonging to the genus *Pseudomonas.* From the examination of isolates from a limited geographical area, Baumann *et al.* (1972) suggested the creation of a new

Table 1

Identification of aerobic, heterotrophic, Gram negative rods – first stage

Motility	Flagella	Oxidase	O-F medium (glucose)	Pigments: Diffusible: Fluorescent	Pigments: Diffusible: Non-fluorescent	Pigments: Non-diffusible	
+	Polar	+	Oxidative or –	Green or –	Blue-green, red, brown or –	– or yellow, red, brown, blue, pink	*Pseudomonas* and allied groups
+	Polar	+	Fermentative	–	– or brown	– or yellow, blue	Vibrionaceae
+	Peritrichous	+	– or oxidative	–	–	–	*Alcaligenes* (Agrobacterium)
+	Peritrichous	+	– or oxidative	–	–	Yellow	*Flavobacterium*
+	Peritrichous	–	Fermentative	–	–	– or yellow, red	Enterobacteriaceae
–	None	–	Fermentative	–	–	–	Enterobacteriaceae
–	None	+ or –	Oxidative or –	–	–	–	*Moraxella*-like *Acinetobacter*
–	None	+ or –	– or oxidative or weak fermentative	–	–	Yellow or orange	*Flavobacterium* *Cytophaga* *Flexibacter*

genus, *Alteromonas*, to accommodate low mol % GC pseudomonads from marine sources. Their only distinction between the genera *Pseudomonas* and *Alteromonas* was the mol % GC of the DNA, a difficult test for routine bacteriology.

In an attempt to resolve the relationships within the group III/IV pseudomonads and between them and groups I and II, Lee *et al.* (1977) subjected 37 strains from a variety of sources to a series of nutritional, biochemical, physiological and antibiotic sensitivity tests, 72 features in all. In addition some aspects of the intermediary metabolism of the strains were investigated using dehydrogenase assays on subcellular fractions (Kersters 1967). The results were used to calculate similarity matrices using various coefficients and dendrograms were constructed using various linkage programmes. A typical dendrogram is shown in Fig. 1. Regardless of the similarity coefficient of linkage programme used the same broad groupings were obtained; five major branches with the high mol % GC forming two (A and B) always distinct from the three of low mol % GC (C, D and E), and only a few strains, usually the same ones, as intermediates. The same pattern emerged when the enzyme data were analysed on their own, and since the two sets of data gave the same groupings real differences between the sets of organisms are indicated. Tests which differentiate the five main branches are given in Table 2.

Lee *et al.* (1977) have suggested that the strains in groups C, D and E having a mol % GC too low for *Pseudomonas* be transferred to the genus *Alteromonas* with the description of the latter genus expanded to incorporate the information gained from their study. Group C includes a named strain of *Alteromonas haloplanktis* (Reichelt & Baumann 1973*a*) viz. MacLeod's strain B-16 (NCMB

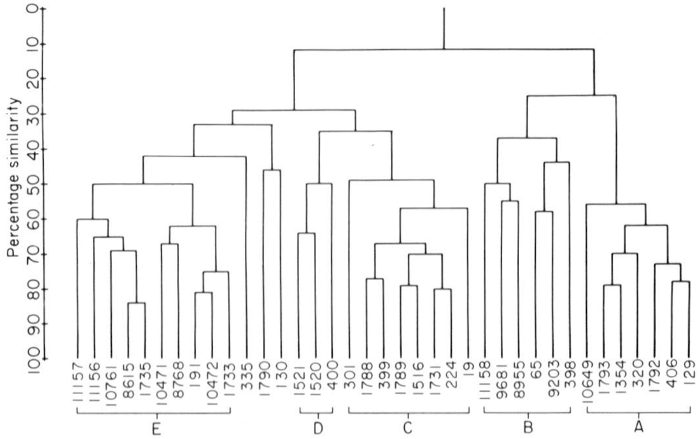

Fig. 1. Dendrogram constructed from a complete linkage analysis of similarities from all data on 37 strains of pseudomonads (after Lee 1973).

Table 2

Characters which differentiate 34 strains of pseudomonads

	A	B	C	D	E
O-F test (glucose)	Oxidative	No acid	No acid	No acid	No acid
H_2S production	1/7*	2/5	–	3/3	11/11
Trimethylamine oxide reduced to trimethylamine	–	–	–	3/3	11/11
DNAase	–	1/5	8/8	3/3	11/11
Arginine dihydrolase	6/7	–	–	–	–
Ornithine decarboxylase	–	–	–	–	9/11
Sensitivity to penicillin (1.5 i.u.)	–	–	–	3/3	–
Sensitivity to chloramphenicol (10 μg)	–	5/5	8/8	3/3	11/11
Carbon sources used for growth: xylose	7/7	–	–	–	–
succinate	7/7	5/5	7/8	–	11/11
malate	7/7	5/5	1/8	3/3	7/11
proline	7/7	5/5	8/8	2/3	–
Na^+ required for growth	–	1/5	8/8	3/3	2/9
Growth in 7.5% NaCl	Variable	Variable	8/8	3/3	1/11
DNA base ratio (range)	58.8–67.3	55.1–68.8	40.1–46.4	41.7–44.7	44.7–54.7
Named species included in study	*Ps. fluorescens* *Ps. putida* *Ps. cepacia* *Ps. fragi*	*Ps. acidovorans* *Ps. testosteroni* *Ps. maltophilia*	*A. haloplanktis* (MacLeod B-16) *Ps. atlantica*		*Ps. putrefaciens* *Ps. rubescens*
Generic designation (Lee *et al.* 1977)	*Pseudomonas*	*Pseudomonas*	*Alteromonas*	*Alteromonas*	*Alteromonas*

–, all strains negative.
* Number of strains positive/number of strains tested.
After Lee (1973).

19; ATCC 19855) previously referred to as a marine pseudomonad or a marine *Pseudomonas* sp. on which many studies on cell membrane structure and function have been carried out (Forsberg *et al.* 1970). This organism is not closely related to *Pseudomonas* and, therefore, it is unwise to extrapolate the results to *Ps. aeruginosa* as has been done (e.g. Meadow 1975). They also propose that the strains in group E, which includes *Ps. putrefaciens* and *Ps. rubescens*, be called *Alteromonas putrefaciens.* This species shows the variation in flagellation referred to in the introduction i.e. it is polar when grown in liquid medium, but peritrichous when grown on solid medium (Fig. 2). The strains in group D are regarded as *Alteromonas* sp. not yet assigned to any named species.

It is well known that some named *Pseudomonas* species give reactions which are different from those given in the description of the genus or the major groupings with which they are associated. Thus *Ps. maltophilia* and some phytopathogenic *Pseudomonas* spp. are oxidase negative; *Ps. cepacia* strains are arginine dihydrolase negative (Stanier *et al.* 1966; Ballard *et al.* 1970) whereas all other species in group A (Table 2) are positive. The presence of DNAase looks promising for the separation of the low mol % GC pseudomonads (*Alteromonas*) of groups C, D and E from the high mol % GC types in group B with one exception – *Ps. maltophilia*, the only high mol % GC strain tested which possessed DNAase.

The main findings of the study of intermediary metabolism was the presence of a tricarboxylic acid cycle and little else. The components of the cycle and substrates metabolically close to it, including some amino acids, were rapidly metabolized. The key enzymes of the Entner-Doudoroff pathway, 6-phosphogluconate dehydrase and 2-keto-3-deoxy-6-phosphogluconate aldolase together with some dehydrogenase enzymes of the pentose phosphate pathway were present in all strains. Thus a negative reaction in the glucose O-F test, given by many strains, does not mean the organism cannot metabolize glucose; they probably do not always use glucose as a principal energy yielding pathway. The Entner-Doudoroff pathway enzymes are induced with gluconate, produced from glucose by simple dehydrogenation. It has recently been shown that the route of glucose metabolism in chemostat cultures of *Ps. aeruginosa* is regulated by the amount of glucose available (Whiting *et al.* 1976). When the culture is growing under nitrogen limitation i.e. with an excess (0.8%) of glucose; most of the glucose is metabolized by extracellular routes to form gluconate and 2-oxogluconate, which accumulate giving an acid reaction. When the culture is glucose limited (0.5% glucose) the metabolism switched to the intracellular route, the Entner-Doudoroff and pentose phosphate pathways were functional and the sugar acids were also metabolized. Whiting *et al.* (1976) suggest that organisms such as *Ps. aeruginosa* and *Ps. fluorescens*, in their natural environment, can sequester glucose as gluconate or 2-oxogluconate which are not so readily utilized by other bacteria which may compete for available glucose. Thus the results of biochemical tests involving carbohydrates must be interpreted with caution, such switches in metabolism may account for some of the erratic and variable results between laboratories as reported by Sneath & Collins (1974).

We conclude that some of the pseudomonad groups of Shewan *et al.* (1960) have proved to be distinct but are rather heterogeneous. Organisms in groups I and II fall into the genus *Pseudomonas* (Buchanan & Gibbons 1974) and into group A (Table 2) and are true *Pseudomonas* spp. and rather heterogeneous in their properties. The organisms in the Shewan *et al.* (1960) groups III and IV are even more heterogeneous, falling into two genera and it is unlikely that organisms similar to those in groups C, D and E (Table 2) would be identified precisely by the scheme in Tables 1 and 2.

3. Vibrionaceae

The family Vibrionaceae – i.e. the facultatively anaerobic, generally oxidase positive, polar flagellate Gram negative rods – is now quite well established. There have been many studies on fairly extensive collections of the various groups within the family (Eddy 1960, 1962; Eddy & Carpenter 1964; Bain *et al.* 1965; Colwell 1970; Citarella & Colwell 1970; Hendrie *et al.* 1970; Baumann *et al.* 1971; Rüger 1972; Reichelt & Baumann 1973*b*; Simidu & Kaneko 1973; Popoff & Véron 1976), but only some of the more recent ones have used nutritional, DNA base ratio or DNA/DNA hybridization studies. However, in some cases the results of these more detailed studies do support the conclusions drawn from the traditional morphological and biochemical tests e.g. Reichelt & Baumann's (1973*b*) nutritional studies on luminous bacteria confirmed the Hendrie *et al.* (1970) groupings based on traditional tests together with some DNA base ratio determinations.

For identifying the genera within the family a fairly simple key based on that of Bain & Shewan (1968) can be used (Table 3). However it is quite evident that there are organisms which resemble the vibrio group in morphology but which do not fit this key e.g. DNA base ratios are outwith the range (Lee *et al.* 1976; Park & bin Jangi 1976). Of the features listed in Table 3 the genus *Aeromonas* has a quite distinct mol % GC. It is also distinguished from *Vibrio* and the luminous bacteria by having a lower optimum NaCl requirement. It is unlikely that any true *Aeromonas* species has ever been isolated from the marine environment, a view supported by Simidu & Kaneko (1973) who found that all the strains isolated from marine fish grouped with *Vibrio parahaemolyticus*/*Vibrio alginolyticus* or with *Photobacterium phosphoreum.* A marine strain isolated by Merkel *et al.* (1964) was named *Aeromonas proteolytica* and Schubert (Buchanan & Gibbons 1974) considers it an anaerogenic subspecies of *Aeromonas hydrophila.* However, Schubert does not include any

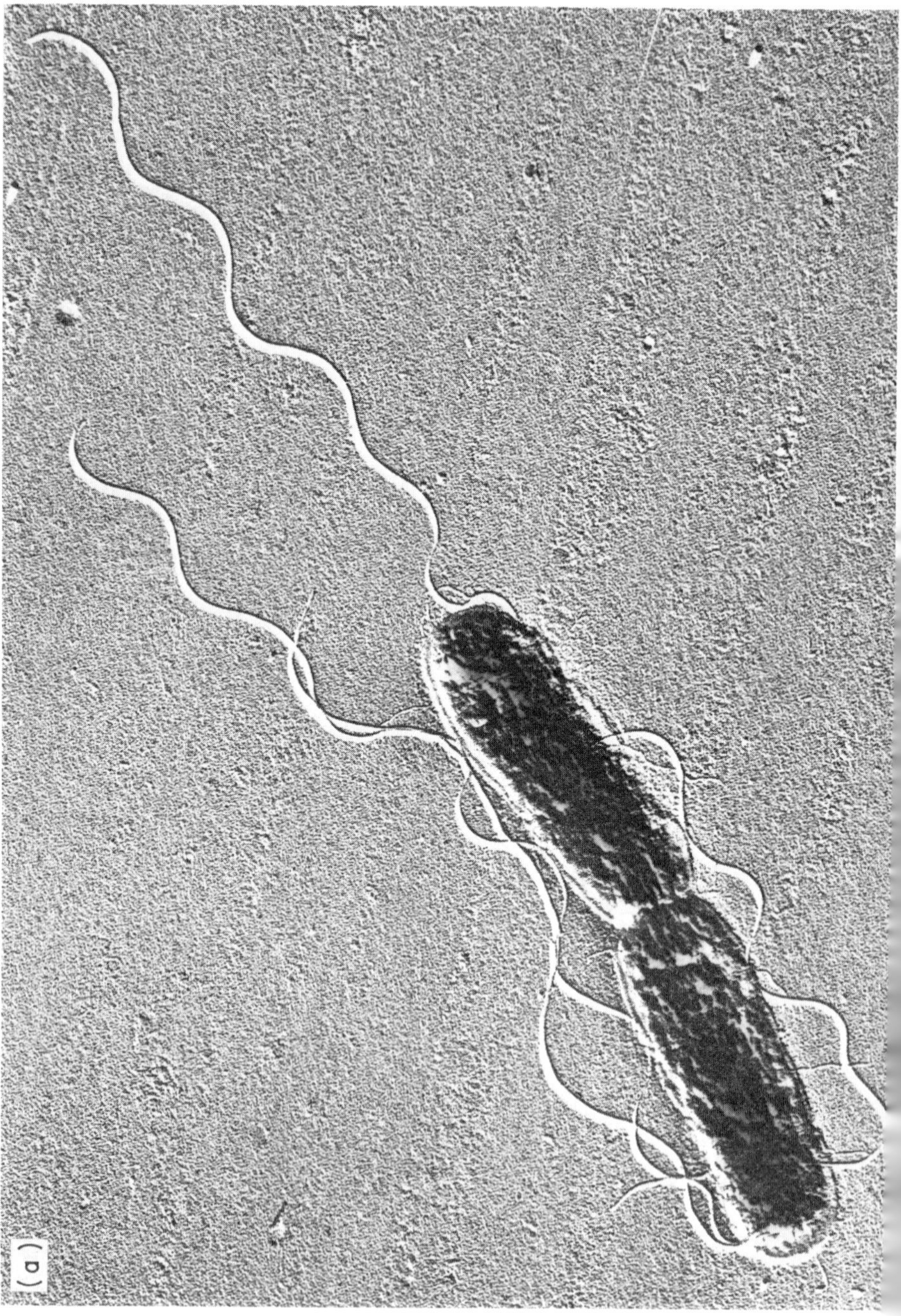
(a)

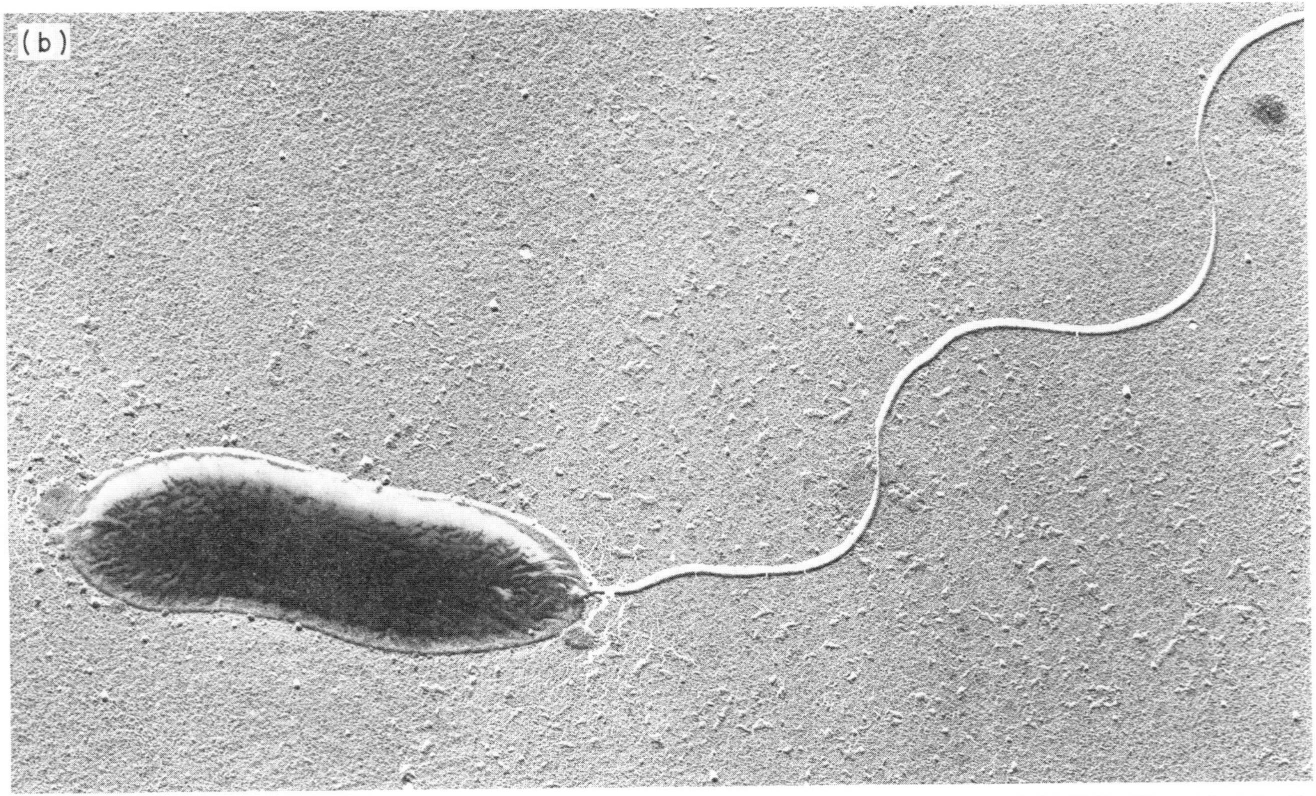

Fig. 2. Electron micrographs of *Alteromonas putrefaciens* grown at 20°C in (a) nutrient agar (× 21,000), (b) nutrient broth (× 28,000).

Table 3

Generic differentiation in the family Vibrionaceae

	Aeromonas	*Vibrio*	*Lucibacterium*	*Photobacterium*	*Plesiomonas*
Flagella	Polar†	Polar or mixed peritrichous*	Mixed peritrichous* or polar	Polar	Polar
Oxidase	+	+	+	d	+
Gas from glucose	d	–	–	d	–
Acid from inositol	–	–	–	–	+
Arginine dihydrolase	+	d	–	+	+
Lysine decarboxylase	–	d	+	+	+
Ornithine decarboxylase	–	d	+	–	+
0/129 sensitivity	–	+	–	+	+
Novobiocin sensitivty	–	+			+
Luminescence	–	d	+	d	–
DNA base ratio (range)	57–63	40–50	45–46	39–42	*ca.* 51

+, over 90% strains positive; –, over 90% strains negative; d, 10–90% strains positive.
* Mixed peritrichous is a thick polar flagellum with many finer lateral flagella.
† One species, *Aeromonas salmonicida*, is non-motile, no flagella.

information on DNA base composition in his description of *Aeromonas* and others have reported that the mol % GC value for '*A. proteolytica*' is much closer to the *Vibrio* (*Beneckea*) group (Baumann *et al.* 1971; McCarthy 1975). Popoff & Véron, (1976) suggest that '*A. proteolytica*' should be excluded from *Aeromonas* and call for the study of further strains to clarify the position. In our opinion it is unlikely that new isolates with characteristics similar to '*A. proteolytica*' would now be assigned to *Aeromonas*; they would be assigned to the *Vibrio* group.

The relationships between the other genera within the family are still a matter of dispute and more studies are required. Baumann *et al.* (1971) proposed that some *Vibrio* spp. of marine origin should be re-allocated to the genus *Beneckea* and Reichelt & Baumann (1973*b*) proposed that *Lucibacterium* should be included in *Beneckea*. The International Committee for Systematic Bacteriology, Subcommittee on the Taxonomy of Vibrios (1975) has rejected the suggestion that *Vibrio parahaemolyticus* should be regarded as *Beneckea* because it is not possible to make an objective assessment of the relationship between *V. parahaemolyticus* and the type species of *Beneckea, B. labra,* since no type of culture of the latter is available. Rüger (1972) suggested that certain marine *Vibrio* spp. be assigned to a new genus, *Marinovibrio*, but from the information in his paper they are still encompassed by the accepted definition of the genus *Vibrio*.

A problem often encountered in classification is that of genera based on the possession of one outstanding, easily recognizable character which tends to obscure the true relationships between strains. This is well demonstrated in this family by the luminous bacteria, which, in the past, some workers have tended to assign to one genus because they emit light, even though they shared many features with other genera. Apart from luminescence, the various species of luminous bacteria have no more in common with each other than they have with some non-luminous species. Liston (1955) found that a group of bacteria from the intestines of fish were identical except that some were luminous while others were not. We believe these are all strains of *Photobacterium phosphoreum*. Similarly, Simidu & Kaneko (1973) reported that light was emitted by only 16% of the strains which grouped with *Ph. phosphoreum* in their numerical analysis. Other species of luminous bacteria are also very similar to non-luminous isolates belonging to other genera (Hendrie *et al.* 1970; Hendrie *et al.* 1971; Reichelt & Baumann 1973*b*).

Strains of *Vibrio alginolyticus*, exhibit varied flagellation dependent on the growth conditions (cf *Alteromonas putrefaciens*). In liquid medium at 37°C they possess only a single polar flagellum (Fig. 3a), but on a solid medium at 20°C numerous peritrichous flagella are demonstrated (Fig. 3b). The International Committee for Systematic Bacteriology, Subcommittee on the Taxonomy of Vibrios (1975) recognized these flagellar morphotypes as infrasubspecific forms

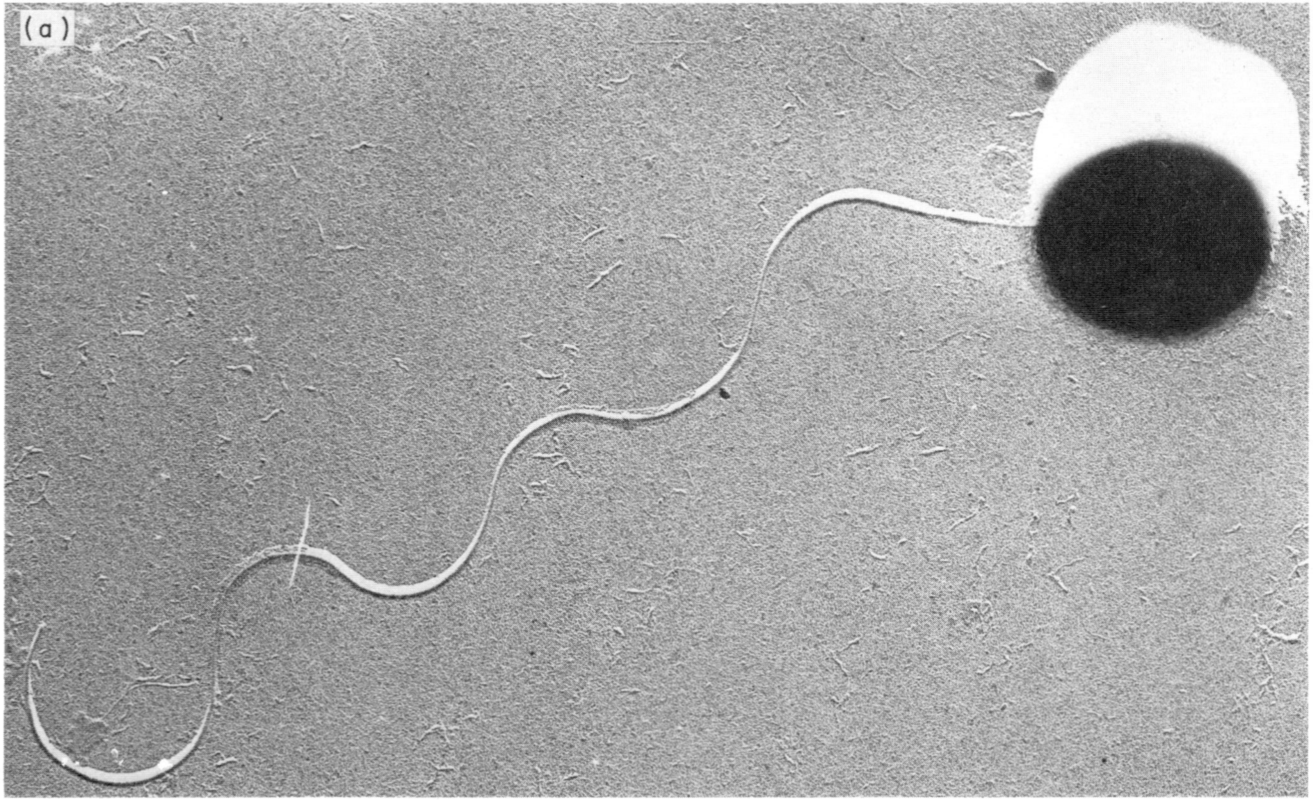

(a)

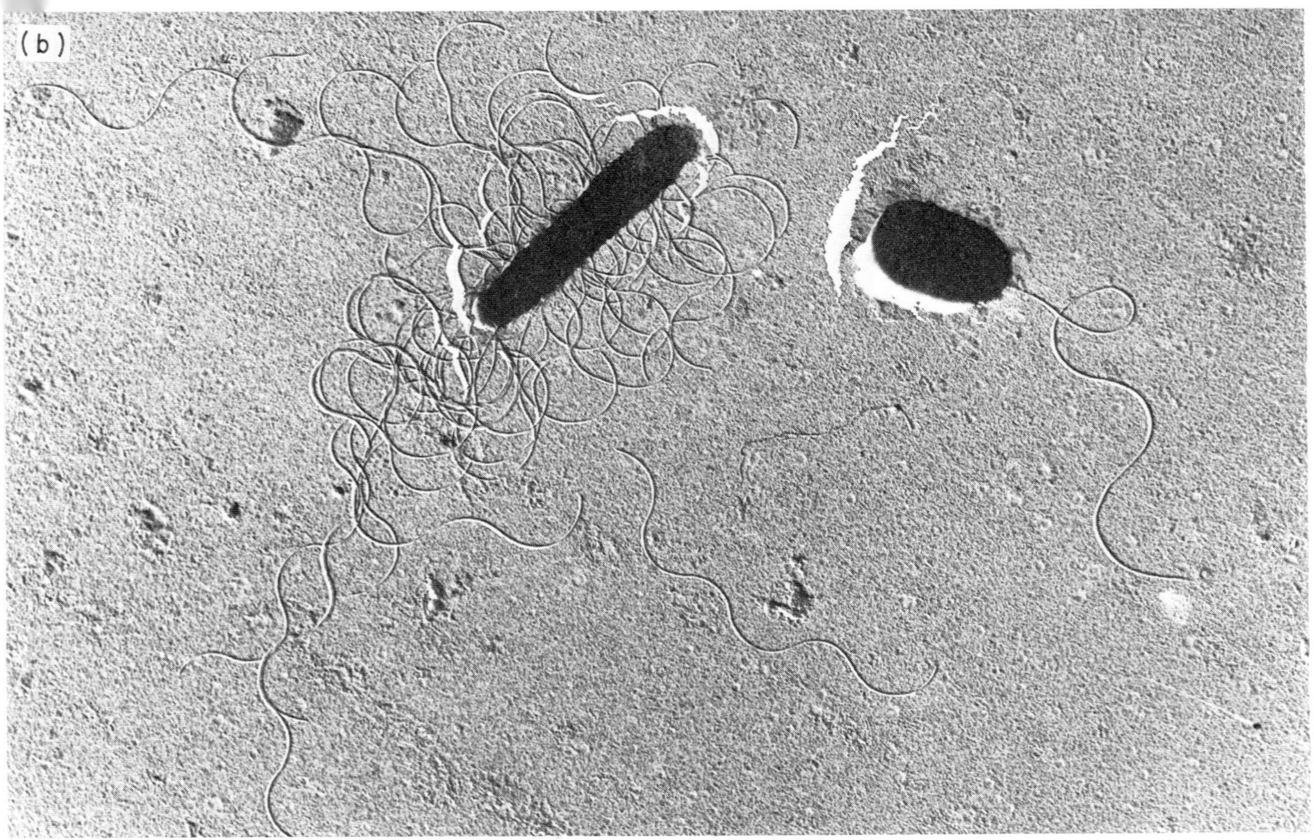

Fig. 3. Electron micrographs of *Vibrio alginolyticus* grown in (a) sea water broth at 37°C (× 35,000), (b) sea water agar at 20°C (× 17,500).

and, therefore, the presence of peritrichous flagella under certain conditions whilst polar flagella only are present in other conditions does not eliminate a strain from a genus described as possessing polar flagella.

4. The *Achromobacter/Alcaligenes* Range

It is not surprising that the *Achromobacter/Alcaligenes* group should contain such a varied collection of strains as it was created to accommodate unreactive colourless rods which did not fit into any other genus (see e.g. Ingram & Shewan 1960). A taxonomic study of over 400 strains, about 120 motile and 280 non-motile, from many different sources has been undertaken (De Ley *et al.* 1967, 1970). The strains motile by means of peritrichous flagella have 58–68 mol % GC, whilst the non-motile strains range from 39 to 47 mol % GC. Almost all our isolates from marine and fish sources fall into the latter category.

The motile organisms, now assigned to *Alcaligenes* (Buchanan & Gibbons 1974) form three subgroups on the basis of their DNA base compositions and complement of enzymes of the Entner-Doudoroff pathway (De Ley *et al.* 1970). Similar groups are evident in similarity diagrams and dendrograms constructed from nutritional and physiological data. It must be emphasized that most organisms gave negative results in over 100 of the 150 tests used to obtain data. Such data on their own are not satisfactory for taxonomic purposes and the groupings obtained must be supported by data from other investigations, as has been done in this case. There are few marine strains in these groups. Baumann *et al.* (1972) assigned some of their strains, from a geographically restricted marine environment, to *Alcaligenes* but not all would be accepted in the genus as defined in *Bergey's Manual of Determinative Bacteriology* (Buchanan & Gibbons 1974) and they were not directly compared with well established *Alcaligenes* species.

A numerical taxonomic study has been carried out on the non-motile strains currently classified as *Moraxella*-like or *Acinetobacter* spp. analogous to the studies of Thornley (1967), Baumann *et al.* (1968*a, b*) and Pagel & Seyfried (1976). The marine strains, which are mostly from fish or northern waters and oxidase positive, form clusters distinct from the oxidase-negative *Acinetobacter* spp. and some of the marker strains included e.g. the *Branhamella/Neisseria* group. Again it must be stressed that only a small proportion, 30 of the 150 tests performed gave usable information.

In transformation tests the *Acinetobacter* spp. could transform the mutants of *Acinetobacter calcoaceticus* used by Juni (1972). The *Moraxella*-like strains could not transform either the *A. calcoaceticus* mutants or the mutant of *Moraxella osloensis* (Juni 1974), so that it is unlikely that these marine strains have any close relationship with the clinically important *Moraxella* spp. Electrophoresis of detergent solubilized cells indicates that the marine *Moraxella*-like strains are all very similar and may represent only one species (unpublished).

The metabolism of the *Achromobacter/Alcaligenes* group is centred on the tricarboxylic acid cycle and energy is obtained from the cycle and compounds chemically close to it (Gibson 1968). This helps to explain the poor results in carbon utilization tests and lack of reactivity in traditional bacteriological tests. Some of the *Moraxella*-like strains can produce acid from glucose and other carbohydrates. It was found that such strains have one enzyme, an aldose dehydrogenase, which can efficiently and completely oxidize aldoses to the aldonic acids, which accumulate (Gibson 1968) (Table 4). The enzyme is different from that described by Hauge (1960) isolated from *Bacterium anitratum* (*Acinetobacter*). Thus in any taxonomic or identification scheme, acid from carbohydrates should be counted as only one feature because only one enzyme of broad specificity is involved. Some *Acinetobacter* spp. can further oxidize some pentonic acids, after a considerable delay (Baumann *et al.* 1968*b*), to α-oxoglutarate and then through the tricarboxylic acid cycle. Thus, much caution is needed in the interpretation of carbohydrate utilization tests.

There is little change from the former procedure in identifying this group of organisms but the nomenclature is modified. The motile strains are assigned to *Alcaligenes* (Buchanan & Gibbons 1974), the oxidase-negative non-motile strains are assigned to *Acinetobacter* and confirmed by use of the transformation assay,

Table 4
Specificity of the aldose dehydrogenase from Moraxella-*like strain NCMB 98*

Substrate	Rate relative to glucose (100)	$K_m \times 10^4$ M*
D-Glucose	100	2.55
D-Galactose	100	8.25
D-Mannose	85	46.25
D-Ribose	100	8.25
D-Xylose	100	12.25
L-Arabinose	100	5.22
D-Arabinose	11	
D-2-Deoxyglucose	100	
D-2-Deoxyribose	70	
D-Glucosamine	34	
D-Fructose	0	
D-Lyxose	0	
L-Lyxose	0	
Sucrose	0	
Lactose	11	
Maltose	27	
Cellobiose	0	

* Average of 4 experiments.

and the oxidase positive non-motile strains are called *Moraxella*-like spp. because they cannot be assigned to the genus *Moraxella* as it is reserved for parasitic strains as it is presently circumscribed (Buchanan & Gibbons 1974). Table 5 shows tests which differentiate the non-motile genera. Some of the former named *Achromobacter* spp. have been allocated to other genera because their properties were originally erroneously described or recorded (Hendrie *et al.* 1974).

5. *Flavobacterium* and Other Yellow Pigmented Rods

Like *Achromobacter*, the genus *Flavobacterium* is not a natural entity (Ingram & Shewan 1960). As defined in *Bergey's Manual of Determinative Bacteriology* (Buchanan & Gibbons 1974), the genus consists of Gram negative bacteria, motile by peritrichous flagella or non-motile without any gliding or swarming motion and with a DNA base ratio in the ranges 63–70 mol % GC and 30–42 mol % GC. Strains exhibiting gliding motion and having DNA base ratios in the lower range are assigned to the genera *Cytophaga* and *Flexibacter.* It is perhaps unfortunate that so many isolates from aquatic sources fall into such ill-defined genera. Using a more exacting microscopic technique than normal, Perry (1973) detected gliding, twitching or flexing in organisms not previously regarded as having this property, including some named *Flavobacterium* spp.

A similarity diagram was constructed by Floodgate & Hayes (1963) from a study of 62 strains of yellow pigmented bacteria of marine origin. The addition of DNA base ratios to this diagram confirms that the pleistrons are quite distinct (Fig. 4). Some of their strains were included by McMeekin *et al.* (1972) in a study of yellow pigmented isolates from poultry, meat products and processing factory water. Fifty-nine strains fell into 16 groups which with Floodgate & Hayes (1963) groupings and DNA base ratios added are shown in Fig. 5. Two groups require comment: group F is identified as *Alteromonas putrefaciens* and NCTC 10016 is one of the strains of *Flavobacterium meningosepticum* used by Owen & Snell (1976) in DNA/DNA hybridization studies which showed that flavobacteria pathogenic to man are closely related to each other but not to other flavobacteria and flexibacteria.

Flavobacteria from cannery cooling water were studied by Bean & Everton (1969) and the work was extended by Byrom (1971) who could place only 108 of the 250 strains in phenons of more than 10 strains. DNA base ratios have been determined on representative strains of each phenon, most of which contain organisms with a wide range of mol % GC (unpublished) (Fig. 6). In addition, Byrom stated that the named *Flavobacterium* spp. he used fell into different pleistrons containing less than 10 strains.

We have studied over 200 strains of flavobacteria from a variety of sources including some from the above studies, from culture collections and isolates

Table 5

Generic identification of non-motile, non-fermentative Gram negative rods

Colony appearance	Morphology	Oxidase	Sensitivity to Penicillin	Sensitivity to Polymyxin B	Gliding motility	Attack on polymers (agar, cellulose or chitin etc)	
Off-white, translucent to opaque, convex	Short stout rods or coccobacilli. Deeply stained	+	+	+	NA	NA	*Moraxella*-like
Off-white, translucent, low convex	Short rods	–	–	+	NA	NA	*Acinetobacter*
Yellow or orange, translucent, convex or low convex, often with a spreading margin	Slender rods of varying length. Longer forms appear to have a flexible axis	+	–	–	+	+	*Cytophaga*
						–	*Flexibacter*
Yellow or orange, translucent, convex, margin entire	Short rods	+ or –	–	+	–	NA	*Flavobacterium*

NA, test not applicable.

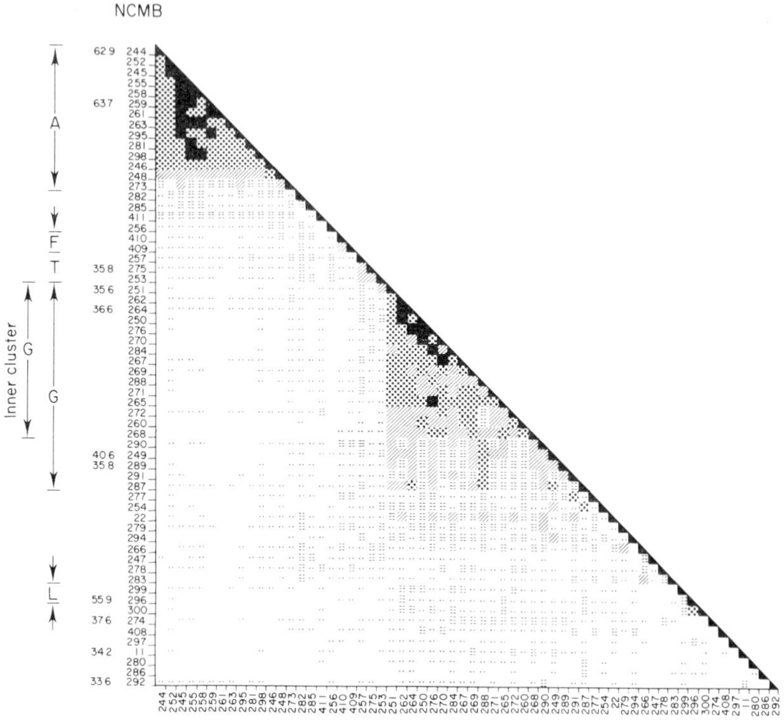

Fig. 4. Modified similarity diagram of Floodgate & Hayes (1963) showing the DNA base ratios (mol % GC) of representative strains from the various pleistons. DNA base ratio values from De Ley & Van Muylem (1963) or our unpublished results.

from iced scampi and carcass meat. They have proved to be very heterogeneous and it was only possible to subject the strains to a battery of 49 tests including motility, oxidase reaction, utilization of single carbon sources and antibiotic sensitivities. A similarity diagram has been constructed from the analysis of the data and an abbreviated form along with some DNA base ratio values is shown in Fig. 7. It can be seen from this figure that, even at a 60% similarity single linkage of positive matches, the phenons are small and usually separate from each other and some phenons contain organisms of widely varying DNA base ratio values. Furthermore, none of the named strains included in the study grouped with any of the other strains even at a 60% similarity level. Among the named strains included was the yellow pigmented *Achromobacter georgiopolitanum* ATCC 25020. In addition to its yellow pigment this organism has other features in common with the *Flavobacterium*/*Cytophaga*/*Flexibacter* complex, e.g. it is insensitive to polymyxin B and sensitive to 0/129, a pattern rarely, if ever, found in *Moraxella*-like or *Acinetobacter* spp. to which a non-motile *Achromobacter* sp. would now be assigned.

Strain	Mol % GC	Floodgate & Hayes group	Number of strains with % S: 61-70	71-80	81-90	91-100
1	32·9					
2						
17			4 (Group A)	4	4	4
18						
3						
12					2	2
6				5		
7					3	
32	37·8					
33						
39			15			2
36			(Group Oi)			2
37						
40	39·3			10	10	
35						2
38						
43						
42	36·4					
44						
24						
25	39·0			3	2	
26						
41						
67			8			
53			(Group Oii)			
68				3	3	
69						
45						
56						
60	35·4		4	4	4	4
65	34·4		(Group S)			
61						
66						
NCMB 249	40·6					
NCMB 251	35·6	G	3	3	3	2
NCMB 264	36·6		(Group G)			
NCIB 9337			2	2	2	
NCIB 9059	32·9					
49	71·3		2	2	2	2
51						
4						
13	69·3			4		
8						
16			6			
NCMB 244	62·9		(Group Yi)			
NCMB 259	63·7	A		2	2	2
23						
48	61·5		3	3		
62			(Group Yii)			
34						
58						
59	47·8		5	5	5	5
64			(Group F)			
63						
46			2	2		
57	39·6					
NCMB 275	35·8	T				
NCMB 296	55·9	L				
NCTC 10016	38·3					

Fig. 5. Modified table of McMeekin *et al.* (1972) showing the DNA base ratio (mol % GC) of representative strains and the relationship to the pleista of Floodgate & Hayes (1963).

From the information that we, and others, have obtained we conclude that rapid accurate identification of many of these yellow pigmented bacteria is difficult. The few features which separate the genera are shown in Table 5, but there is no published satisfactory key for rapid identification beyond this level. the DNA/DNA hybridization studies of Owen & Snell (1976) suggest that the

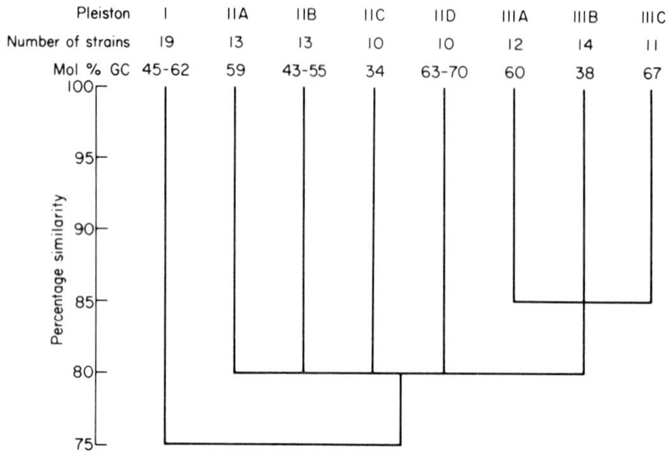

Fig. 6. Modified dendrogram of Byrom (1971) showing the range of DNA base ratio values of representative strains from each division.

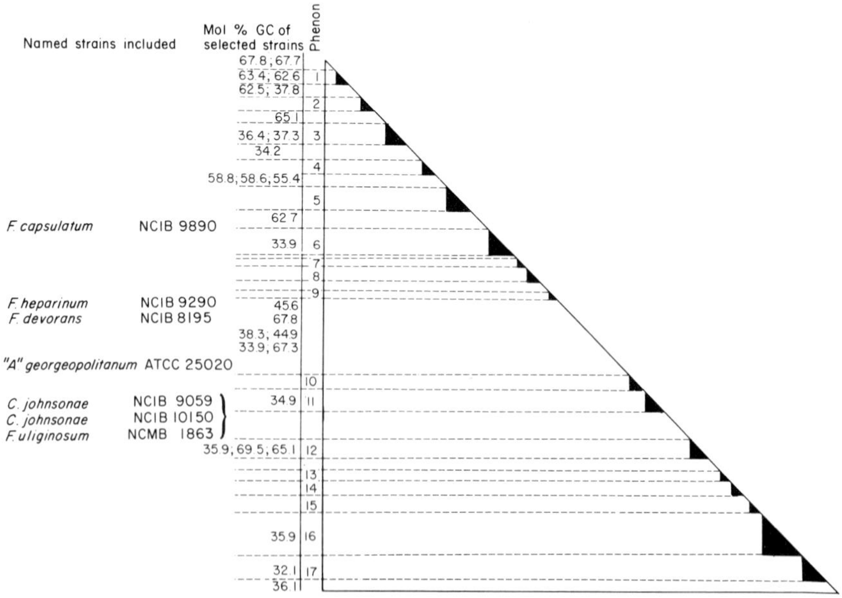

Fig. 7. Simplified similarity (S) diagram constructed from a single linkage analysis of positive matches only from over 200 yellow pigmented rods. Phenons 1–17 were selected at 60%S. The position of named marker strains and the DNA base ratios of representative strains has been added.

medically important *Flavobacterium* spp. are genetically distinct from the non-medical types, but it is difficult to separate them phenotypically. Transformation assays with the mutant of *Acinetobacter calcoaceticus* (Juni 1972) indicates that the oxidase-negative *Flavobacterium* spp. are not related to *Acinetobacter*. Clearly there is a need for more work on this group of organisms to elucidate the situation. The situation in *Bergey's Manual of Determinative Bacteriology* (Buchanan & Gibbons 1974) is far from satisfactory and more work is needed to evaluate the relationships and/or differences between the low mol % GC species of *Flavobacterium* and the *Cytophaga/Flexibacter* group.

6. Conclusions

It is clear that many of the organisms described by Shewan *et al.* (1960) and identified by their scheme did not fit comfortably into the genera existing then or now. The difficulties encountered in identifying such strains were, and remain real. While some can be identified with confidence many will continue to masquerade under some inappropriate name until new species or even genera are established to accommodate them.

The statement that the identification scheme had no taxonomic pretensions and expressed no phylogenetic relationships (Shewan *et al.* 1960) has been amply justified but it does raise the question of the relationships between some of our groups of aquatic bacteria. It is evident from the results discussed that many of the organisms in different groups show similarities to each other. They give a uniformly negative response in many bacteriological tests; they look alike; they are similar in DNA base composition (40–50 mol % GC), in intermediary metabolism and in their ecological distribution. It seems reasonable, therefore, to suggest that they may be related.

These bacteria, characteristically survive for long periods in low nutrient and in adverse environments. Cells which die are likely to spill their contents into the environment and this 'debris' would be available to the surviving cells. It is known that many cells can take up large pieces of DNA (e.g. as in the transformation assays referred to earlier, see also Carlile & Skehel 1974) and such processes can affect the evolution of the cells. If this happens in aquatic environments, then the overall effect would be a convergence of evolution in bacteria in the habitat so that they become more alike. Thus many of the aquatic bacteria may be more related than presently supposed.

7. Acknowledgements

The electron micrographs were prepared by W. Hodgkiss. DNA base ratio determinations were carried out by D. G. McLeod and I. D. Ogden.

8. References

ALLEN, R. D. & BAUMANN, P. 1971 Structure and arrangement of flagella in species of the genus *Beneckea* and *Photobacterium fischeri. Journal of Bacteriology* **107**, 295–302.

BAIN, N. & SHEWAN, J. M. 1968 Identification of *Aeromonas, Vibrio* and related organisms. In *Identification Methods for Microbiologists*, Part B, eds Gibbs, B. M. & Shapton, D. A. London & New York: Academic Press.

BAIN, N., TAYLOR, M. E. & SHEWAN, J. M. 1965 Some preliminary data on a study of the *Vibrio* and allied genera. In Reports from the Conference on the Taxonomy of Bacteria held in the Czechoslovak Collection of Microorganisms, J. E. Purkyně University, Brno on September 29th–October 1st, 1965. *Spisy Přírodovědecké Fakulty University J. E. Purkyně, Brno*, Serie K **35**, 298–299.

BALLARD, R. W., PALLERONI, N. J., DOUDOROFF, M., STANIER, R. Y. & MANDEL, M. 1970 Taxonomy of the aerobic pseudomonads: *Pseudomonas cepacia, P. marginata, P. alliicola* and *P. caryophylli. Journal of General Microbiology* **60**, 199–214.

BAUMANN, L., BAUMANN, P., MANDEL, M. & ALLEN, R. D. 1972 Taxonomy of aerobic marine eubacteria. *Journal of Bacteriology* **110**, 402–429.

BAUMANN, P., DOUDOROFF, M. & STANIER, R. Y. 1968*a* Study of the *Moraxella* group. I. Genus *Moraxella* and the *Neisseria catarrhalis* group. *Journal of Bacteriology* **95**, 58–73.

BAUMANN, P., DOUDOROFF, M. & STANIER, R. Y. 1968*b* A study of the *Moraxella* group. II. Oxidase-negative species (genus *Acinetobacter*) *Journal of Bacteriology* **95**, 1520–1541.

BAUMANN, P., BAUMANN, L. & MANDEL, M. 1971 Taxonomy of marine bacteria: the genus *Beneckea. Journal of Bacteriology* **107**, 268–294.

BEAN, P. G. & EVERTON, J. R. 1969 Observations on the taxonomy of chromogenic bacteria isolated from cannery environments. *Journal of Applied Bacteriology* **32**, 51–59.

BUCHANAN, R. E. & GIBBONS, N. E. (eds) 1974 *Bergey's Manual of Determinative Bacteriology*, 8th Edn. Baltimore: Williams & Wilkins Co.

BYROM, N. A. 1971 The Adansonian taxonomy of some cannery flavobacteria. *Journal of Applied Bacteriology* **34**, 339–346.

CARLILE, M. J. & SKEHEL, J. J. (eds) 1974 *Evolution in the Microbial World.* Symposium 24 of the Society for General Microbiology. Cambridge: Cambridge University Press.

CITARELLA, R. V. & COLWELL, R. R. 1970 Polyphasic taxonomy of the genus *Vibrio*: polynucleotide sequence relationships among selected *Vibrio* species. *Journal of Bacteriology* **104**, 434–442.

COLWELL, R. R. 1970 Polyphasic taxonomy of the genus *Vibrio*: numerical taxonomy of *Vibrio cholerae, Vibrio parahaemolyticus*, and related *Vibrio* species. *Journal of Bacteriology* **104**, 410–433.

DE LEY, J. & VAN MUYLEM, J. 1963 Some applications of deoxyribonucleic acid base composition in bacterial taxonomy. *Antonie van Leeuwenhoek* **29**, 344–358.

DE LEY, J., BAIN, N. & SHEWAN, J. M. 1967 Taxonomy of the *Achromobacter* and allied species. *Nature, London* **214**, 1037–1038.

DE LEY, J., KERSTERS, K., KHAN-MATSUBARA, J. & SHEWAN, J. M. 1970 Comparative D-gluconate metabolism and DNA base composition in *Achromobacter* and *Alcaligenes. Antonie van Leeuwenhoek* **36**, 193–207.

EDDY, B. P. 1960 Cephalotrichous, fermentative Gram-negative bacteria: the genus *Aeromonas. Journal of Applied Bacteriology* **23**, 216–249.

EDDY, B. P. 1962 Further studies on *Aeromonas.* I. Additional strains and supplementary biochemical tests. *Journal of Applied Bacteriology* **25**, 137–146.

EDDY, B. P. & CARPENTER, K. P. 1964 Further studies on *Aeromonas.* II. Taxonomy of *Aeromonas* and C27 strains. *Journal of Applied Bacteriology* **27**, 96–109.

FLOODGATE, G. D. & HAYES, P. R. 1963 The Adansonian taxonomy of some yellow pigmented marine bacteria. *Journal of General Microbiology* **30**, 237–244.

FORSBERG, C. W., COSTERTON, J. W. & MACLEOD, R. A. 1970 Quantitation, chemical characteristics, and ultrastructure of the three outer cell wall layers of a Gram-negative bacterium. *Journal of Bacteriology* **104**, 1354–1368.

GIBSON, D. M. 1968 Comparative carbohydrate metabolism of some Gram-negative bacteria belonging to the *Achromobacter* and related genera. Ph.D. Thesis, University of Aberdeen.

HAUGE, J. G. 1960 Kinetics and specificity of glucose dehydrogenase from *Bacterium anitratum*. *Biochimica et Biophysica Acta* **45**, 263–269.

HENDRIE, M. S., HODGKISS, W. & SHEWAN, J. M. 1964 Considerations on organisms of the Achromobacter-Alcaligenes group. *Annales de l'Institut Pasteur de Lille* **15**, 43–59.

HENDRIE, M. S., HODGKISS, W. & SHEWAN, J. M. 1970 The identification, taxonomy and classification of luminous bacteria. *Journal of General Microbiology* **64**, 151–169.

HENDRIE, M. S., HODGKISS, W. & SHEWAN, J. M. 1971 Proposal that *Vibrio marinus* (Russell 1891) Ford 1927 be amalgamated with *Vibrio fischeri* (Beijerinck 1889) Lehmann and Neumann 1896. *International Journal of Systematic Bacteriology* **21**, 217–221.

HENDRIE, M. S., HOLDING, A. J. & SHEWAN, J. M. 1974 Emended descriptions of the genus *Alcaligenes* and of *Alcaligenes faecalis* and proposal that the generic name *Achromobacter* be rejected: status of the named species of *Alcaligenes* and *Achromobacter*. *International Journal of Systematic Bacteriology* **24**, 534–550.

HERBERT, R. A., HENDRIE, M. S., GIBSON, D. M. & SHEWAN, J. M. 1971 Bacteria active in the spoilage of certain sea foods. *Journal of Applied Bacteriology* **34**, 41–50.

HILL, L. R. 1966 An index to deoxyribonucleic acid base compositions of bacterial species. *Journal of General Microbiology* **44**, 419–437.

HUGH, R. & LEIFSON, E. 1953 The taxonomic significance of fermentative versus oxidative metabolism of carbohydrates by various Gram negative bacteria. *Journal of Bacteriology* **66**, 24–26.

INGRAM, M. & SHEWAN, J. M. 1960 Introductory reflections on the *Pseudomonas-Achromobacter* group. *Journal of Applied Bacteriology* **23**, 373–378.

INTERNATIONAL COMMITTEE ON SYSTEMATIC BACTERIOLOGY. SUBCOMMITTEE ON THE TAXONOMY OF VIBRIOS 1975 Minutes of Closed Meeting, 3rd September, 1974. *International Journal of Systematic Bacteriology* **25**, 389–391.

JUNI, E. 1972 Interspecies transformation of *Acinetobacter*: genetic evidence for a ubiquitous genus. *Journal of Bacteriology* **112**, 917–931.

JUNI, E. 1974 Simple genetic transformation assay for rapid diagnosis of *Moraxella osloensis*. *Applied Microbiology* **27**, 16–24.

KERSTERS, K. 1967 Rapid screening assays for soluble and particulate bacterial dehydrogenases. *Antonie van Leeuwenhoek* **33**, 63–72.

KING, E. O., WARD, M. K. & RANEY, D. E. 1954 Two simple media for the demonstration of pyocyanin and fluorescin. *Journal of Laboratory and Clinical Medicine* **44**, 301–307.

KOVACS, N. 1956 Identification of *Pseudomonas pyocyanea* by the oxidase reaction. *Nature, London* **178**, 703.

LEE, J. V. 1973 Some comparative biochemical and physiological studies on selected Gram-negative bacteria. Ph.D. Thesis, University of Aberdeen.

LEE, J. V., DONOVAN, T. J. & FURNISS, A. L. 1976 The taxonomy of some oxidase-negative Vibrio-like aquatic bacteria. *Journal of Applied Bacteriology* **41**, iv–v.

LEE, J. V., GIBSON, D. M. & SHEWAN, J. M. 1977 A numerical taxonomic study of some *Pseudomonas*-like marine bacteria. *Journal of General Microbiology* **98**, 439–451.

LISTON, J. 1955 A group of luminous and non-luminous bacteria from the intestine of flatfish. *Journal of General Microbiology* **12**, i.

LOCKHART, W. R. & LISTON, J., editors 1970 *Methods for Numerical Taxonomy.* Bethesda, Md: American Society for Microbiology.

MCCARTHY, D. H. 1975 *Aeromonas proteolytica* – a halophilic aeromonad? *Canadian Journal of Microbiology* **21**, 902–904.

MCMEEKIN, T. A., STEWART, D. B. & MURRAY, J. G. 1972 The Adansonian taxonomy and the deoxyribonucleic acid base composition of some Gram negative, yellow pigmented rods. *Journal of Applied Bacteriology* **35**, 129–137.

MANDEL, M. 1966 Deoxyribonucleic acid base composition in the genus *Pseudomonas*. *Journal of General Microbiology* **43**, 273–292.

MEADOW, P. M. 1975 Wall and membrane structures in the genus *Pseudomonas*. In *Genetics and Biochemistry of Pseudomonas,* eds Clarke, P. H. & Richmond, M. H. London & New York: John Wiley & Sons.

MERKEL, J. R., TRAGANZA, E. D., MUKHERJEE, B. B., GRIFFIN, T. D. & PRESCOTT, J. M. 1964 Proteolytic activity and general characteristics of a marine bacterium, *Aeromonas proteolytica* sp. n. *Journal of Bacteriology* **87**, 1227–1233.

OWEN, R. J. & SNELL, J. J. S. 1976 Deoxyribonucleic acid reassociation in the classification of flavobacteria. *Journal of General Microbiology* **93**, 89–102.

PAGEL, J. E. & SEYFRIED, P. L. 1976 Numerical taxonomy of aquatic Acinetobacter isolates. *Journal of General Microbiology* **95**, 220–232.

PARK, R. W. A. & BIN JANGI, S. 1976 Surface water vibrios – a mixed bunch. *Journal of Applied Bacteriology* **41**, v–vi.

PERRY, L. B. 1973 Gliding motility in some non-spreading flexibacteria. *Journal of Applied Bacteriology* **36**, 227–232.

POPOFF, M. & VÉRON, M. 1976 A taxonomic study of the *Aeromonas hydrophila–Aeromonas punctata* group. *Journal of General Microbiology* **94**, 11–22.

QUIGLEY, M. M. & COLWELL, R. R. 1968 Proposal of a new species *Pseudomonas bathycetes. International Journal of Systematic Bacteriology* **18**, 241–252.

REICHELT, J. L. & BAUMANN. P. 1973*a* Change of the name *Alteromonas marinopraesens* (ZoBell and Upham) Baumann *et al.* to *Alteromonas haloplanktis* (ZoBell and Upham) comb. nov. and assignment of strain ATCC 23821 (*Pseudomonas enalia*) and strain c-A1 of De Voe and Oginsky to this species. *International Journal of Systematic Bacteriology* **23**, 438–441.

REICHELT, J. L. & BAUMANN, P. 1973*b* Taxonomy of the marine luminous bacteria. *Archiv für Mikrobiologie* **94**, 283–330.

RHODES, M. E. 1965 Flagellation as a criterion for the classification of bacteria. *Bacteriological Reviews* **29**, 442–465.

ROSYPAL, S. & ROSYPALOVÁ, A. 1966 Genetic, phylogenetic and taxonomic relationships among bacteria as determined by their deoxyribonucleic acid base composition. *Folia – Facultatis Scientiarum Naturalium Universitatis Purkynianae Brunensis* **VII** (3), 1–90.

RÜGER, H.-J. 1972 Taxonomic studies on marine bacteria from North Sea sediments: genus *Vibrio*. Proposal for combining marine vibrio-like bacteria in a new genus *Marinovibrio. Veröffentlichungen des Institut für Meeresforschung in Bremerhaven* **13**, 239–254.

SHEWAN, J. M., HODGKISS, W. & LISTON, J. 1954 A method for the rapid differentiation of certain non-pathogenic asporogenous bacilli. *Nature, London* **173**, 208–209.

SHEWAN, J. M., HOBBS, G. & HODGKISS, W. 1960 A determinative scheme for the identification of certain genera of Gram-negative bacteria, with special reference to the Pseudomonadaceae. *Journal of Applied Bacteriology* **23**, 379–390.

SIMIDU, U. & KANEKO, E. 1973 A numerical taxonomy of *Vibrio* and *Aeromonas* from normal and diseased marine fish. *Bulletin of the Japanese Society of Scientific Fisheries* **39**, 689–703.

SNEATH, P. H. A. 1957*a* Some thoughts on bacterial classification. *Journal of General Microbiology* **17**, 184–200.

SNEATH, P. H. A. 1957*b* The application of computers to taxonomy. *Journal of General Microbiology* **17**, 201–226.

SNEATH, P. H. A. 1972 Computer taxonomy. In *Methods in Microbiology*, **Vol. 7A** eds Norris, J. R. & Ribbons, D. W. London & New York: Academic Press.

SNEATH, P. H. A. & COLLINS, V. G., editors 1974 A study in test reproducibility between laboratories: report of a Pseudomonas Working Party. *Antonie van Leeuwenhoek* **40**, 481–527.

SOKAL, R. R. & SNEATH, P. H. A. 1963 *Principles of Numerical Taxonomy.* San Francisco & London: W. H. Freeman & Co.

STANIER, R. Y., PALLERONI, N. J. & DOUDOROFF, M. 1966 The aerobic pseudomonads: a taxonomic study. *Journal of General Microbiology* **43**, 159–272.

THORNLEY, M. J. 1967 A taxonomic study of *Acinetobacter* and related genera. *Journal of General Microbiology* **49**, 211–257.

WHITING, P. H., MIDGLEY, M. & DAWES, E. A. 1976 The role of glucose limitation in the regulation of the transport of glucose, gluconate and 2-oxogluconate, and of glucose metabolism in *Pseudomonas aeruginosa. Journal of General Microbiology* **92**, 304–310.

Nitrogen Assimilation in Marine Environments

R. A. HERBERT AND C. M. BROWN

Department of Biological Sciences, University of Dundee, Scotland

AND

S. O. STANLEY

Dunstaffnage Marine Laboratory, Oban, Argyll, Scotland

CONTENTS

1. Introduction

NITROGEN is an essential element required by all biological systems, and consequently its availability has frequently been implicated as one of the factors limiting primary productivity in aquatic environments (Gerloff & Skoog 1957; Thomas 1970; Skelef *et al.* 1971). It is therefore important to establish the forms in which nitrogen is available, the mechanisms by which they are transformed, the organisms involved and the dynamics of these processes *in situ.*

The principal forms of nitrogen occurring in aquatic environments, excluding dissolved nitrogen gas, are ammonia (NH_4–N), nitrate (NO_3–N), nitrite (NO_2–N) and organic N. Due to complex interactions the levels of NH_4–N, NO_3–N and NO_2–N fluctuate considerably both seasonally and, more importantly, even within a 24 h period. Considerable caution must therefore be exercised when interpreting inorganic N levels in aquatic environments. In general, nitrogen levels are highest during winter and early spring due to increased precipitation and land drainage and decline sharply during middle and late summer as a result of assimilation by phyto- and zoo-plankton.

Oligotrophic waters are considered to be those with inorganic N levels $<1 \times 10^{-6}$ M nitrogen, and eutrophic systems those which exceed this level

(Vollenwieder 1968). In oceans, inorganic N levels are normally low and correspond to those observed in oligotrophic fresh waters. Sverdrup *et al.* (1946) quoted levels of NO_3–N ranging between 0.07–3.0 x 10^{-6} M N; NO_2–N, 0.007–0.25 x 10^{-6} M N and NH_4–N, 0.03–0.25 x 10^{-6} M N. However, inorganic nitrogen levels vary both geographically and with depth. Where upwelling occurs, such as the Peru current, surface NO_3–N levels of 1.4–2.1 x 10^{-6} M N have been reported (Wooster *et al.* 1965; Eppley *et al.* 1970). Lower values have been observed in the Sargasso Sea, off Bermuda (Riley 1957) and Sagami Bay, Japan (Miyazaki *et al.* 1973). In Loch Etive, north west Scotland, Solorzano & Grantham (1975) recorded a NO_3–N maximum of 0.4 x 10^{-6} M N in winter and early spring which fell below detectable levels later in the year concurrent with the development of a phytoplankton bloom. Nitrite usually occurs at low concentrations in surface oceanic waters but may reach appreciable levels (0.2 x 10^{-6} M N) in anoxic waters, largely it is thought, as a result of the activities of denitrifying bacteria (Thomas 1966; Fiadeiro & Strickland 1968). Ammonia also presents an important source of nitrogen in ocean waters although the concentrations appear to fluctuate widely, probably due to rapid turnover by the planktonic population (Goering *et al.* 1964; Eppley *et al.* 1971). In the English Channel NH_4–N levels of 0.2 x 10^{-6} M N have been reported, and these may rise in coastal inshore water to 0.9 x 10^{-6} M N (Cooper 1933). In the west coast sea lochs of Etive and Eil, levels of 0.04–0.13 x 10^{-6} M N have been recorded.

In estuarine environments the levels of NO_3–N, NO_2–N and NH_4–N vary considerably depending upon the input of agricultural run-off and domestic sewage. In the Tay Estuary preliminary data show NO_3–N levels ranging from 0.17–0.3 x 10^{-6} M N and NH_4–N levels of 0.2–0.49 x 10^{-6} M N.

A major source of inorganic nitrogen available in aquatic environments, fresh water or marine, is dissolved nitrogen gas. It has been estimated (Murray *et al.* 1969) that the saturation concentrations of dissolved N_2 gas are in the range 7.8–16.4 x 10^{-4} M N depending upon the salinity of the water. Biological nitrogen fixation thus represents a potentially important source of inorganic nitrogen in oligotrophic environments such as the oceans. This communication is concerned with the bacterial assimilation of inorganic nitrogen, i.e. nitrogen fixation, nitrate reduction and the conversion of ammonia to amino acids in marine and estuarine environments. These mechanisms are discussed in relation to the known physiological constraints known to operate and include field data from three distinct sampling sites, the Tay Estuary and two deep sea lochs in Argyll, Scotland.

2. Physiological Aspects of Nitrogen Assimilation

Essentially, nitrogen assimilation can be considered to be the reduction of atmospheric nitrogen and nitrate to ammonia and the conversion of ammonia to glutamine and glutamate which serve as nitrogen donors in most biosynthetic reactions. The physiological background of these processes and their ecological implications in aquatic environments have been reviewed recently (Brown *et al.* 1974; Brown & Johnson 1976) and will not be considered here in detail.

Nitrogen fixation is the reduction of atmospheric nitrogen to ammonia and requires a source of reducing power (usually reduced ferredoxin or an equivalent e^- donor) and a source of energy (ATP). Photosynthetic micro-organisms satisfy these requirements by photoreduction of the electron donor and photophosphorylation, but heterotrophic bacteria must generate both by the metabolism of organic carbon sources. Since marine and estuarine waters are not usually rich in organic carbon significant nitrogen fixation in these environments might well be restricted to photosynthetic micro-organisms and those heterotrophs existing epiphytically on macrophytes or attached to detritus. Pshenin (1963) reported that in the Black Sea *Azotobacter* sp. were found associated with the macro-algae *Phylophora* and *Ulva* and in sediments due to precipitation of detritus. A further physiological complication is the fact that the enzyme nitrogenase, from all sources studied, is an oxygen-sensitive complex and nitrogen fixation must therefore be an anaerobic process. For this reason facultative anaerobes fix only under anaerobic conditions and aerobes and the oxygen-producing blue-green algae have developed mechanisms for maintaining the sites of nitrogen fixation anaerobic. In general, fixation occurs most efficiently at low oxygen tensions, and for *Azotobacter* sp. the mechanism involved has been termed 'respiratory protection' (Postgate 1971), oxygen being removed by a very high rate of respiration. This can occur only when significant quantities of carbon are available and the implications during growth in nutrient poor environments such as the sea or an estuary are obvious. In sediments which are low in available carbon and oxidized the same arguments apply but in those which are anoxic or nearly so then nitrogen fixation by a range of heterotrophs becomes possible although the requirements for reductant and ATP still apply. Laboratory studies have shown that the ability to synthesize the enzyme nitrogenase, from all sources studied, is an oxygen-sensitive complex and of bound nitrogen, especially NO_2–N and NH_4–N. In experiments with *Azotobacter chroococcum* (Drozd *et al.* 1972) and *Klebsiella pneumoniae* the presence of detectable quantities of NH_4–N in the cultures coincided with the repression of nitrogenase synthesis. In the marine environment NH_4–N levels are rarely likely to be high enough to be repressive (Stewart 1971), but for example, the situation in polluted regions of an estuary remains to be assessed. When the levels of fixed nitrogen are low then it would seem probable that they are assimilated simultaneously with the NH_4–N produced by nitrogen fixation.

The ability to assimilate nitrate is widespread in heterotrophic bacteria (Payne 1973) and involves reduction via nitrate to ammonia, catalysed by nitrate and nitrite reductases. These reactions are endergonic and Syrett (1954) calculated that the AF value for the conversion of NO_3–N to NO_2–N was

22.3 kcal.mol^{-1} and for NO_2–N to NH_4–N, 59.3 kcal.mol^{-1}. In the light of these calculations it is not surprising that micro-organisms utilize NH_4–N in preference to NO_3–N and in many instances the presence of sufficiently high concentrations of NH_4–N results in the cessation of nitrate uptake and the repression of nitrate and nitrite reductases. Nevertheless, since NO_3–N levels are normally much higher than NH_4–N in non-polluted waters it is assumed that nitrate is the principal fixed nitrogen source.

Whilst the NH_4–N levels are normally low the significance of ammonia may be in its rapid turnover rather than its low standing concentration. In a marine pseudomonad (Brown *et al.* 1975) the uptake of NO_3–N ceased if the NH_4–N concentration exceeded more than 1 mM but at lower concentrations both nitrogen sources were used simultaneously. The saturation constants for the uptake of NO_3–N and NH_4–N and the maximum uptake rates for both substrates were equivalent and it was suggested that when present at high concentrations NH_4–N had a direct effect on nitrate uptake. The nitrate reductase of this organism used NADH as electron donor and required FAD and a cation for maximal activity. Unlike the algal enzymes which require nitrate to be present for activity to be developed the bacterial enzyme is synthesized in the absence of nitrate provided that the culture concentrations of NH_4–N and glutamate are low. This can be achieved readily in a nitrogen-limited chemostat culture whereupon nitrate and ammonia were utilized simultaneously and nitrate reductase activity was the highest recorded. Cultures grown on an excess of NH_4–N or glutamate, however, do not show detectable levels of nitrate reductase. In no experimental system containing an excess of NO_3–N as sole nitrogen source was NO_3–N, NO_2–N or NH_4–N detected intracellularly in appreciable concentrations (Brown *et al.* 1972, 1973, 1975). This indicates that either the uptake of NO_3–N or its reduction to NO_2–N is intrinsically a rate-limiting step and that all cultures with NO_3–N as sole nitrogen source are physiologically nitrogen limited. Whilst it is possible to demonstrate the preferential utilization of NH_4–N in the laboratory, it is unlikely to occur in the sea or an unpolluted estuary where the NH_4–N levels are probably not sufficiently high to repress the synthesis of nitrate reductase. The ability to take up and reduce NO_3–N is therefore expected to be commonly expressed in marine and estuarine environments not subject to pollution.

The mechanisms of ammonia utilization have been studied in a number of marine and fresh water bacteria (Brown *et al.* 1972, 1973). When the concentration of NH_4–N is low then its assimilation proceeds via the ATP-requiring route involving the enzyme glutamine synthetase (GS) and glutamate synthase (GOGAT) (Fig. 1). The significance of this system lies in the low K_m (0.25 mM) for NH_4–N of bacterial glutamine synthetase and the ability therefore of these enzymes to 'scavenge' ammonia when present at low concentrations. Since, as discussed earlier, growth on NO_3–N produces a system

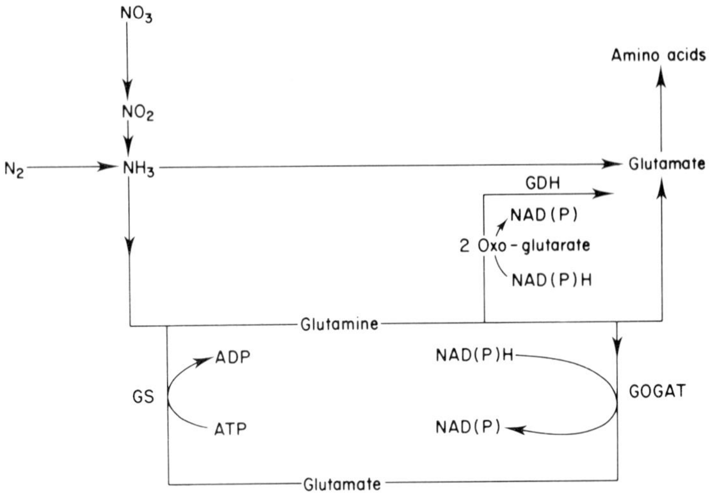

Fig. 1. The pathways of ammonia assimilation in heterotrophic micro-organisms.

similar physiologically to that observed with low NH_4–N levels, then in aquatic environments low in either NH_4–N and/or NO_3–N it is likely that NH_4–N will be assimilated via glutamine synthetase and glutamine synthase. In the unlikely event of NH_4–N being present at high concentrations assimilation proceeds via the classical ATP-independent glutamate dehydrogenase (Fig. 1), which possesses relatively high K_m values for NH_4–N (10–22 mM) and synthesis of glutamine synthetase and glutamate synthase is repressed. The synthesis of glutamate in non-polluted marine and estuarine environments containing NO_3–N and NH_4–N is therefore likely to be an energy-dependent process. Ammonia assimilation during nitrogen fixation probably also proceeds via glutamine synthetase and glutamate synthase. This is certainly true of *Clostridium pasteurianum* and a *Clostridium* sp. isolated from a Loch Etive sediment which synthesize glutamine synthetase and glutamate synthase constitutively and do not contain glutamate dehydrogenase. Information on *Desulfovibrio* sp. isolated from marine sediments shows that it possesses a glutamate dehydrogenase but no data are yet available on glutamine synthetase and glutamate synthase in these organisms.

3. Ecological Aspects of Nitrogen Fixation

Biological nitrogen fixation is important since it represents an input of new nitrogen into an aquatic ecosystem. Consequently the process has attracted considerable attention, especially in relation to the process of lake eutrophication. Nitrogen fixation, the reduction of atmospheric nitrogen to ammonia appears, on present evidence, to be restricted to certain bacterial groups and the blue-green algae. Quantitatively, blue-green algae are considered to contribute most of the fixed nitrogen in aquatic environments. Since they are not dependent upon a fixed carbon source they are widely distributed, occurring in fresh, brackish and marine waters. Nearly all strains of filamentous heterocystous blue-green algae have been shown to fix nitrogen whereas the ability appears to be more restricted within the filamentous non-heterocystous and unicellular forms (Fogg *et al.* 1973). Whilst blue-green algae may be the dominant organisms involved in nitrogen fixation (Dugdale *et al.* 1964; Goering *et al.* 1966) recent data have shown that low rates of nitrogen fixation, presumably bacterial, occur in the aphotic and anoxic zones of several lakes (Brezonik & Harper 1969; Keirn & Brezonik 1971). Free-living heterotrophic bacteria which may occur in aquatic environments and have been shown unequivocally to fix nitrogen include purple, green and purple non-sulphur bacteria, *Azotobacter* spp., *Clostridium* spp., *Bacillus* spp., *Desulfovibrio* spp., *Klebsiella* spp. and methane-oxidizing bacteria (Postgate 1974).

Relatively few data are available on nitrogen fixation in estuarine and marine environments. Yet, Hardy & Holstein (1972) estimate that of the 175×10^6 tonnes fixed per annum, 20% occurs in the oceans. Stewart (1971) cites evidence of nitrogen fixation for 18 blue-green algae belonging to the genera *Anabaena, Calothrix, Michrochaete, Nodularia, Nostoc, Rivularia* and *Trichodesmium.* Many studies have implicated *Trichodesmium* sp. as being the most important blue-green alga involved in nitrogen fixation in tropical seas (Dugdale *et al.* 1961, 1964; Goering *et al.* 1966; Taylor *et al.* 1973). Stewart (1965, 1967) studied *in situ* nitrogen fixation by epilithic algae on the Scottish coast and by sand-dune slack algae on the east coast of England using the stable isotope ^{15}N as tracer. *Calothrix* sp. dominated the Scottish shore and were estimated to contribute 2–5 g $N.m^{-2}$ $year^{-1}$, equivalent to 41% of the total annual nitrogen budget whereas in the dune slack *Nostoc* sp. were dominant and contributed *ca.* 20% of the annual nitrogen input. High fixation rates, 4.6 mg $N.m^{-2}.day^{-1}$ have also been reported from a Long Island salt marsh by Whitney *et al.* (1975) although they give no indication of the organisms involved.

The significance of heterotrophic nitrogen fixation in the marine environment is as yet unclear. Pshenin (1963) was the first to demonstrate the potential of heterotrophic nitrogen fixation in the Black Sea. He observed *Azotobacter* sp. at all depths even in the presence of hydrogen sulphide. Maximum counts ($24 \times 10^3 .ml^{-1}$) occurred in summer and he proposed that the carbon source used by the *Azotobacter* sp. was moribund phytoplankton. There have been no subsequent reports of marine *Azotobacter* sp. although several attempts have been made to isolate them (Wynn-Williams & Rhodes 1974; Herbert 1975). Since marine and estuarine waters are not usually rich in organic carbon it is unlikely that *Azotobacter* sp. even if present would fix significant quantities of nitrogen. The occurrence of nitrogen-fixing, anaerobic bacteria in the marine environment has been well documented. Photosynthetic bacteria are frequently isolated from salt marshes and mass developments of these organisms have been reported from the coasts of Germany, Denmark, the northern Black Sea and eastern seaboard of the USA (Trüper 1970). However, the significance of these organisms in fixing nitrogen remains unknown. Likewise *Desulfovibrio* and *Clostridium* spp. are frequently isolated from anoxic sediments yet their contribution to the nitrogen budget remain obscure.

Brooks *et al.* (1971) provided the first real evidence of *in situ* bacterial nitrogen fixation in estuarine sediments and succeeded in isolating a *Clostridium* sp. capable of fixing nitrogen. They observed low but positive rates of fixation in the surface sediments, as measured by acetylene reduction, and estimated that this contributed a net input of 37 μg $N.cm^{-2}.year^{-1}$. More recently Werner and coworkers (1974) isolated *Klebsiella* sp. and *Aerobacter* sp. capable of fixing nitrogen from marine sediments and demonstrated nitrogen fixation in a model ecosystem. Whilst heterotrophic nitrogen fixation in sediments is probably not important as a nitrogen source to the overlying waters because of the low rates involved it may nevertheless play a significant role in the microbial activity of the sediments.

4. Heterotrophic Nitrogen Fixation in Loch Etive, Loch Eil and Kingoodie Bay Sediments

Heterotrophic bacterial nitrogen fixation in estuarine and marine sediments has been studied at three distinct sites on the east and west coasts of Scotland (Fig. 2). The estuarine study area, Kingoodie Bay in the Tay Estuary, is composed of fine silts and muds overlying sand. In this region of the estuary considerable changes in salinity occur depending upon tidal state and at low tide the sediments are exposed for several hours. Whilst the sediments are aerobic at the surface they rapidly become anaerobic, black, and rich in hydrogen sulphide with depth. Loch Etive and Loch Eil are two deep sea lochs situated on the west coast of Scotland. Both are deep bodies of water and whilst salinities of surface waters may vary due to fresh water input, the water immediately overlying the deeper sediments is permanently marine. The sediments are anoxic beneath the surface and show only slight variations in temperature during the year. Loch Etive receives no industrial and little agricultural effluent whereas Loch Eil receives industrial waste from the Fort William pulp mill at a rate of *ca.* 1 g of $cellulose.m^{-2}.day^{-1}$. The principal physical and chemical characteristics of the three sites are summarized in Tables 1 and 2. Nitrate levels are lower in the Kingoodie sediments than in the west coast sea lochs but the reverse occurs with

Fig. 2. Location of the sampling sites on the east and west coasts of Scotland.

ammonia. Total C and N levels in the Tay are significantly lower than those found in Loch Etive and Loch Eil.

Heterotrophic nitrogen bacteria belonging to the genera *Azotobacter, Clostridium* and *Desulfovibrio* can be isolated from sediments from each of the study areas. However, laboratory studies with several *Azotobacter* isolates have shown that they neither grow nor fix nitrogen in the presence of 0.4 M NaCl although they do grow, albeit with a long lag phase, if NH_4–N or NO_3–N is added to the growth medium. Since all the sediments are anoxic it is doubtful that, even if *Azotobacter* sp. could tolerate marine conditions, significant quantities of nitrogen would be fixed. Under these conditions anaerobic nitrogen-fixing bacteria would be expected to predominate. Significant numbers of nitrogen-fixing clostridia are found in Loch Eil and Loch Etive (Tables 3 and 4), but like the *Azotobacter* sp., none will fix nitrogen or grow in the presence of 0.4 M NaCl. *Clostridium* sp. are rarely found in the Kingoodie Bay sediments but those that have been isolated will fix nitrogen with or without 0.4 M NaCl, i.e. they are euryhaline (Herbert 1975). However their apparently low numbers preclude any significant role in heterotrophic nitrogen fixation in the Tay. In Loch Etive the data suggest that the clostridia are present as spores, since counts from heated sediments are similar to those from unheated samples. No nitrogen-fixing facultative anaerobes have been isolated from either Loch Eil or Loch Etive but they occur in low numbers in the Tay (Table 5). None of the strains isolated, identified as *Klebsiella* and *Aerobacter* spp., will fix nitrogen in

Table 1

Physical characteristics of water at the sampling sites at Loch Eil, Loch Etive and the Tay Estuary

Characteristic	Sampling area				
	Loch Eil		Loch Etive		Tay Estuary (Kingoodie Bay)
	E24	E70	E6	E24	
Temperature (°C)	8–12	7.5–11.5	8–11	8–11	3–17
Salinity (per thousand)	28–30	28–30	ND	ND	2–16.5
pH at 4 cm	7.35–7.59	7.04–7.20	7.42–7.96	7.25–7.73	7.4–8.1
E_h (mV) depth	+199 to −37	−101 to −190	+179 to −23	+144 to −182	+100 to −200
O_2 ($mg.l^{-1}$)	7.04–11.04	7.36–10.56	ND	ND	7.28–9.86
Depth (m)	28	54	55	23	1–3

ND, not determined.

Table 2

Chemical characteristics of sediments at the sampling sites at Loch Eil and the Tay Estuary

Inorganic nitrogen in 0–5 cm depth of sediment	Sampling area: Loch Eil E24	Sampling area: Loch Eil E70	Sampling area: Tay Estuary (Kingoodie Bay)
NO_3 –N (μg)	1.1–5.08	1.1–1.8	0.17–0.3
NO_2 –N (μg)	0.22–3.12	0.12–1.59	0.0–0.03
NH_4 –N (μg)	1.3–5.0	1.1–2.5	0.2–0.49
Total C (%)	4.31–6.51	3.75–7.98	2.6
Total N (%)	0.36–0.55	0.28–0.62	0.2

the presence of salt (Herbert 1975). The dominant bacteria found in the sediments from all the sampling sites are sulphate-reducing bacteria belonging to the genus *Desulfovibrio*. The majority of these isolates fix nitrogen, as measured by acetylene reduction in the presence of salt and, more significantly, maximum rates occur at salt concentrations of 0.2–0.4 M NaCl depending upon the strain (Herbert 1975). Particularly high counts occur at Loch Eil stations E24 and E70 (Table 4) which may reflect the input of cellulose into this loch. *Desulfovibrio* counts in Loch Etive and Kingoodie Bay (Tables 3 and 5) are significantly lower than those found in Loch Eil.

Nitrogen fixation rates, as measured by the acetylene reduction method (Stewart *et al.* 1968) have been determined in sediment samples from each sampling site. Low but detectable rates have been observed at all sampling

Table 3

Most probable numbers of Desulfovibrio *spp.,* Clostridium *spp. and facultative anaerobes from Loch Etive sediments*

Bacterial group	NaCl content of medium*	Sampling site: Etive 6 (count.g^{-1} dry wt)	Sampling site: Etive 24 (count.g^{-1} dry wt)
Desulfovibrio spp.	+0.4 M NaCl	10^3-10^4	10^2-10^3
	–NaCl	10^2	10^2
Clostridium spp.	+0.4 M NaCl	0	0
	–NaCl	10^3-10^4	10^3
Facultative anaerobes	+0.4 M NaCl	0	0
	–NaCl	0	0

* Nitrogen-free media used.

Table 4

Most probable numbers of Desulfovibrio *spp.*, Clostridium *spp. and facultative anaerobes from Loch Eil sediments*

Bacterial group	NaCl content of medium*	Control	Sampling site		
			EIL 24	EIL 70	EIL 2
Desulfovibrio spp.	+0.4 M NaCl	10^3-10^4	10^3-10^4	10^4-10^6	10^4-10^5
	−NaCl	10^2-10^3	10^2-10^3	10^3-10^4	10^3-10^4
Clostridium spp.	+0.4 M NaCl	0	0	0	0
	−NaCl	ND	ND	ND	ND
Facultative anaerobes	+0.4 M NaCl −NaCl	ND	ND	ND	ND

ND, not determined.
* Nitrogen-free media used.

stations (Table 6) when incubated anaerobically. These rates are similar to those reported by Brooks *et al.* (1971) for the Wacassassa Estuary, Florida. Whilst highest rates of acetylene reduction occur at Eil 70 where the highest counts of *Desulfovibrio* sp. were recorded, the rates are lower than might have been expected. In Loch Etive station E6, the acetylene reduction rate is only marginally lower than Eil 70 yet the *Desulfovibrio* count is only 1–10% of that at Eil 70. Conversely, a high *Desulfovibrio* count at Eil 24 does not yield a correspondingly greater acetylene reduction rate than at other sampling sites where the count is 100-fold lower. These data suggest that one or more factors may well be limiting bacterial nitrogen fixation in Loch Eil sediments. Carbon

Table 5

Most probable numbers of Desulfovibrio *spp.,* Azotobacter *spp. and anaerobes from Kingoodie Bay sediments*

Bacterial group	NaCl content of medium*	Sampling site	
		1	2
Desulfovibrio spp.	+0.4 M NaCl	10^3	10^3
	−NaCl	10^2	10^2
Azotobacter spp.	−0.4 M NaCl	10^2-10^3	10^2
	+NaCl	0	0
Obligate/facultative anaerobes	+0.4 M NaCl	10^2	10^2
	−NaCl	10^2-10^3	10^3

* Nitrogen-free media used.

Table 6

Anaerobic nitrogen fixation rates, as measured by acetylene reduction, in Loch Eil, Loch Etive and Kingoodie Bay sediments

Sampling site	nmoles of $C_2H_4.g^{-1}$ dry wt. h^{-1}
Loch Eil	
E24	0.13
E70	0.24
E12	0.13
Loch Etive	
E6	0.21
E24	0.11
Tay Estuary	
Site 1	0.074
Site 2	0.144

limitation has frequently been implicated as limiting nitrogen fixation and there have been several reports of stimulation of nitrogen fixation following the addition of exogenous carbon sources (Keirn & Brezonik 1971; Maruyuma *et al.* 1974; Herbert 1975). Addition of 10% (w/v) glucose to Kingoodie Bay sediments stimulates nitrogen fixation rates by a factor of 2.5 under anaerobic conditions (Herbert 1975). However, no stimulation of nitrogen fixation occurs when exogenous carbon, whether glucose, cellulose or lactate, is added to Loch Eil sediments suggesting that some other factor(s) presently unknown may be limiting. Certainly, at stations E70 and Eil 24 sulphide levels are high. Using an ion-specific electrode (Orion probe) mean sulphide levels of between 10.6 and 39.6 $mg.l^{-1}$ have been recorded in Loch Eil sediments. These high levels may well be inhibitory and are currently being investigated.

5. Effects of Salinity on Nitrogen Assimilation

In the open sea or in a deep sea loch nitrogen assimilation occurs at a relatively constant salinity and temperature variations are usually small. Nutrient concentrations will clearly vary considerably on a seasonal basis as well as due to the mixing of different bodies of water. All these changes will have a profound effect on the microbial populations and their metabolic activities. The estuarine environment is more complex because considerable changes in salinity and temperature also occur. Salinity variations, in particular, exert an effect upon nitrogen assimilation since the intracellular pools of amino acids vary in a consistent manner with increasing or decreasing NaCl concentration. Variations of this type are known to occur in a wide range of bacteria (Tempest *et al.* 1970; Brown & Stanley 1972) and are of physiological significance because, depending upon the nutrient status of the culture and the salinity, the pool amino acids may account for between 5 and 20% of the total amino nitrogen. The principal pool components involved in these changes are glutamate, the product of ammonia assimilation, and proline which is derived directly from glutamate. The pool content of a marine strain of *Pseudomonas* grown in a chemostat under nitrogen limitation was 0.4 $mg.l^{-1}$ at 0.2 M NaCl and 2.4 $mg.l^{-1}$ at 0.5 M NaCl, while in a carbon-limited system the pool sizes were 3.9 $mg.l^{-1}$ at 0.2 M NaCl and 14.0 $mg.l^{-1}$ at 0.5 M NaCl (Table 7). These pools appear to have several

Table 7

Influence of NaCl on the amino acid pools of the marine pseudomonad PL-1 in carbon limited chemostat culture

NaCl concentration (M)	Pool amino acids ($mg.l^{-1}$)
0.2	3.9
0.4	6.9
0.6	14.3
0.8	18.2

functions (Stanley & Brown 1976) including the supply of precursors for protein synthesis, essential metabolites, reserves of nitrogen (but not, apparently, carbon), some involvement in osmoregulation and as a mechanism for balancing the intracellular electric charge. The response of the pools to changes in salinity are rapid and reproducible (Stanley & Brown 1974, 1976). An increase in NaCl from 0.2 to 0.5 M results in a three-fold increase in pool glutamate concentrations within 30 min in chemostat cultures growing at a dilution of 0.1 h^{-1}. This response was dependent upon the availability of a nitrogen source (NO_3–N or NH_4–N) and is due to synthesis of glutamate *de novo*. In media containing low concentrations of nitrate or ammonia the response is slower and appears linked to nitrogen availability. The significance of these pools in osmoregulation is established by the fact that even in a nitrogen-limited environment the pool amino acids vary with salinity. Neither salinity nor Na^+ or K^+ influence the synthesis of the enzymes involved in the conversion of NO_3–N or NH_4–N to glutamate. Stanley & Brown (1976) proposed that the rapid increase in the synthesis of glutamate on increasing the salinity of the growth medium was due to increased permeability of the bacteria to NO_3–N and NH_4–N. A decrease in salinity results in a rapid and specific adjustment of pool composition with some excess amino acids being excreted. This occurs without any detectable loss in culture viability. It is not known at present whether these effects, observed with the *Pseudomonas* sp., are applicable to other organisms.

As discussed earlier, nitrogen fixation by *Azotobacter, Klebsiella* and *Aerobacter* isolates is inhibited by the presence of salt in the growth medium.

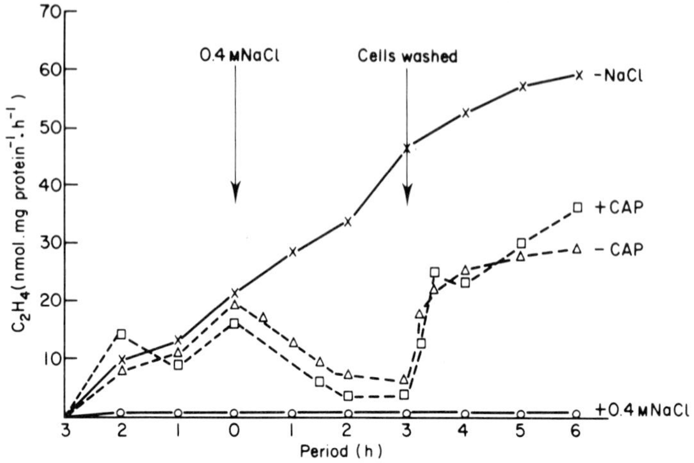

Fig. 3. Effect of 0.4 M NaCl on nitrogen fixation rate of *Azotobacter* isolate K-1.

The effect of salt on nitrogen fixation by a strain of *Azotobacter agilis* is reversible (Fig. 3). When 0.4 M NaCl is added to a nitrogen-fixing culture of *Azotobacter K1*, fixation is inhibited almost immediately but rapidly returns after the cells are centrifuged, washed and transferred to salt-free and nitrogen-free growth medium. Activity returns rapidly both in the presence and absence of chloramphenicol indicating that synthesis of nitrogenase *de novo* is not required for the return of nitrogen-fixing ability. When fixed inorganic nitrogen sources, i.e. NO_3–N or NH_4–N are added to the nitrogen-free growth medium, growth occurs in the presence of 0.4 M NaCl albeit with a greatly increased lag period yet the total cell yield is ultimately the same as in salt free media. Whilst *Azotobacter* sp. do not grow or fix nitrogen in the presence of 0.4 M NaCl they are able to survive, and should low levels of bound inorganic nitrogen be available they will grow slowly.

6. Conclusions

From the data presented in this paper the following conclusions can be made.

1. Low rates of nitrogen fixation occur in sediments from Loch Etive, Loch Eil and Kingoodie Bay.
2. *Desulfovibrio* spp. appear to be the dominant organisms involved since not only are they quantitatively the major component of the anaerobic microflora, they comprise the only group of nitrogen-fixing bacteria so far isolated which function under marine conditions.

3. Carbon availability may well be limiting nitrogen fixation in the Kingoodie sediments since exogenous carbon stimulates higher rates of acetylene reduction, whereas in Loch Eil exogenous carbon causes no stimulation of activity and other factors may be limiting, e.g. sulphide inhibition.
4. In marine and estuarine environments where NH_4–N levels are low, assimilation proceeds via the energy-dependent glutamine synthetase–glutamate synthase pathway which has a high affinity for NH_4. This is the price the organism has to pay for scavenging trace quantities of inorganic nitrogen.
5. Whilst changes in salinity may exert a profound effect on micro-organisms there is good evidence to show that heterotrophic bacteria have evolved mechanisms for rapid adaptation to the changed conditions.

7. Acknowledgements

Investigations on the physiology of nitrogen assimilation were supported by grants BRG/647 from the Science Research Council and GR3/2042 from the Natural Environment Research Council. The River Tay Project is financed by grant GR3/2729 from the Natural Environment Research Council.

Results quoted for Loch Eil are data from the Loch Eil project undertaken jointly by members of the Scottish Marine Biological Association and the Universities of Dundee and Strathclyde.

8. References

BREZONIK, P. L. & HARPER, G. L. 1969 Nitrogen fixation in some anoxic lacustrine environments. *Science, New York* **164**, 1277–1279.

BROOKS, R. H., BREZONIK, P. L., PUTNAM, H. D. & KEIRN, M. A. 1971 Nitrogen fixation in an estuarine environment: the Wacassassa on the Florida gulf coast. *Limnology and Oceanography* **16**, 701–710.

BROWN, C. M. & JOHNSON, B. 1976 Inorganic nitrogen assimilation in aquatic microorganisms. *Advances in Aquatic Microbiology* **1**, 1–113.

BROWN, C. M. & STANLEY, S. O. 1972 Environment mediated changes in the cellular content of the 'pool' constituents and their associated changes in cell physiology. *Journal of Applied Chemistry and Biotechnology* **22**, 363–389.

BROWN, C. M., MACDONALD-BROWN, D. S. & STANLEY, S. O. 1972 Inorganic nitrogen metabolism in marine bacteria: nitrogen assimilation in some marine pseudomonads. *Journal of the Marine Biological Association of the United Kingdom* **52**, 793–804.

BROWN, C. M., MACDONALD-BROWN, D. S. & STANLEY, S. O. 1973 The mechanisms of nitrogen assimilation in pseudomonads. *Antonie van Leeuwenhoek* **39**, 89–98.

BROWN, C. M., MACDONALD-BROWN, D. S. & MEERS, J. L. 1974 Physiological aspects of microbial inorganic nitrogen metabolism. *Advances in Microbial Physiology* **11**, 1–52.

BROWN, C. M., MACDONALD-BROWN, D. S. & STANLEY, S. O. 1975 Inorganic nitrogen metabolism in marine bacteria: nitrate uptake and reduction in a marine pseudomonad. *Marine Biology* **31**, 7–13.

COOPER, L. H. M. 1933 Chemical constitments of biological importance in the English Channel. Pt 1. phosphate, silicates, nitrate, nitrite, ammonia. *Journal of the Marine Biological Association of the United Kingdom* **18**, 677–728.

DROZD, J. W., TUBB, R. S. & POSTGATE, J. R. 1972 A chemostat study of the effect of fixed nitrogen sources on nitrogen fixation, membranes and free amino acids in *Azotobacter chroococcum. Journal of General Microbiology* **73**, 221–232.

DUGDALE, R. C., MENZEL, D. W. & RYTHER, J. H. 1961 Nitrogen fixation in the Sargasso Sea. *Deep Sea Research* **7**, 298–300.

DUGDALE, R. C., GOERING, J. J. & RYTHER, T. H. 1964 High nitrogen fixation rates in the Sargasso Sea and Arabian Sea. *Limnology and Oceanography* **9**, 507–510.

EPPLEY, R. W., PACKARD, T. T. & MAC ISSAC, J. J. 1970 Nitrate reductase in Peru current phytoplankton. *Marine Biology* **6**, 195–199.

EPPLEY, R. W., RODGERS, J. N., MCCARTAY, J. J. & SOURNIA, A. 1971 Light/dark periodicity in nitrogen assimilation of the marine phytoplankton *Skeletonema costatum* and *Coccolithus huxleyi* in N-limited chemostat culture. *Journal of Phycology* **7**, 150–154.

FIADEIRO, M. & STRICKLAND, J. D. H. 1968 Nitrate reduction and the occurrence of a deep sea nitrogen maximum in the ocean off the west coast of South America. *Journal of Marine Research* **26**, 187–201.

FOGG, G. E., STEWART, W. D. P., FAY, P. & WALSBY, A. P. 1973 *The Blue Green Algae*. New York & London: Academic Press.

GERLOFF, G. C. & SKOOG, F. 1957 Nitrogen as a limiting factor for the growth of *Microcystis aeruginosa* in Southern Wisconsin Lakes. *Ecology* **38**, 556–561.

GOERING, J. J., DUGDALE, R. C. & MENZEL, D. W. 1964 Cyclic diurnal variations in the uptake of ammonia and nitrate by photosynthetic organisms in the Sargasso Sea. *Limnology and Oceanography* **9**, 448–451.

GOERING, J. J., DUGDALE, R. C. & MENZEL, D. W. 1966 Estimates of *in situ* rates of nitrogen uptake by *Trichodesmium* sp. in the tropical Atlantic Ocean. *Limnology and Oceanography* **11**, 614–620.

HARDY, R. W. F. & HOLSTEIN, R. D. 1972 Global nitrogen cycling, pools, evolution, transformations, transfers and research needs. In *The Aquatic Environment: Microbial Transformations and Water Quality Management Implications*, eds Ballentine, R. K. & Guarraia, L. J. Washington D.C.: Environmental Protection Agency.

HERBERT, R. A. 1975 Heterotrophic nitrogen fixation in shallow estuarine sediments. *Journal of Experimental Marine Biology and Ecology* **18**, 215–225.

KEIRN, M. A. & BREZONIK, P. L. 1971 Nitrogen fixation by bacteria in Lake Mize, Florida and in some lacustrine sediments. *Limnology and Oceanography* **16**, 720–731.

MARUYUMA, Y., SUZUKI, T. & OTOBE, K. 1974 Nitrogen fixation in the marine environment. The effect of organic substrates on acetylene reduction. In *Effect of the Ocean Environment on Microbial Activities*, eds Colwell, R. R. & Morita, R. Y. Baltimore: University Park Press.

MIYAZAKI, T., WADA, E. & HATTORI, A. 1973 Capacities of shallow waters of Sagami Bay for oxidation and reduction of inorganic nitrogen. *Deep Sea Research* **20**, 571–577.

MURRAY, C. M., RILEY, J. P. & WILSON, T. R. S. 1969 The solubility of gases in distilled water and sea water. I. Nitrogen. *Deep Sea Research* **16**, 297–310.

PAYNE, W. J. 1973 Reduction of nitrogenous oxides by microorganisms. *Bacteriological Reviews* **37**, 409–452.

POSTGATE, J. R. 1971 Relevant aspects of the physiological chemistry of nitrogen fixation. In *Microbes and Biological Productivity*, eds Hughes, D. E. & Rose, A. H. Symposium of the Society for General Microbiology. Cambridge: Cambridge University Press.

POSTGATE, J. R. 1974 New advances and future potential in biological nitrogen fixation. *Journal of Applied Bacteriology* **37**, 185–202.

PSHENIN, L. N. 1963 Distribution and ecology of *Azotobacter* in the Black Sea. In *Symposium on Marine Microbiology*, ed. Oppenheimer, C. H. Springfield, Illinois: C. C. Thomas.

RILEY, G. A. 1957 Phytoplankton of the north central Sargasso Sea. *Limnology and Oceanography* **2**, 252–270.

SKELEF, G., OSWALD, W. J. & GOLUEKE, G. C. 1971 Assaying algal growth with respect to nitrate concentration by a continuous flow turbidostat. *Proceedings of the 5th International Congress on Water Pollution Research* **111**, 25–34.

SOLORZANO, L. & GRANTHAM, B. 1975 Surface nutrients, chlorophylls and phaeopigment in some Scottish sea lochs. *Journal of Experimental Marine Biology and Ecology* **20**, 63–76.

STANLEY, S. O. & BROWN, C. M. 1974 Influence of temperature and salinity on the amino acid pools of some marine pseudomonads. In *Effect of Ocean Environment on Microbial Activities.* eds Colwell, R. R. & Morita, R. Y. Baltimore: University Park Press.

STANLEY, S. O. & BROWN, C. M. 1976 Inorganic nitrogen metabolism in marine bacteria: the intracellular free amino acid pools of a marine pseudomonad. *Marine Biology* **(in** press).

STEWART, W. D. P. 1965 Nitrogen turnover in marine and brackish environments. I. Nitrogen fixation. *Annals of Botany* **20**, 229–239.

STEWART, W. D. P. 1967 Nitrogen turnover in marine and brackish environments. II. Use of ^{15}N in measuring nitrogen fixation in the field. *Annals of Botany* **31**, 385–407.

STEWART, W. D. P. 1971 Nitrogen fixation in the sea. In *Fertility of the Sea*, ed. Costlow, J. D. London: Gordon & Breach.

STEWART, W. D. P., FITZGERALD, G. P. & BURRIS, R. H. 1968 Acetylene reduction by nitrogen fixing blue-green algae. *Archiv für Mikrobiologie* **68**, 336–348.

SVERDRUP, H. A., JOHNSON, M. W. & FLWMING, R. H. 1946 *The Oceans, their Physics, Chemistry and General Biology*. New York: Prentice Hall.

SYRETT, P. J. 1954 Ammonia and nitrate assimilation by green algae. In *Autotrophic Micro-organisms*, eds Fry, B. A. & Peel, J. L. Symposium of the Society for General Microbiology. Cambridge: Cambridge University Press.

TAYLOR, B. F., LEE, C. C. & BUNT, J. S. 1973 Nitrogen fixation associated with the marine blue-green alga *Trichodesmium*, as measured by the acetylene reduction technique. *Archiv für Mikrobiologie* 88, 205–212.

TEMPEST, D. W., MEERS, J. L. & BROWN, C. M. 1970 Synthesis of glutamate in *Aerobacter aerogenes* by a hitherto unknown route. *Biochemical Journal* **117**, 405–407.

THOMAS, W. H. 1966 On denitrification in the north eastern tropical Pacific Ocean. *Limnology and Oceanography* **11**, 393–414.

THOMAS, W. H. 1970 Effect of ammonia and nitrate concentration on chlorophyll increases in natural tropical Pacific phytoplankton populations. *Limnology and Oceanography* **15**, 386–394.

TRÜPER, H. C. 1970 Culture and isolation of phototrophic sulphur bacteria from the marine environment. *Helgoländer Wissenschaftliche Meeresuntersuchungen* **20**, 6–10.

VOLLENWIEDER, R. A. 1968 Scientific fundamentals of the eutrophication of lakes and flowing waters with particular reference to nitrogen and phosphorus as factors in eutrophication. DAS/CS1/6827. Paris: Organisation Economic Cooperation and Development.

WERNER, D., EVANS, H. J. & SEIDLER, R. J. 1974 Facultative anaerobic N-fixing bacteria from the marine environment. *Canadian Journal of Microbiology* **20**, 59–64.

WHITNEY, D. E., WOODWELL, G. M. & HOWARTH, R. W. 1975 Nitrogen fixation in Flax Pond. A Long Island salt marsh. *Limnology and Oceanography* **20**, 640–643.

WOOSTER, W. S., CHOW, T. J. & BARRETT, J. 1965 Nitrate distribution in Peru current waters. *Journal of Marine Research* **23**, 210–221.

WYNN-WILLIAMS, D. D. & RHODES, M. E. 1974 Nitrogen fixation in seawater. *Journal of Applied Bacteriology* **37**, 203–216.

Microbiological Aspects of Drinking Water Supplies

M. HUTCHINSON* AND J. W. RIDGWAY

Water Research Centre,
Medmenham Laboratory, Henley Road,
Medmenham, Marlow, Buckinghamshire, England

CONTENTS

1. Objectives of Water Supply Microbiology

THE MOST ESSENTIAL FEATURE of a water supply is sanitation or the prevention of infection, because the water cycle represents an obvious mode of transmission of enteric disease in the community. This was recognized as early as 1854 by John Snow, who, in his researches into the mode of propagation of cholera, demonstrated (by removal of the pump handle in Broad Street, London) that the infective agent was conveyed by water. In 1856, Budd related the incidence of typhoid fever to polluted water supplies. Both observations predated the laboratory identification of the causative agents by some 30 years.

In those early days contamination of water supplies was recognized by the presence of certain chemical substances and as early as 1858 regular chemical analysis of the water supply for London first began. The recognition by Pasteur

* Present address: Pynes Laboratory of the South West Water Authority, Upton Pyne, Exeter, Devon, England.

and others that micro-organisms were the cause of disease naturally led to the development of bacteriological examination as being a valuable supplement to chemical analysis, from the public health aspect. In 1885 Percy Frankland started the first routine examination of water in London using gelatine plate counts and in 1891 enunciated the concept that organisms characteristic of sewage must be identified to provide evidence of potentially dangerous pollution. From that time forwards it has been the aim of the water bacteriologist to devise methods to detect and enumerate faecal bacteria in water. Amongst the early pioneers figure such names as Klein, Houston and MacConkey. Houston recognized the three main groups of indicator organisms which are still in use to this day, notably coliforms, faecal streptococci and gas-forming clostridia. He argued that because of their origin in the excrement of man, animals and birds, their detection in water signified faecal pollution. This concept is as true today as it was then and although methods may have changed, it is still the criterion by which water samples are routinely assessed.

Water supply bacteriology has many aspects, but that of public health is of paramount importance and concerns the following.

1. Detection and assessment of the degree of faecal pollution in a potential source of supply in order to design a suitable method of treatment.
2. Assessment of the performance of various stages of water treatment.
3. Confirmation of hygienic safety of the final water entering supply.

   It is unwise to rely solely on a single line of defence to prevent microbial pollution reaching the consumer and accordingly the following strategies are used for ground waters: (a) protection of the aquifer from pollution by drawing up safety zones; (b) proper well construction to prevent contamination from surface waters; (c) disinfection. For surface waters the strategies are: (a) to limit the pollution at source; (b) use of adequate number of appropriate water treatment processes for the type of water; (c) final disinfection.
4. Demonstration by regular bacteriological surveillance that the quality of water is maintained throughout distribution. Again protection is provided in several ways, i.e.: (a) distribution systems soundly constructed to specifications laid down in Codes of Practice; (b) constructed with materials and fittings which have been tested to ensure that they are satisfactory mechanically and will not lead to contamination of the water or lead to the growth of micro-organisms; (c) maintenance of a disinfectant residual through the distribution system.

Other aspects of microbiology are becoming increasingly important not only to improve the quality of the product but also to achieve an economic and trouble-free system of supply. On the one hand this would include the use of biological processes for water treatment, and on the other, discouragement of those microbial growths which lead to a deterioration in water quality or give rise to special problems in supply. These aspects will be dealt with later.

2. Micro-organisms Associated with Water Supply

The bacteria of particular interest for water supply purposes are listed in Table 1. Apart from pathogens, other organisms important to the water supply bacteriologist are those that can be used as indicators of faecal pollution. Ideally these should: (a) be present in greater numbers than any pathogen; (b) be unable to proliferate in water to any extent; (c) be more resistant than any pathogen to the water environment and disinfection processes; (d) give simple and characteristic reactions enabling unambiguous identification in culture.

Although the importance of *Escherichia coli* is irrefutable, the value of coliforms as indicator organisms has been questioned (Dutka 1973), however, they are still widely used throughout the world. It is claimed that faecal streptococci have certain advantages in that: (a) they do not multiply in water (Dutka 1973); (b) there is correlation between their presence and that of faecal coliforms (*E. coli*) (Litsky *et al.* 1955); (c) they die off more slowly than coliforms, especially in run-off from agricultural areas (Geldreich 1970; Evans & Owens 1972); (d) they have a greater potential for survival in potable water than *E. coli* and so verify the pollution of a distribution system remote from the point of contamination (Burman 1961); (e) certain species show a measure of specificity for a particular host thus permitting characterization of the source of pollution (Holden 1970). However, faecal streptococci are not ideal in that they are less abundant than *E. coli* in faeces and may grow on vegetation and insects, especially *Streptococcus faecalis liquefaciens* (Geldreich *et al.* 1964; Geldreich 1970).

The anaerobic spore former, *Clostridium perfringens* is valuable for identifying remote pollution because the spores invariably survive chlorination better than vegetative cells. However, they are removed to some extent by treatment and numbers decline through the distribution system (*Anon.* 1968). In observations on distribution systems the Water Research Centre has found clostridial spores in 33% of samples from dead ends, which suggests that they sediment out because, being anaerobes, it is unlikely that they can grow in this situation. The ability of many coliforms to grow under natural conditions in the tropics has led workers to study alternative organisms, including bifidobacteria (Evison & James 1975). Preventing the entry of pathogenic organisms and the indicators of faecal pollution into potable water supplies does not preclude the detection of large numbers of other micro-organisms. Many of the organisms listed in Table 1 have been found by the Water Research Centre in distribution

Table 1

Micro-organisms associated with water supply

Organism	Disease
Water-borne disease	
Bacterial	
Vibrio cholerae	Cholera
Salmonella typhi	Typhoid fever
Salm. paratyphi A and B	Paratyphoid fever
Salm. montevideo B	Salmonellosis
Salm. typhimurium	Salmonellosis
Shigella sonnei	Bacillary dysentery
Shig. dysenteriae	Bacillary dysentery
Shig. flexneri	Bacillary dysentery
Escherichia coli serotypes	Enteritis
Leptospira icterohaemorrhagia	Leptospirosis
Protozoan	
Entamoeba hystolytica	Amoebic dysentery
Naegleria spp.	Amoebic meningoencephalitis
Hartmanella spp.	Amoebic meningoencephalitis
Acanthamoeba spp.	Amoebic meningoencephalitis
Viral	
Hepatitis A virus	Viral hepatitis type A
Enteroviruses (polio, coxsackie A and B and echo)	Respiratory tract infections Non-bacterial enteritis (role of water route uncertain)
Adenoviruses	Respiratory tract infections Non-bacterial enteritis (role of water route uncertain)
Reoviruses	Respiratory tract infections Non-bacterial enteritis (role of water route uncertain)
Parvoviruses	Respiratory tract infections Non-bacterial enteritis (role of water route uncertain)

Intestinal bacteria indicative of faecal pollution.
E. coli, coliforms (*Citrobacter, Enterobacter, Klebsiella*), *Streptococcus faecalis* (typically human), *Strep. faecium* (man and animals), *Strep. durans* and *equinus* (strictly non-human), *Clostridium perfringens*, Bifidobacteria and *Pseudomonas aeruginosa*

Other bacteria
Fluorescent pseudomonads, *Flavobacterium, Achromobacterium, Chromobacterium, Proteus, Bacillus, Corynebacterium, Mycobacterium, Arthrobacter, Serratia,* Micrococci (including *Staphylococcus aureus*), *Vibrio, Aeromonas, Hyphomicrobium, Nocardia, Desulfovibrio*, Thiobacilli, Brevibacteria, *Caulobacter, Leptothrix, Crenothrix, Gallionella. Actinomyces*, yeasts and fungi

systems, but only in a few instances is the significance of these organisms known. They may contribute towards problems related to water quality by causing depletion of oxygen, production of undesirable tastes and odours, corrosion, dirty water, visible growth on, and deterioration of, material used in water supply and spoilage in food processing and other manufacturing industries using the public water supply (Bean & Everton 1969). A more detailed discussion of many of these organisms in relation to water supply is to be found in *Water Treatment and Examination* (Holden 1970).

3. Treatment

The objectives of water treatment are to reduce the levels of turbidity, and of chemical and microbiological pollution by a series of unit processes. For surface waters the number and type of processes depends on the nature of the pollutant and its concentration.

The most important stage of water treatment, now universally applied, is disinfection and accordingly most other processes serve to condition the water for final disinfection. These include removal of: (a) ammonia and organic nitrogenous compounds which react with certain disinfectants, notably chlorine, and thereby reduce efficiency; (b) organic compounds which make an additional demand on the disinfectant used, whether chemical, such as chlorine, bromine, iodine, chlorine dioxide and ozone, or physical such as u.v. radiation; (c) turbidity which may shield cells physically against disinfection.

Originally the treatment required was specified on the basis of the level of bacterial pollution in the source water in terms of coliform counts (*Anon.* 1946, 1961), but with auxilliary treatment, especially prechlorination, it is possible to treat satisfactorily waters having a far greater level of faecal pollution (*Anon.* 1961). Similar guidelines are now being considered by the European Economic Community which has proposed gradings for waters with the following contents of micro-organisms indicating faecal pollution (*Anon.* 1975*a*).

	A1	A2	A3	A4
Total coliforms (number.100 ml^{-1})	50	5000	50,000	greater than this level
Faecal coliforms (number.100 ml^{-1})	20	2000	20,000	greater than this level
Faecal streptococci (number.100 ml^{-1})	20	1000	10,000	greater than this level
Salmonellae	not present in 5 *l*	not present in 1 *l*		greater than this level

The recommended treatments for these categories are listed below.

- A1 Simple rapid filtration and disinfection.
- A2 Prechlorination, coagulation, flocculation, rapid filtration and final chlorination.
- A3 Breakpoint chlorination, coagulation, flocculation, settlement, filtration, treatment with activated carbon or ozone and final chlorination.
- A4 Would require some form of preliminary storage or biological pretreatment to remove ammonia and/or some other form of advanced water treatment.

The efficiency of various unit processes have been assessed in terms of the removal of bacteria and viruses (*Anon.* 1961; Robeck *et al.* 1962) and in general each stage may be expected to remove 50–90% or even 99% of the bacterial and viral load. Therefore, water treatment processes can be designed to deal with the expected level of pollution. Although all treatments have some effect on the microbiological quality of water, only certain processes are microbiological in nature. Table 2 summarizes typical combinations of treatment processes used in the UK and it is proposed to discuss these individually in terms of microbiology.

(a) *Storage*

Although storage of surface waters is principally used for reasons other than treatment, many valuable chemical, biological and microbiological changes occur. These are important in terms of reduction not only of certain types of organic pollution and bacterial numbers in general, but of viruses, pathogenic bacteria and faecal organisms in particular (Poynter & Stevens 1975).

Normally storage is for several weeks but Geldreich (1970) has shown that storage for only seven days will ensure removal of 75–99% of faecal organisms. Factors which operate include u.v. radiation, natural flocculation and sedimentation, competition for available nutrients, production of biocidal waste products by various micro-organisms, predation by protozoa and possible parasitic bacteria of the genus *Bdellovibrio* (Fry & Staples 1976). Removal of bacteria and viruses is usually greatest in the summer months.

Overall, storage confers benefits but it may also give rise to occasional problems. These include: (a) solution of iron and manganese following stratification and overturn in deep reservoirs (Bernhardt 1967); (b) pollution from gulls and other water fowl which are attracted to large bodies of water for roosting at certain periods of the year (Fennel *et al.* 1974). Water fowl may produce an equivalent level of faecal pollution per day per head of population as man (Geldreich 1966) and a significant proportion of this directly pollutes the body of water. Moreover, gulls may carry salmonellae (Williams and Richards 1976). (c) The profuse growth of algae at certain times of the year can lead to an increase in numbers of coliforms particularly in the later stages of treatment where algal cells are filtered off and decay, releasing organic matter which serves as substrate for the bacteria (Burman 1961).

It has also been shown that death of many algal cells in a reservoir promotes the growth of actinomycetes which can give rise to undesirable tastes and odours in the water (Silvey 1953, 1954). Also, the algae themselves may produce taste and odours in water or cause serious blockage of filters used in water treatment.

(b) *Pretreatment*

To reduce the chemical and mechanical load on subsequent stages of water treatment, the following pretreatment processes may be used.

Table 2

Typical combinations of treatment processes used in the UK

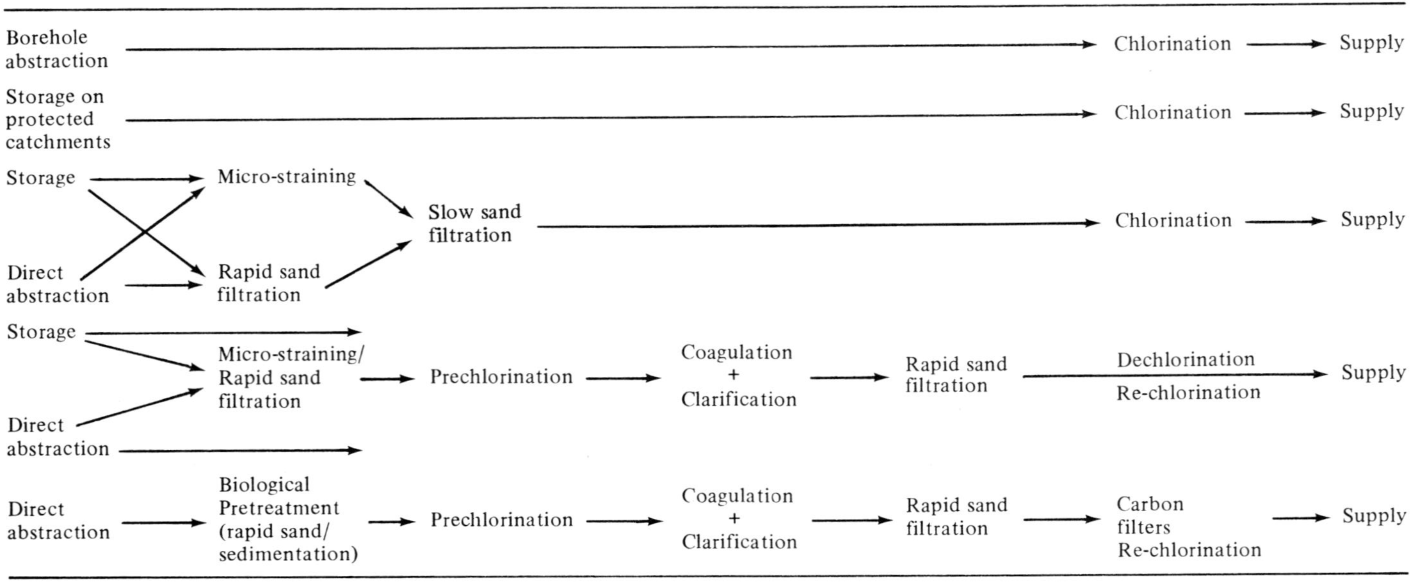

(i) *Micro-strainers*

These are simple mechanical filters in the form of a revolving drum with a mesh size of 23, 35 or 65 μm through which the water passes, the larger particulate matter, including algae, being strained off and continuously backwashed to waste. Bacterial removal is virtually non-existent (Boucher 1967) although a zoogloeal slime may form and blind the filter screen. In order to overcome the problem u.v. lamps are sometimes placed over the filter surface to prevent such growths.

(ii) *Roughing filters*

By using coarse sand and rapid rates of filtration of up to 12 $m.h^{-1}$ filamentous algae and large suspended solids may be removed thereby reducing the load on slow sand filters. With continued operation organic matter is metabolized and a biological slime forms on the surfaces of the sand particles. This biological film is active in the removal of ammonia and soluble iron and manganese, and a measure of bacterial removal is also achieved (Table 3).

Table 3

Biological pretreatment for the removal of ammonia, nitrite and bacteria

River	Treatment	Percentage removal			
		NH_3	NO_2^-	Coliforms	37°C plate count
Avon†	Filter	50–100	50–100	55–85	50–80
Trent††	Filter	50–100	50–100	50	50
Severn*	Sedimentation tank	100	100	80	60
Severn‡	Sedimentation tank	90	90	–	–
Trent††	Sedimentation tank	75	–	50	50

–, no data available.
* Data from Milliner *et al.* (1972).
† Data from Pugh (1949).
†† Data from *Anon.* (1972).
‡ Data from Parker (1972).

(iii) *Biological sedimentation*

By critical control of flow in horizontal and upflow tanks it is possible to hold a biofloc seeded on river silt, sand or alum floc in suspension. By presenting a large surface area of biologically active floc to the water rapid removal of ammonia is achieved, Milliner *et al.* (1972) report complete removal of 0.83 $\mu g.ml^{-1}$ of NH_3–N in river water with less than 30 min retention and in pilot plant studies with River Thames water, levels of 2.0 $\mu g.ml^{-1}$ of NH_3–N have been completely removed with 4 min retention (Short 1973). In addition to ammonia removal the bacterial load is also reduced to a limited extent (Table 3), and some removal of organic matter including natural tastes and odours has been recorded (Lewis 1966).

This treatment has been taken one stage further by following the aerobic nitrification process in which there is production of nitrate and depletion of oxygen, by an anaerobic phase of biological sedimentation in which the nitrate is reduced to nitrogen gas. This is achieved by the addition of methanol as a suitable but inexpensive organic substrate. The reduction of 1 $\mu g.ml^{-1}$ of NO_3–N requires 2.5 to 3.0 $\mu g.ml^{-1}$ of methanol but if oxygen is present then an additional 0.9 $\mu g.ml^{-1}$ of methanol is required for removal of each $\mu g.ml^{-1}$ of dissolved oxygen by biological activity (Water Research Centre, unpublished).

(c) *Slow sand filtration*

This is the classical biological treatment for water purification. By filtration of water at a slow rate (0.1 $m.h^{-1}$) through fine sand (0.2–0.4 mm diam.) a biologically active filter mat (Schmutzdecke) forms in the top few centimetres which not only removes suspended matter and bacteria but also modifies the character of the water biochemically. The ability to remove micro-organisms is due very largely to microbial competition and predation by protozoa and invertebrates and is clearly shown by deterioration in the quality of the effluent when the temperature falls below 4°C. Although the Schmutzdecke is very active, filtration also takes place throughout the sand bed and further chemical changes affect total organic carbon and nitrogen and the removal of natural taste and odours, but there is little effect on the more refractory humic compounds found in certain coloured upland waters.

Ammonia is normally oxidized completely to nitrate but under low temperature conditions serious breakthrough of ammonia can occur resulting in more expensive and less effective disinfection by chlorination.

Other problems which occur with slow sand filters include: (a) clogging of filters following algal blooms; (b) putrefactive decomposition of algae, producing ammonia and undesirable tastes and odours, depleting oxygen and causing excessive growths of bacteria including coliforms; (c) lack of dissolved oxygen can produce solubilization of iron from the filter sand leading to excessive growths of iron bacteria within the underdrains of the filters, sometimes causing blockage (Hornby 1954); (d) spore-forming bacteria may become established in the filter bed and spores may pass through in relatively large numbers thereby permitting chlorine-resistant organisms to enter supply (*Anon.* 1973); (e) as with reservoirs, large uncovered slow sand filters attract seagulls and thereby lead to serious pollution (*Anon.* 1954, 1966*b*).

Problems of this type, the vast amount of space that the filters occupy and their labour-intensive cleaning and maintenance have lost slow sand filtration some of its popularity. However, compared with modern chemical coagulation and rapid sand filtration they can still effectively decrease the organic content of waters going to supply and also the risks of biological aftergrowths in the distribution system (Houghton 1970). When used with roughing filters, it is now possible to increase the filtration rates from 0.1 to 0.4 $m.h^{-1}$, with no significant decline in efficiency.

(d) *Chemical coagulation*

In order to remove colloidal matter and suspended solids it may be necessary to add chemical coagulants such as ferrous or aluminium sulphate which form a positively charged floc of ferric or aluminium hydroxide. This coagulates negatively charged colloids such as clay, humic compounds and micro-organisms. Specially designed contact chambers achieve a gentle circulation of the water and encourage aggregation and settlement of the floc, leaving a sludge ready for disposal. Although settlement may be very effective this process is usually followed by rapid sand filtration for maximum removal of floc.

(e) *Rapid sand filtration*

Unlike slow sand filtration in which straining occurs largely in the surface layers, with rapid sand filtration removal occurs by adsorption of foreign matter coagulated with alum or iron floc, on to sand surfaces throughout the depth of the filter. To achieve filtration in depth, sand of a larger diameter is used (0.5–2.0 mm) and this is sometimes covered with a coarser layer of anthracite to form a graded preliminary filtration layer without the risk of blinding the sand filter. Anthracite is used because of its lower specific gravity so that during backwashing, when the filter is fluidized, this coarser medium will remain on top of the filter.

With prolonged filtration the floc penetrates deeper into the bed until a stage is reached when breakthrough of the floc begins or the head required to drive water through the filter becomes too great. It is then necessary to clean the filters by passing water in the reverse direction at such a rate as to expand the sand bed by 50% and at the same time allowing the wash water to run to waste. The cleaning process may be further improved by use of compressed air or mechanical rakes to stir the sand.

Although rapid sand filtration is strictly physicochemical in nature there is evidence that rapid sand filters, either fed by gravity or in pressure shells, can become infected with micro-organisms. Table 4 shows the effect of consecutive backwashings on a pressure sand filter used for swimming pool water. After repeated backwashings in the presence of 2.0 $\mu g.ml^{-1}$ of free chlorine, it was not possible to achieve completely satisfactory bacteriological results. There was

Table 4

The effect of backwashing on the bacterial numbers present in the effluent from a pressure sand filter

Sample	2 day, 37°C plate ct.ml^{-1}	Coliforms MPN.100 ml^{-1}	*E. coli* MPN.100 ml^{-1}
Drain down			
After 3 min	1.3×10^5	180+	0
After 10 min	3.8×10^5	25	3
Backwash 1 (1.5 m^3.sec^{-1})			
After 2 min	5.6×10^4	0	0
Clean condition	5.0×10	0	0
Backwash 2 (1.5 m^3.sec^{-1})			
After 2 min	3.9×10^2	0	0
Clean condition	3.5×10^3	0	0
Backwash 3 (4.0 m^3.sec^{-1})			
After 1 min	8.8×10^3	1.0	0
After air scour for 1 min	1.2×10^5	90	0
Clean condition	9.2×10^2	1.0	0

evidence that this filter was introducing contamination into the pool, at least in excess of amounts previously recommended for swimming pools (*Anon.* 1951). This condition had arisen as a result of ineffective backwashing which had failed to fluidize the sand bed completely with the result that mud balls rich in organic matter had formed in stagnant areas of the filter bed, and these could not be broken up by backwashing and air scour. The remedy lay in complete renewal of the sand in the filter shell.

(f) *Activated carbon filters*

These have been introduced into water treatment for the final removal of residual organic pollutants in water, including taste and odour compounds. The principle is strictly one of chemical adsorption and in that sense their useful life is limited. However, in practice, granular carbon filters are still operative and producing satisfactory improvement of water quality after several years of continuous operation. This is believed to be due largely to the microbes which develop; thus, the filter is both biological and chemical. Further evidence of this is that carbon filters can contribute large numbers of bacteria to the water (Table 5).

This effect is most noticeable where carbon filtration is preceded by a biocidal form of pretreatment such as prechlorination or softening of the water with lime. This may arise by the selection of relatively few physiological types resistant to disinfection, but subsequently, competitive pressures on the indi-

Table 5

Changes in bacterial counts through packed carbon filters in a pilot plant treating a highly polluted river water

Water treatment process	3 day, 22°C plate ct.ml^{-1}	Carbon filter: Before	Carbon filter: After
Basic treatment*	Mean	1360	237
	Max	5400	780
	Min	15	6
Caustic soda softening plus basic treatment	Mean	403	1210
	Max	2260	6800
	Min	0	0
Basic treatment*	Mean	1590	521
	Max	6400	5000
	Min	80	4
Prechlorination plus basic treatment	Mean	21	3820
	Max	280	12,000
	Min	0	1

* Ferric sulphate coagulant plus filtration.

vidual are reduced. These organisms are then able to multiply freely. With many different physiological types entering a carbon filter competition is much greater and restrains proliferation of any one species.

This effect has serious repercussions in distribution systems because in the presence of a declining residual in the carbon filter where prechlorination is practised, bacterial types will be selected by their resistance to chlorine. These same types then pass out of the filter to the final stage of disinfection and a large proportion of them survive, although they may be chlorine damaged, and enter the distribution system where they appear to multiply even in the presence of a nominal chlorine residual (Table 6).

(g) *Disinfection*

This is the most important stage of water treatment on which the safety of the supply depends. However, it has limitations largely brought about by the quality of water presented for disinfection, hence several stages of water treatment are used to condition the water, physically, chemically and microbiologically, for final disinfection.

In addition to temperature, the water quality factors influencing disinfection include organic compounds which react with the disinfectant, pH, ammonia, together with inorganic and organic particles that may shield cells physically.

Various disinfectants have been used for water supply including chlorine,

Table 6

Bacterial numbers throughout a water supply system employing activated carbon filtration

Stage of treatment	Chlorination practice	Chlorine residual $\mu g.ml^{-1}$		30°C, 7 day plate ct.ml^{-1}
		Free	Total	
Raw river water	–	–	–	10,700
–	Prechlorination	–	–	–
Post coagulation	–	7.8	9.4	2
Post filtration		0.4	5.8	2
Post carbon filters	Dechlorination	0.0	0.82	5500*
	Final disinfection chlorine plus ammonia for 15 h contact			
Final water		0.0	1.07	690*
End of distribution		0.0	0.05	10,600

* Chlorine-damaged bacteria.

chlorine dioxide, iodine, ozone, potassium permanganate, silver and u.v. radiation on the grounds that each has certain advantages under adverse environmental conditions.

(i) *The effect of temperature*

Disinfection of micro-organisms is a chemical reaction and accordingly the rate is affected by prevailing temperature. It is often necessary to increase the disinfection dose when temperatures are low. A complication can occur with warmer conditions in summer when it may also be necessary to increase the level of disinfectant because more disinfectant is lost to the environment in chemical side reactions and is not therefore available for disinfection.

(ii) *The effect of pH*

Halogen disinfectants which undergo hydrolysis and dissociation in water are affected by pH because the level of undissociated hypohalous acid present is predominantly the bactericidal agent.

$$X_2 + H_2O \rightleftharpoons HX + HOX$$

$$HOX \rightleftharpoons OX^- + H^+$$

Being uncharged, the hypohalous acid molecule is able to penetrate cell wall boundaries more easily. From published work on disinfection of vegetative bacteria, viruses, amoebic cysts and bacterial spores, Hall (1973) has calculated the disinfection level,

$$D = \int [\mathrm{HO\,Cl}]\ \mathrm{d}t\ (\mathrm{mol.min^{-1}.l^{-1}}).$$

These range from *ca.* 1×10^{-6} for 99% kill of vegetative cells and bacteriophage; 20×10^{-6} for 99% kill of viruses; 110×10^{-6} for 90% kill of amoebic cysts and 200×10^{-6} for 10% kill of bacterial spores.

(iii) *Effect of ammonia*

The above levels of disinfection are required over and above all other reactions. The halogens chlorine and bromine react with ammonia and organic nitrogenous compounds to form mono-, di- and trichloramines and bromamines. The species and levels formed depending on the concentration of reactants, pH and reaction time in that overall removal of ammonia may be achieved by the 'breakpoint' reaction (Palin 1950; Johnson & Overby 1970). The removal of ammonia by breakpoint reactions occurs with a ratio of halogen to NH_3–N by weight of 10 : 1 for chlorine and 20 : 1 for bromine and in that sense is wasteful of disinfectant with the result that other processes of ammonia removal are often used in water treatment (Short 1973).

Various philosophies exist concerning disinfection by chlorine. It is often argued that monochloramine is not an effective disinfectant and in that sense only free chlorine residuals should be used. However, many systems operate with apparent success using marginal chlorination in which the chlorine residual formed is predominantly monochloramine. In other circumstances, ammonia is added to the water to stabilize the chlorine residual as monochloramine during distribution. However, it should be realized that in water treatment, chlorine is added to water containing natural ammonia and bacteria and in that sense the chlorine has an equal opportunity of reacting with bacteria as with ammonia, and in experimental studies there is little difference observed between chlorine alone and chlorine reacting with ammonia *in situ* but both are far superior to preformed chloramines (Fig. 1). Other disinfectants do not react with ammonia (Fig. 2) and so possess certain advantages in situations where significant levels of ammonia are present as for example in river waters during cold weather when nitrification is minimal.

(iv) *Effect of organic matter*

Oxidizing disinfectants also react with organic matter in water, humic and fulvic acids being predominant. These represent a further demand on the available disinfectant and suppress the level of disinfection (Fig. 3). In order to satisfy this disinfectant demand during treatment and thereby promote the

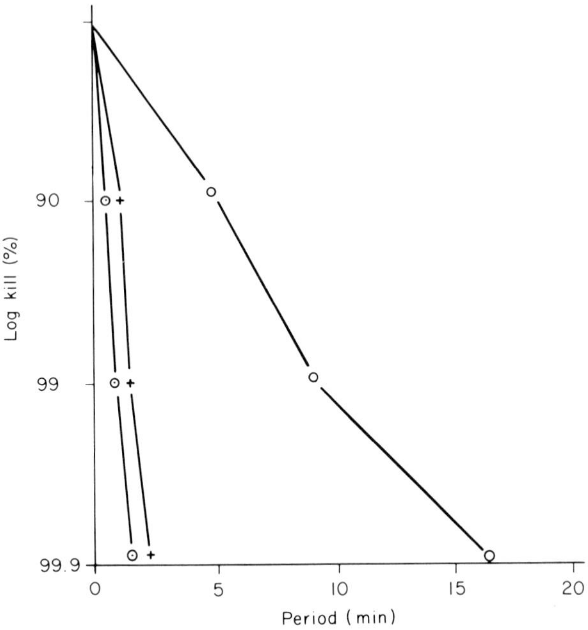

Fig. 1. The percentage kill of *E. coli* at pH 7.5 and 5°C using: ⊙, 0.03 μg.ml^{-1} available chlorine, +, 0.03 μg.ml^{-1} chlorine + 0.03 μg.ml^{-1} NH_3–N, chloramine formation *in situ*; ○, 0.1 μg.ml^{-1} available chlorine as chloramine.

chances of maintaining a chlorine residual during distribution, prechlorination is often practised and this together with coagulation and filtration are effective in removing much of the organic matter.

The effect of humic colour on disinfection by u.v. radiation is also significant in that the radiation is absorbed and does not affect the cells to the same extent. Other substances present in water may also dissipate the energy, for example, inorganic particles. The deleterious effect of both humic colour and clay suspension on the destruction of *E. coli* is shown in Table 7, and when these components are used together the effect is approximately additive.

(h) *Ground waters*

The infiltration of polluted surface waters through the soil cover constitutes an efficient process of water treatment. Two distinct phases of infiltration may be recognized namely one in the essentially aerobic, unsaturated zone and the other in the saturated zone which may be partially or wholly depleted of oxygen. Removal of organic pollution occurs in the unsaturated zone and can result in the formation of an effective biological filter. Whilst there are probably no

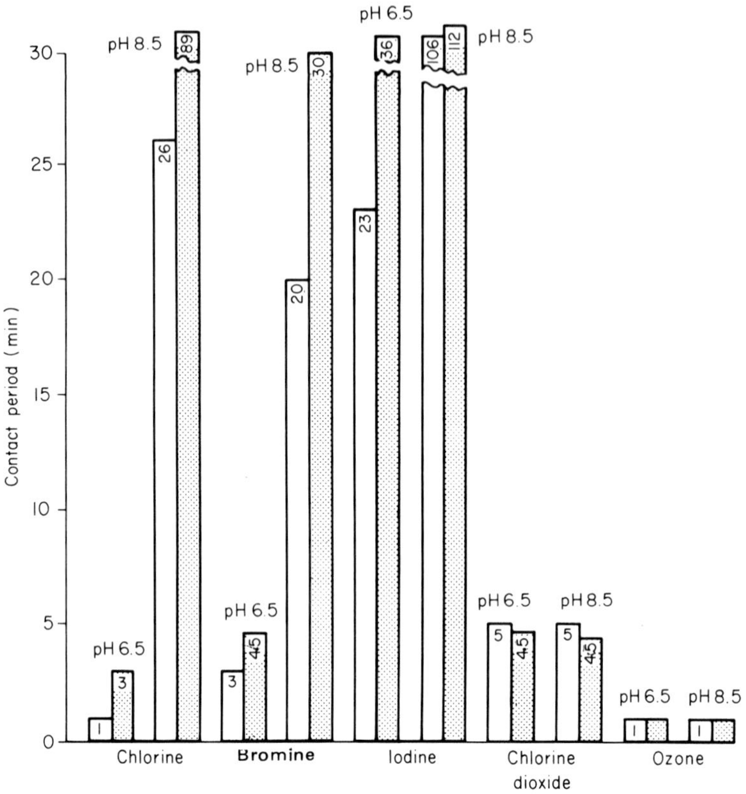

Fig. 2. Comparison of disinfectants at a concentration of 0.02 μg.ml^{-1} for 99% destruction of *E. coli* at 5°C and pH 6.5 and 8.5 in the presence of [open box], 0.03 μg.ml^{-1} (typical of a borehole level); [dotted box], 0.3 μg.ml^{-1} (typical level for a river) of ammonia nitrogen.

specific mechanisms for selective removal of pathogens or indeed any other bacteria, the majority are unsuited to this highly competitive microbial environment and evidence suggests a limit of travel of *ca.* 3 m in depth (Romero 1970). Wells and boreholes are accordingly lined for at least this depth to prevent contamination.

In the saturated zone, mechanisms of removal are limited to adsorption and in these circumstances organisms may be carried passively in the ground water flow; microbiological penetration has been observed in up to 30 m of horizontal travel (Romero 1970). For this reason hydrological evidence is sought on the strata concerned, and the direction and rate of ground water flow in order to site the borehole away from obvious sources of pollution such as contaminated surface water, septic tanks, agricultural wastes and refuse tips. In any event a minimum zone of protection of 30 m is essential.

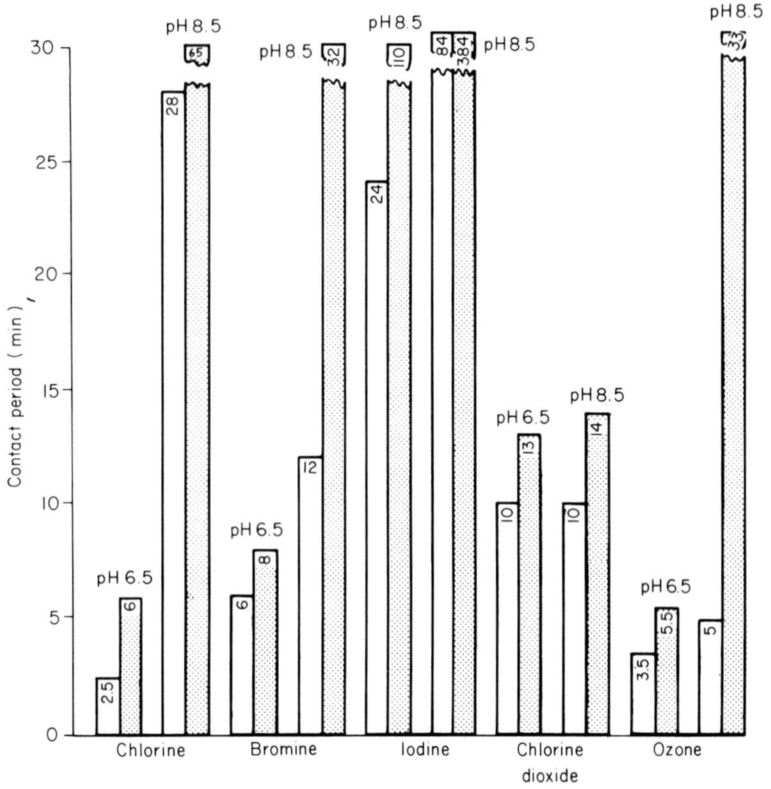

Fig. 3. Comparison of disinfectants at a concentration of 0.02 μg.ml^{-1} for 99% destruction of *E. coli* at 5°C and pH 6.5 and 8.5 in the presence of: [ ], 5 hazen units; [dotted], 25 hazen units of organic colour.

Table 7

Combined effect of humic colour and turbidity of River Thames water on destruction of E. coli *by u.v. light (110 μW.cm^{-2} at 253 nm)*

Humic colour (°Hazen)	Turbidity*	Time for 90% kill of *E. coli* (sec)
–	–	34
5	–	51
25	–	170
–	50	115
5	50	175
25	50	260

* Expressed as μg.ml^{-1} of bentonite.

In those circumstances where fissured strata are involved the extent of travel is unlimited and the degree of purification virtually non existent. These sources are obviously suspect and should be monitored regularly both chemically and bacteriologically for evidence of pollution, but the greatest safeguard of all lies with adequate chlorination (Allen & Geldreich 1975).

Although pathogenic and pollution indicator organisms are of paramount importance (Hutchinson 1974*a*), in ground water supplies other organisms are important. Their beneficial activities include degradation of man-made pollutants such as pesticides, herbicides and detergents, and general mineralization including the cycling of the essential elements, nitrogen, phosphorus and sulphur. However, when taken to extremes by excessive pollution, the same processes can give rise to problems often associated with ground waters. These include: (a) depletion of dissolved oxygen; (b) reduction of nitrate to nitrite or ammonia; (c) reduction of sulphate to sulphide giving rise to offensive odours and growth of filamentous sulphur bacteria; (d) reaction of sulphide with iron to form an insoluble precipitate which can restrict ground water flow; (e) mobilization of iron from soil under conditions of reduced oxygen tension only to be oxidized and precipitated in other regions of the aquifer either by chemical or microbiological means.

Iron bacteria are frequently present in ground waters and in particular those subject to a degree of organic pollution. This was evident in the episode described by Taylor (1960) where disposal of burnt sugar and biscuit wastes in a flooded gravel pit resulted in profuse growths of filamentous iron bacteria in a nearby well. Similarly, Brighton (1958) reported that the fermentation of ensilaged pea haulms polluted a chalk borehole via a swallow hole. This produced profuse growths of filamentous iron bacteria (*Leptothrix* and *Crenothrix*) and also filamentous sulphur bacteria (Beggiatoa) growing at the expense of hydrogen sulphide generated by the enhanced activity of sulphate-reducing bacteria. The combined activities of all of these organisms severely restricted the yield of the borehole and rendered the water unpalatable.

The increasing demand for potable water supplies has lead to the concept of replenishing ground waters by artificial recharge (*Anon.* 1971*c*). However, the dangers of this approach must be clearly recognized if irretrievable contamination of the aquifer is to be avoided and only recharge of a fully treated water should be permitted. The reasoning behind this is not solely from a microbiological health aspect but because organic pollution might promote growths which could cause blockage and reduce the yield of the aquifer.

4. Problems Related to Distribution of Water

If water could be distributed to the consumer in the condition in which it is produced following treatment, the tasks of the water supply scientist would be relatively simple. Unfortunately during distribution significant physical, chemical and microbiological changes occur. These include: (a) microbial aftergrowths; (b) microbiologically induced chemical changes; (c) contamination.

(a) *Microbial aftergrowths*

(i) *Source of organism*

Not all bacteria in water are indigenous: some are derived from soil and sewage and a proportion of these will survive the water treatment process. Other organisms gain access from the treatment processes themselves or at the time of assembly or repair of pipelines. Service reservoirs and water towers may add further pollution either as the result of structural defects or simply through contact of the water with air and the walls of such structures. It has been shown that up to $1.5 \times 10^6 .ml^{-1}$ bacteria may be recovered from condensation on the ceiling of water towers (Thofern & Speh 1974).

(ii) *Factors controlling numbers of bacteria in supply*

The multiplication of organisms within the system will depend on many factors such as: (a) level and type of inorganic and organic nutrients including growth factors; (b) environmental factors such as temperature, redox potential and pH; (c) whether an adequate level of residual disinfectant is maintained throughout the system.

All of these factors can be related in that with increased temperature there is a more rapid decline of chlorine residual due to reaction with organic matter present. Accordingly, the biocidal effect is less but with the increased temperature microbiological growth is greater.

It is not surprising therefore that an increase in bacterial numbers occurs through the distribution system and in certain cases for which the hydraulic flows are known, the increases correlate roughly with the detention time (Fig. 4). Similarly, growth coincides with the decline in chlorine residual and in that sense it is difficult to decide at which concentration the residual ceases to be effective. In this particular system, the chlorine appears to fail to control growth from the time the water enters supply, and as the residual falls, more and more bacteria develop.

These observations were made in the field but a control experiment was set up at the same time in which the water was held in three Winchester bottles at 22°C in the dark and these were monitored daily for total residual chlorine and yeast extract agar plate counts after incubation for 7 days at 22°C. The three bottles were for: (a) monitoring chlorine; (b) monitoring bacterial numbers; (c) monitoring bacterial numbers after initial dechlorination.

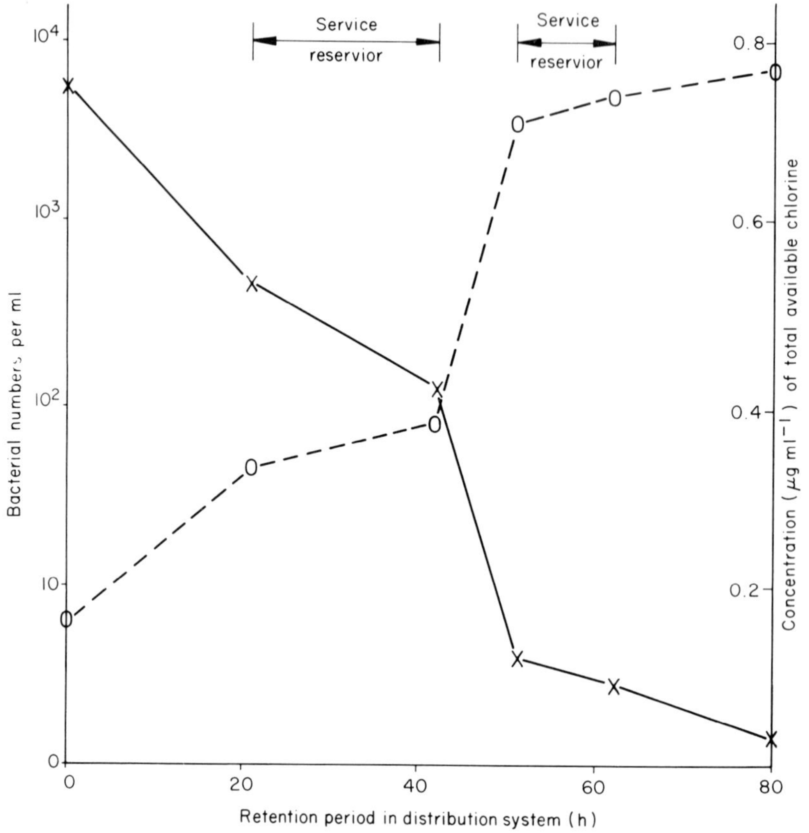

Fig. 4. The variation with retention time in a distribution system of: O, bacterial numbers; X, total available chlorine.

The results for these three systems are given in Fig. 5 which shows that without the chlorine demand of the distribution system, the chlorine residual remained fairly constant after an initial decline, and bacterial numbers were suppressed. However, in the dechlorinated sample, the number of bacteria rose to a level comparable to that observed in the distribution system but in the latter case this occurred in 3 days at 10°C compared with 6 days for the bottle sample. Similar experiments for other waters are summarized in Table 8.

In system (1), a high quality chalk borehole supply, the chlorine residual was maintained throughout the distribution system and accordingly bacterial numbers were controlled. This was confirmed in the bottle test. System (2) was the soft, moderately polluted river water described previously but here the counts, after 3 days detention, give the same picture. System (3) was a soft

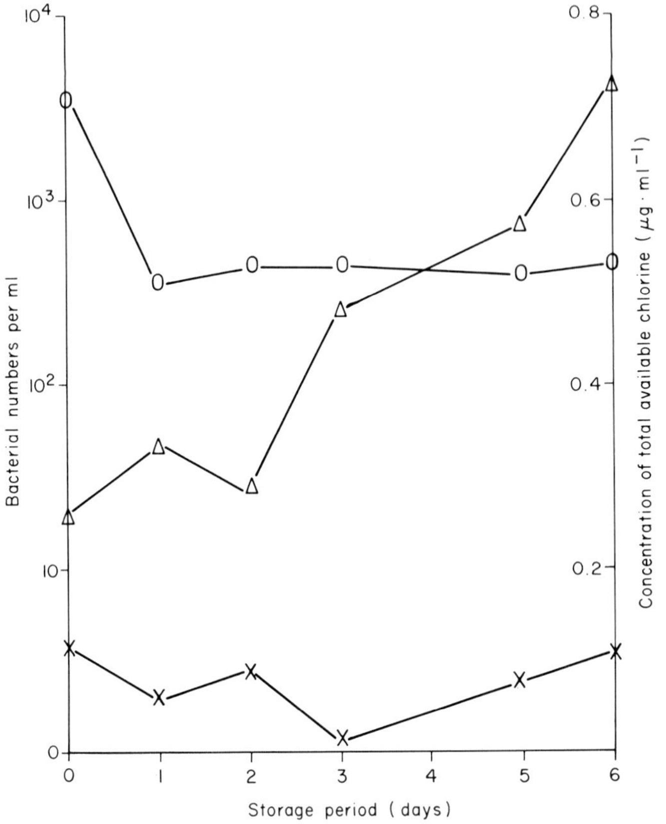

Fig. 5. The effect of storage time on the numbers of bacteria in a fully treated potable water stored both in the presence and absence of chlorine: ○, total chlorine residual; X, bacterial numbers in the presence of chlorine; △, numbers in the absence of chlorine.

upland water with significant organic colour which rapidly destroyed any residual chlorine in supply. However, despite this loss, counts did not rise appreciably confirming that refractory humic compounds are of limited nutritional value to these bacteria. System (4) was an impounded river water with high levels of sewage pollution. In system (4a), conducted in the prescribed manner, growth was rapid in both bottles, but numbers were controlled somewhat in the distribution system by applying further chlorination at an intermediate stage of supply. In system (4b), water was taken from the system at a stage corresponding to 3 days retention and this was incubated in a bottle for an extra 3 days. Further increases in this case confirm that with declining chlorine residual, bacteria can multiply to the nutritive potential of the system.

In certain cases, the bacterial increases observed through a distribution

Table 8

Comparison of bacterial numbers in distributed waters after retention for 3 days in distribution or incubation in large glass bottles (7 day plate counts at 22°C on YF agar)

System	Bottle incubation for 3 days at 22°C		After 3 days retention in distribution	Total chlorine residual in distribution after 3 days	Total organic carbon in the water entering supply
	With chlorine	Dechlorinated			
	$No.ml^{-1}$	$No.ml^{-1}$	$No.ml^{-1}$	$\mu g.ml^{-1}$	$\mu g.ml^{-1}$
Borehole (1)	<0.1	4.4×10^2	1.0	0.09	0.2
Polluted, soft river water (2)	1.5	4.0×10^2	7.4×10^3	0.03	3.0
Soft upland catchment (3)	<1	1.6×10^1	5.1×10^1	<0.02	2.0
Sewage polluted impounded river water (4)					
Water entering supply (a)	9.3×10^2	5.3×10^4	2.8×10^2	0.08	5.0
Water from supply after 3 days retention (b)	7.5×10^3	1.5×10^4	–	–	–

system cannot be accounted for on the basis of retention and likely generation times. This suggests that the site of activity occurs largely on surfaces within the pipe. Any apparent deficiency in organic carbon in potable water is probably redressed by the continuous flow of water bringing fresh nutrients to the laminar layer of near zero velocity at the surface of the pipe. This gives rise to a biological slime layer which may account for important changes to the pipeline and water conveyed through it.

(iii) *Sources of energy*

Three distinct sources of energy are available to bacteria in distribution systems: (a) inorganic compounds; (b) soluble organic matter; (c) particulate or colloidal organic matter.

(a) In water supply, energy sources for chemoautotrophs include ferrous iron, hydrogen and reduced sulphur compounds, all derived from the process of corrosion. Other autotrophic transformations include oxidation of manganese, ammonia and nitrate. It has been calculated, assuming a biological efficiency of 5%, that to produce 1 g of organic carbon from carbon dioxide by iron bacteria would require the oxidation of 448 g of ferrous iron. This potential is demonstrated by the chemoautotrophic *Thiobacillus ferrooxidans* which in culture will utilize amounts of ferrous iron as large as 3 $\mu g.ml^{-1}$ (Hutchinson *et al.* 1966). It is unlikely that this obligate acidophile will grow under neutral conditions in water supply. However, *Gallionella* is assumed to be autotrophic and operates in water supply systems to a limited extent (Wolfe 1964). The other important energy source is hydrogen generated in the corrosion process, which may be utilized by sulphate-reducing bacteria. Similarly the sulphide produced by this reaction may itself be oxidized autotrophically by species of thiobacilli. It may be argued therefore that corrosion of iron pipelines is an essential driving force of many biological processes which occur in water supply.

(b) The level of dissolved organic matter entering water supply will differ with the type of treatment, biological processes achieving a greater measure of mineralization of oxidizable organic compounds. The levels and type of organic compounds also differ according to source. A high quality chalk borehole water may have a level of total organic carbon as low as 0.2 $\mu g.ml^{-1}$ whereas a river derived supply may be 10 to 20 times this value and highly coloured upland catchment waters higher still. Table 9 summarizes the nature of these compounds.

This would suggest that these waters differ mainly in the levels of refractory humic and fulvic acids and accordingly bacterial growth does not necessarily relate to the total organic carbon value. This difference in bacterial response for different types of water and water treatment is illustrated in Table 10 in which the final bacterial level achieved once the chlorine residual is lost is largely related to the type of water.

Table 9
Types of organic compounds found in various water supplies

	Total organic carbon ($\mu g.ml^{-1}$)	Carboxylic acids ($\mu g.ml^{-1}$)	Non-polar compounds ($\mu g.ml^{-1}$)	Polar compounds ($\mu g.ml^{-1}$)
Borehole	0.9	0.3	0.3	0.3
Stored river water	2.5	2.0	0.3	0.2
River water	4.7	3.2	1.2	0.3

(c) The remaining source of organic carbon present in water supplies is particulate, and includes algae and variously flocculated organic materials which may pass through treatment, or even arise within the distribution system itself by adsorption on hydrated ferric oxide floc formed in the pipelines as a result of corrosion.

(b) *Microbiologically induced chemical changes in supply*

(i) *Corrosion*

Because water supply pipelines have traditionally been made of iron the effect of corrosion is very important. To the water industry the principal concern is not so much for the loss in service life due to corrosion of the external surfaces of buried pipelines, but the effect on the interior. This is because the products of corrosion give rise to deposits and nodules which interfere with the carrying capacity of the main and also affect the quality of the water conveyed. Corrosion is an electrochemical process in which iron goes into solution at the anode and hydrogen forms at the cathode from dissociation of water causing the cell to become polarized; thus, for corrosion to continue, depolarization of the cathode is essential. Initially this is achieved by oxygen in the water but the accumulation of iron oxides at the site of corrosion changes the situation. In water supply corrosion rarely ceases at this stage because other reactions, termed anaerobic corrosion, are initiated microbiologically. The overall mechanism for anaerobic corrosion according to Wolzogen-Kuhr & van der Vlugt (1934) may be summarized as follows.

Anodic reaction	$4Fe \rightarrow 4Fe^{2+} + 8e.$
Dissociation of water	$8H_2O \rightarrow 8H^+ + 8OH^-.$
Cathode reaction	$8H^+ + 8e \rightarrow 8H.$
Bacterial sulphate reduction	$SO_4^{2-} + 8H \rightarrow S^{2-} + 4H_2O.$
	$4Fe + SO_4^{2-} + 4H_2O \rightarrow FeS + 3Fe(OH)_2 + 2OH.$

Table 10

Effect of type of water, treatment and chlorine residuals in controlling bacteria in distribution

Type of water	Treatment										Raw water	After treatment		Water entering distribution			Intermediate stage of distribution				End of distribution			
	Impounded	Prechlorination	Coagulation	Rapid filtration	Pressure filtration	Slow sand filtration	pH adjustment	Chlorination	Ammoniation	Contact tank	No.ml^{-1}	Bacterial count (no.ml^{-1})	Chlorine residual (ug.ml^{-1}) free total	Bacterial count (no.ml^{-1})	Chlorine residual (ug.ml^{-1}) free	total	Bacterial count (no.ml^{-1})	Chlorine residual (ug.ml^{-1}) free	total	Distance travelled (km)	Bacterial count (no.ml^{-1})	Chlorine residual (ug.ml^{-1}) free	total	Distance travelled (km)
Underground																								
chalk borehole								+	+	+	4		–	81	0.0	0.35	72	0.0	0.01	2.8	460	0.0	0.0	6.1
sandstone borehole								+			79		–	4	0.23	0.23	–	–	–		180	0.06	0.06	20.6
mine water								+			83		–	1	0.32	0.33	0	0.24	0.26	3.2	10	0.08	0.08	5.5
Upland catchment																								
							+	+	+	+	212		–	104	0.0	0.3	11	0.0	0.15	10.1	11	0.0	0.0	21.4
	+		+		+		+	+			2170	54	–	3	0.0	0.35	16	0.0	0.2	9.0	1070	0.0	0.1	14.3
	+		+	+			+	+		+	660	390	–	11	0.40	0.50	530	1.05	0.14	32.2	1220	0.01	0.04	42.0
	+		+		+		+	+		+	720	310	–	133	0.01	0.21	580	0.0	0.03	3.2	1310	0.0	0.01	10.1
	+					+		+			190	60	–	60	0.0	0.00	93	0.0	0.0	14.5	190	0.0	0.0	16.9
River derived																								
						+		+		+	211,000	130	–	38	0.83	0.83	9	0.1	0.25	5.5	4700	0.02	0.10	8.0
	+	+	+	+				+	+	+	10,700	2	5.6	690	0.02	1.07	420	0.0	0.40	9.7	10,600	0.0	0.05	27.4
	+		+	+			+	+		+	7000	1700	–	26	1.10	1.55	8	0.35	0.70	2.4	2500	0.0	0.10	27.0
			+	+				+		+	10,100		–	35	0.35	0.40	57	0.30	0.35	2.1	6000	0.0	0.00	5.3

Bacteria counted on YE agar after 7 days at 30°C.

Horvath & Solti (1959) showed from polarization curves in the presence and absence of cultures of sulphate-reducing bacteria that organisms brought about cathodic depolarization. Booth & Tiller (1960, 1962*a,b*) working with different species of sulphate reducers showed that this ability was related to hydrogenase activity: *Desulfovibrio vulgaris* (Hildenborough strain) isolated from a corroded pipe had a high hydrogenase activity resulting in marked cathodic depolarization whereas with the hydrogenase-negative *Desulfotomaculum orientis* no effect was demonstrable. In the corrosion of central heating exchangers, hydrogenase-positive strains of the thermophile *Desulfotomaculum nigrificans* have been shown to be active. The hydrogenase-positive strains utilize the hydrogen for energy but require organic matter for cell growth. In theory any bacterium able to utilize free hydrogen should be capable of cathodic depolarization and Postgate (1963), reporting the work of Cook, claims unequivocal corrosion of iron by nitrate reducing strains of *Pseudomonas, Bacillus* and *Micrococcus.*

Micrococcus denitrificans is reported to remove 70 mg of Fe from steel wool in five days compared with 31 mg in ten days for a sterile control system. Booth *et al.* (1968) found no evidence of cathodic depolarization with hydrogen or methane bacteria. Although important in the early stages, cathodic depolarization is not the only mechanism involved in corrosion. Secondary aspects arising directly or indirectly from microbial activity include the following.

1. The formation of hydrogen sulphide which is frequently evident inside corrosion nodules and which maintains a high e.m.f. in the Fe–H corrosion cell.
2. Ferrous sulphide which acts as a cathodic depolarization agent and is presumably reduced to hydrogen sulphide.
3. The presence of ferrous sulphide and hydrogen sulphide under localized conditions of high pH resulting from the corrosion process, produces the polysulphides and elemental sulphur encountered in corrosion nodules (Bunker 1940). Elemental sulphur also stimulates the cathodic process, presumably by its reduction to sulphide.
4. As corrosion proceeds, deposits form as a film or tubercle which can promote corrosion by creating conditions of differential aeration, the oxygen-deficient interior becoming anodic to the rest of the pipe. It is claimed that iron bacteria contribute to the formation of this condition. In the oxygen-deficient interior of such tubercles sulphate reducers may be found though it is possible that when the condition reaches an advanced state, the whole becomes somewhat fossilized and the site of sulphate reduction moves out towards the surface of the nodule where nutrients are more freely available.
5. Other mechanisms of corrosion have been postulated including oxidation of elemental sulphur and sulphide to sulphuric acid by species of *Thiobacillus* under aerobic conditions. This mechanism may operate on the exterior of

pipelines under conditions of fluctuating ground water levels in which aerobic and anaerobic conditions are produced alternately (Wormwell 1973). It is unlikely to operate inside pipelines since the concentration of acid produced would be constantly removed.

6. The phenomenon of acid corrosion due to the combined activities of sulphate-reducing bacteria producing sulphides under anaerobic conditions and their oxidation by thiobacilli in an aerobic part of the system is well documented for steel and concrete sewers (Parker 1945; Boon & Lister 1975).

Ideally, the water industry would like to control corrosion of the interior of pipelines; however, few mechanisms of control are truly effective. The methods adopted are presented here.

1. Aeration of oxygen-deficient water supplies. Many systems with severe problems of discoloured water are supplied from boreholes deficient in dissolved oxygen (<20% saturation). Aeration should suppress the growth of sulphate-reducing bacteria, but in practice these measures may not be completely effective unless supported by a preliminary programme of mains scraping to destroy the corrosion nodules and expose the sulphate-reducing bacteria.
2. The introduction of chlorination to control sulphate-reducing bacteria is suspect for the same reasons.
3. Cathodic protection is said to control microbiologically induced corrosion by increasing the level of polarization and at the same time raising the pH value at the surface of the pipe. However, it is not easily applied to the interior of a pipeline.
4. Other treatments include the use of active silica on polyphosphates to stabilize corrosion products and thereby avoid consumer complaints. There is no evidence to suggest that the action is on microbiological processes.
5. The only sure treatment is by complete scraping of the interior of the pipeline and effective relining with bitumen, epoxy or polyester resin or cement.

(ii) *Involvement of micro-organisms in the production of 'dirty water'*

The release of ferrous iron by corrosion of iron mains and its possible oxidation to the insoluble ferric state on contacting aerated water in the pipeline, can have serious effects on the quality of water supplied to the consumer. In one oxygen-deficient supply with minimum flow rates, a level of 11.9 $mg.l^{-1}$ of soluble (ferrous) iron has been found though under normal conditions of pH and dissolved oxygen the oxidation of ferrous iron is a rapid chemical reaction. In situations where oxygen is limiting the iron bacteria may establish themselves in a position relative to the oxygen gradient and ferrous iron so that they may catalyse the reaction. Baas-Becking & Wood (1956) have isolated these organisms from water with pH values ranging from 4.6 to 8.0 and E_h values from +200 to +500 mV.

Iron bacteria frequently encountered in water mains include *Gallionella* and filaments of the *Sphaerotilus – Leptothrix* group, but rarely in large numbers so their involvement in the oxidation of iron resulting from corrosion may be limited compared with their more common involvement with iron oxidation in boreholes and springs.

MaCrae & Edwards (1972) have demonstrated that many bacteria, including *Caulobacter, Micrococcus, Ps. fluorescens, Mycobacterium phlei, E. coli, Klebsiella* spp. and *Corynebacterium* spp., when incubated with a ferric iron sol, will passively adsorb hydrated ferric oxides. Victoreen (1969) has demonstrated that this condition can give rise to discoloured water in supply.

Hydrated ferric oxides arising from corrosion also complex with organic matter including humic and fulvic acids remaining after treatment and ultimately coagulation is achieved in which viable bacteria become entrained. These may then grow on the humic compounds or other adsorbed organic matter. This occurs in the stagnated area adjoining the pipe wall giving rise to a chemical–biological slime layer. These deposits contain between 2 and 10% of organic matter.

Other workers believe that certain bacteria have the ability to utilize the organic fraction of iron–humic complexes leading to a secondary precipitation of iron around the cells (Drake 1965; Clark *et al.* 1967).

Bays (1975) is of the opinion that microbial populations, pseudomonads in particular, are directly responsible for enhancing the process of corrosion of iron mains by changing the pH and E_h in the immediate environment of the pipe wall. Problems occurred with the use of an organically polluted surface water particularly during the summer when bacterial numbers in the distribution system were large. By increasing the level of chlorination and minimizing the detention period in supply he has succeeded in controlling bacterial numbers and reduced the incidence of consumer complaints towards the periphery of the distribution system.

(iii) *Oxidation of manganese*

As with iron, many water supplies contain low levels of reduced manganese which is soluble and difficult to remove. However, in pipelines manganese is often precipitated as its oxide and micro-organisms may be involved (Tyler & Marshall 1967*a,b*).

(iv) *Depletion of oxygen*

In many distribution systems there is a measurable decline in the level of dissolved oxygen in the water which is in part due to corrosion but also to the growth of micro-organisms. The level of oxygen depletion is related to the chlorine residual and flow rate because retention is greater towards the periphery of the system. This also correlates with the fact that with reduced flow rate particulate matter including corrosion products with adsorbed organic matter, will sediment to a greater extent and considerable biological growth will be associated with these deposits.

(v) *Nitrate and sulphate reduction*

With the limitation of oxygen, certain bacteria utilize nitrates or sulphates (see Section 4bi) as terminal hydrogen acceptors in respiration. This has repercussions on water quality as reduction of nitrate to give nitrite in the water could give rise to the condition of methaemaglobinaemia in infants. For this reason statutory limits for nitrate (and nitrite) are laid down for water supplies. The risk from nitrate reduction is far greater from the microbial population within the intestine than that which is likely to occur in a water supply because the formation of nitrite in a water supply is rare, reduction normally proceeding to nitrogen gas. Under certain conditions nitrate may be reduced to ammonia which in the presence of chlorine can lead to undesirable tastes and odours (see below). Reduction of sulphate to hydrogen sulphide is largely related to anaerobic corrosion.

(vi) *Tastes and odours*

The presence of sulphides in water supplies is unacceptable to the consumer and rarely is the condition allowed to develop. Programmes of flushing usually introduce sufficient oxygen to suppress sulphide formation, but in waters deficient in oxygen aeration may be necessary. Reduction of nitrate to ammonia may occur at some specific location in a distribution system where the redox conditions are favourable. Chlorine in the distributed water may be in excess of this (>10 parts chlorine: 1 part NH_3–N by weight) and would lead to the formation of nitrogen trichloride which, even at a concentration of 0.01 $\mu g.ml^{-1}$, is very odourous resembling TCP (trichlorophenol) and at these extremely low levels is difficult to measure (Victoreen 1974). Justification for this is largely based on the observations that: (a) phenols are never found in waters which are the subject of complaints; (b) TCP-like odours occur only with borehole supplies of impeccable quality but which are deficient in dissolved oxygen; (c) the odours are not due to the residual chlorine itself for with these systems very low levels of free chlorine are applied; (d) complaints tend to arise in a particular area towards the periphery of the system suggesting that at a certain point sufficient ammonia may have been produced by reduction of nitrate with which the chlorine reacts; (e) an analogous situation may arise as the result of chemical reduction of nitrate by fresh copper surfaces in new plumbing installations.

The growth of micro-organisms in pipe sediments could in itself give rise to taste and odours. Notable in this respect are various fungi and actinomycetes (Burman 1965; Bays *et al.* 1970). *Streptomyces* spp. produce metabolic byproducts having the characteristic musty-earthy odours for which the specific compounds geosmin (Gerber & Lechevalier 1965) and methyl*iso*borneol (Piet *et al.* 1972) have been identified.

(c) *Contamination*

(i) *Cross-connection and back siphonage*

Contamination is possible when pumps or private supplies of water which may be polluted are connected to the mains supply. Danger of pollution exists when reduced pressures occur in the public supply as a result of heavy draw-off at times of peak demand, for fire fighting or serious leaks in water mains. Under these conditions pollutants can be drawn in. Many episodes of enteric disease have occurred for this reason (Jebb 1965). In a recent survey of the risk of back siphonage in Britain (*Anon.* 1971*a*) 52 out of 119 water undertakings questioned reported a total of 120 incidents of which: (a) 78 cases were in industrial, 22 in domestic, 15 in farm and 4 in hospital supplies, and 1 occurred during mains repair; (b) 82 were the result of non-compliance with byelaws; and 42 of these related to cross-connections between the public water supply and another source or system; 16 cases were due to failure of non-return valves, and 13 to temporary hose connections; (c) of the 120 incidents only 3 came to light on the evidence of illness amongst consumers, the remainder were identified by changes in taste, colour, temperature of water supplied or by routine inspection.

It was suggested that many other instances had probably occurred but passed unnoticed.

In an investigation of back siphonage risks in domestic properties (Gilfillan 1971), 26 water undertakings each inspected 50 properties for compliance with *Model Water Byelaws* (*Anon.* 1966*a*). Bacteriological tests were made on the water drawn from cold water storage cisterns and taps fed directly from the service pipe to assess deterioration in quality with storage. In was found that 61% of properties were at 'risk' with respect to contravening at least one criterion of the byelaw as far as this concerns cross-connection and back siphonage. These figures relate to properties of all ages whereas age-group figures were: <5 years 36%; 5–15 years 63%; >15 years 84%. If temporary connections to taps fed directly from the service pipes were included as a risk the overall figure would rise to 85%. For the purpose of bacteriological examination 37°C plate counts were made on water drawn directly from the service pipe (a) and water drawn from the storage cistern (b). As a measure of deterioration, the ratio of b/a was calculated and fell into the following categories.

Ratio b/a	Percentage of results in each category
0–1	46
1–10	39
10–100	11
100–1000	4
>1000	0

The majority of results (85%) fall within the range 0–10 signifying no change in water quality. However, ratios in excess of this suggest deterioration with storage. Deterioration was also correlated with uncovered cisterns and cisterns with a low number of outlets suggesting a low turn over. Coliform determinations were made for 426 properties and in 17 instances these were positive: coliforms were present in 5 cases for the service pipe sample and 11 cases after storage. In one case coliforms, including *E. coli*, were present in both samples. Only one example contravened the requirements in Report 71 (*Anon.* 1969). The distribution of coliforms in covered and non-, or inadequately, covered cisterns were comparable. It was concluded that in domestic properties there was a finite risk of back siphonage. However, this hazard was considered to be low because the contamination was unlikely to be harmful to health but could give rise to complaint by consumers (class 3 risk).

Class 2 installations include applications such as hose connections to domestic taps, domestic washing machines, dishwashers, car wash plant and dental chairs. These must be protected by an air gap or non-return valve of class A reliability whereas for class 1 risks only, an air gap would suffice. These include fire fighting appliances, connections for industrial, agricultural, hospital and laboratory purposes, bidets and haemodialysis machines, mobile tankers, gully emptiers and sewer cleaning machines together with hose connections to taps in non-domestic premises. Safe-guarding against back siphonage is largely a matter of education and strict enforcement of byelaws not only because this affects plumbing in factories in which there is a greater opportunity for cross-connection, but also in houses where the 'do-it-yourself' enthusiast might not appreciate the potential danger.

(ii) *Materials of construction*

Although some measure of growth of micro-organisms on low levels of organic matter present in potable water is inevitable, it is becoming increasingly evident that many common materials used in water supply can promote excessive growth of micro-organisms. This can lead to a deterioration in the quality of water by promoting growths of indicator bacteria, visible growths and turbidity and the production of taste and odours. As part of quality control in relation to consumer complaints the Metropolitan Water Board, London has for many years pioneered the investigation of materials used in water installations. This has included jointing materials and sealants, greases for tap and valve spindles, pump and valve gland packings, tallow and soldering fluxes, lubricants, tap washers and various plastic materials. (*Anon.* 1968, 1973; Burman & Colbourne 1976). Originally, concern was with the growth of coliforms on leather washers used on taps (*Anon*. 1916; Pratt 1965) and hemp yarn used for caulking run lead joints. Both situations gave rise to difficulty in obtaining satisfactory bacteriological samples in supply. Taste and odour problems have also been associated with various plumbing materials, particularly under conditions of stagnation and localized warming which favour the growth of fungi and actinomycetes (Burman 1965). Conditions of this type are probably responsible for a great deal of the deterioration in water quality encountered in supply and may explain the chance occurrence of coliforms and possibly *Ps. aeruginosa* in supply. There is evidence to suggest that the industry should be more concerned about the latter organism which is potentially pathogenic (Hunter & Ensign 1947; Cross *et al.* 1966).

A special feature of this problem occurs with the disinfection of all newly laid and repaired mains by the water undertaking in which the main is usually charged with a large dose of chlorine (20 μg.ml^{-1}) as laid down in regulations (*Anon.* 1967). In spite of these measures, most water undertakings in Britain experience difficulty in obtaining satisfactory bacteriological results following the operation. An investigation of the problem by the WRC has shown that the lubricants used in assembly of push-fit joints in pipelines promote the growth of coliforms within the joint space, and in this remote situation chlorination is ineffective. The solution to the problem lies in replacement of the soap- and castor-oil-based lubricants by a water soluble, biocidal lubricant. This aids assembly of the pipe joint and also exerts a biocidal action from within the joint space, thereby avoiding the problems of penetration of biocide into the joint space (Hutchinson 1974*b*).

On the basis of water supply byelaws which contain clauses prohibiting the consumer from any practice which would lead to waste, undue consumption, misuse or contamination of the water, the National Water Council Approvals Board now requires that all materials and fittings for use in contact with potable water should be tested to ensure that they do not contaminate the water supply, contamination being interpreted as any deterioration in quality of the water, irrespective of whether it is injurious to health or not. (Concern is not solely for microbiological, but also for toxicological and organoleptic aspects.) Waste can be the result of consumers flushing the tap to get rid of odourous, bad tasting and dirty looking water before consumption.

(iii) *Biodeterioration*

A related problem concerns the resistance of materials to biodeterioration. There is evidence that certain constructional materials in a water environment will deteriorate through the activities of micro-organisms and lose their essential mechanical properties. This has occurred with certain pipe jointing cements based on sulphur which is oxidized by thiobacilli. The effect is loss of sealant and localized attack of the pipe joint by the acid produced (Duecker *et al.* 1948; Frederick & Starkey 1948). Rubber-like sealants of polyurethane or polysulphide used with expansion joints of concrete water retaining structures have been observed to deteriorate (*Anon.* 1970*a*; Appleton 1973). With polyurethanes the attack is probably the direct result of fungi on the polymer, but with polysulphides deterioration is more likely to be the result of depolymerization brought about by hydrogen sulphide generated in the environment through the activities of sulphate-reducing bacteria. This is understandable for many sewage treatment processes, but in the case of a service reservoir the attack was confined to the floor of the tank which had a covering of sediment favourable to sulphate reduction.

Deterioration has also been observed with rubber sealing rings used both in water and sewage pipelines (Leeflang 1963; Hutchinson *et al.* 1975). The organism responsible for rubber deterioration is a species of *Nocardia*, forming orange-pink colonies, which has been found in numbers of 10^5–$10^7.g^{-1}$ of deteriorated rubber (Hutchinson *et al.* 1975). It has been found that deterioration in a pipeline may be identified by the presence of these organisms rather than by expensive removal of whole pipe sections. With distance travelled along the pipeline, there is an increasing opportunity for the water to pick up significant numbers of these organisms from each pipe joint. Analysis is by membrane filtration of the water and incubation of filters on the medium of Orchard & Goodfellow (1974). By this technique the characteristic orange-pink Nocardia associated with deterioration have been found in numbers as great as $200 \times 100\ ml^{-1}$ in mains where there is evidence of rubber deterioration. Deterioration of natural rubber sealing rings may occur after only 4 years exposure and certain seals have lost over half of their mass after 18 years exposure and are no longer serviceable. However, it appears that not all rubbers behave similarly because WRC has examined examples of natural rubber rings which are still satisfactory after 75 years service.

(iv) *Growths related to storage*

As mentioned previously certain constructional materials can promote excessive growth of micro-organisms, as has been found for certain cold water cisterns constructed from glass-reinforced polyester. This growth occurs at the expense of soluble organic compounds leached from the resin. Manufacturers, now aware of this condition, are endeavouring to solve the problem. However, not all slime growths result from constructional materials. Poynter & Mead (1964) showed that a common feature of many slime growths in header tanks was the presence of volatile organic solvents in the environment which condense on the cool surfaces. This condition occurs frequently, although not exclusively, in industrial premises such as those of printers, paint and cosmetic industries and brewing and wine and spirit trades. In domestic properties this can occur where home brewing and wine making is practised, often in the airing cupboard adjacent to the cold water cistern. Another common occurrence is in domestic premises over a hairdressing salon where presumably similar factors apply. These jelly-like slimes may consist of a variety of encapsulated bacteria, actinomycetes and fungi. On occasions fungi predominate, on others bacteria or combinations. The condition is extremely difficult to erradicate without siting the cold water header tank outside the premises or by sealing the tank completely and venting it to the exterior.

(v) *Animal growths*

Where microbiological growths become established, the condition may rapidly change into a complete aquatic ecosystem with the appearance of various forms of animal life. In distribution systems, Collingwood (1966) and Smalls & Greaves (1968) have shown that the presence of animals can be a major aesthetic problem in water supply. These animals not only survive in water mains but actively multiply. Certain species of chironomid can even reproduce parthenogenetically in the larval state without recourse to its natural, free-flying, adult state (Williams 1974). These animals must feed on organic matter present in the water which enters either in a particulate or soluble state. If one accepts that material passing a 0.45 μm filter is soluble, then something like 90% of organic matter in water after treatment is soluble and unavailable to animal species until it is converted into a particulate state by bacteria, fungi or actinomycetes. Therefore to prevent soluble organic matter becoming available to animals, bacterial growth must be controlled by maintaining an adequate chlorine residual throughout the distribution system. Of course with surface waters other forms of particulate organic matter are present, notably algae. Collingwood (1970) has estimated that the organic contribution from this source, assuming 50% of the dry weight of diatoms is silica, would vary from 25 to 1000 g for a modest daily supply of 5000 m^3. However, not all of this organic carbon is retained because some must inevitably pass through the system unchanged.

5. Water Quality, Standards and Surveillance

Over the years a great deal has been written on the bacteriological examination of water supplies and standards, both nationally and internationally. The procedures adopted for ensuring compliance with these standards are equally varied.

In the UK water undertakings have a statutory obligation to provide 'wholesome water' and should be prepared to provide evidence of this quality to local authorities through their health officers.

An independent check may be made by health department staff operating through the Public Health Laboratory Service, but obviously both co-operate in the best interests of the community.

Recommendations have been drawn up which although not carrying statutory authority, nevertheless should be complied with. These are set out in report No. 71 – The Bacteriological Examination of Water Supplies (*Anon.* 1969). This document has evolved with time in the light of advances in water bacteriology and covers the following aspects.

1. To help engineers in the management of their systems and interpretation of results.
2. The importance of satisfactory sample collection.
3. To obtain consistency, media and test procedures are fully specified.
4. Differentiation between water entering supply and water from within the distribution system since it is now appreciated that the quality of water is likely to deteriorate during transit.
5. For standards to be truly effective it is essential to specify the number of samples and the frequency of sampling.
6. Originally, plate counts at 22°C and 37°C were believed to have important hygienic significance, but with progressive reports this status has retracted. Now they are only mentioned in the context of monitoring water treatment and the general cleanliness of the distribution system.
7. Water quality standards are based on coliforms and *E. coli*, which ideally should be absent in 100 ml but nevertheless may be permitted in 5% of samples per year provided
   –No sample >10 coliforms.100 ml^{-1}
   –No sample >2 *E. coli.* 100 ml^{-1}
   –No sample 1 or 2 *E. coli* out of a total coliform count of –3 or more .100 ml^{-1}
   –coliforms not detected in two consecutive samples.

It is appreciated by engineers and water scientists that the above standards are realistic, nevertheless they feel that when coliforms are detected in a supply then it is better to take positive action to identify and eliminate the cause rather than accept their presence philosophically. Accordingly the water industry operates towards the quality goal of 'absence' of coliforms in 100 ml samples so that any 'spurious' results arising from such causes as failure to effect proper disinfection of new structures or growth of coliforms on materials, can be detected immediately and action taken.

Although originally important, bacteriological plate counts are no longer considered in many standards (*Anon.* 1970*b*, 1971*b*). However, they have recently been reintroduced in the USA (*Anon.* 1975*b*) on the grounds that high plate counts at 35° (>500/ml) can interfere seriously with the detection of coliforms (Geldreich *et al.* 1972).

Certain Authorities, WHO and US-Environmental Protection Agency, and to some extent the UK, are prepared to accept that bacteriological examination of water supplies can, in part, be replaced by monitoring for an adequate chlorine residual (>0.2 μg.ml^{-1}).

Recent statistics suggest that a large proportion of incidences of disease occur during distribution from such causes as cross-connections and back siphonage (Craun & McCabe 1973). No amount of bacteriological surveillance could ever be expected to detect these chance occurrences. Therefore, they must be covered with by Codes of Practice and Specification for plumbing systems, as is done for example, in *Model Water Byelaws* and *Safeguards to be Adopted in the Operation and Management of Waterworks* (*Anon.* 1966*a*, 1967).

These regulations, together with bacteriological standards embodying a stipulation as to the minimum number of samples to be examined for each system according to the population served, together with improved water treatment facilities and universal adoption of disinfection, inspires a measure of confidence in public water supply. This is confirmed by the low incidence of enteric disease attributable to water supplies in the UK.

Having largely resolved the health aspect, the next objective is improvement of water quality by better management of water resources, improved water treatment, better operational management of water distribution systems and recognition of the fact that improvements could be made in the choice of materials which come into contact with water. To this end, the important role that bacteriology can play should be fully realized.

6. References

ALLEN, M. J. & GELDREICH, E. E. 1975 Bacteriological criteria for groundwater quality. *Ground Water* **13**, 45–51.

ANON. 1916 *Twelfth Report on Research Work.* London: Metropolitan Water Board.

ANON. 1946 *Manual of Recommended Water-sanitation Practice.* United States Public Health Service, Publication No. 525. Washington: U.S.G.P.O.

ANON. 1951 *Purification of the Water of Swimming Baths.* Ministry of Housing and Local Government. London: H.M.S.O.

ANON. 1954 *Thirty-Sixth Report on Results of Bacteriological, Chemical and Biological Examination of the London Waters for 1953–54.* London: Metropolitan Water Board.

ANON. 1961 *Effectiveness of Water Treatment Processes.* United States Public Health Service, Publication No. 898. Washington: U.S.G.P.O.

ANON. 1966*a* *Model Water Byelaws.* Ministry of Housing and Local Government. London: H.M.S.O.

ANON. 1966*b* *Forty-Second Report on Results of Bacteriological, Chemical and Biological Examination of the London Waters for 1965–66.* London: Metropolitan Water Board.

ANON. 1967 *Safeguards to be Adopted in the Operation and Management of Waterworks.* Ministry of Housing and Local Government. London: H.M.S.O.

ANON. 1968 *Forty-Third Report on Results of Bacteriological, Chemical and Biological Examination of the London Waters for 1967–68.* London: Metropolitan Water Board.

ANON. 1969 *Reports on Public Health and Medical Subjects No. 71. The Bacteriological Examination of Water Supplies.* Department of Health and Social Security. London: H.M.S.O.

ANON. 1970*a Forty-Fourth Report on Results of Bacteriological, Chemical and Biological Examination of the London Waters for 1969–70.* London: Metropolitan Water Board.
ANON. 1970*b European Standards for Drinking Water.* Geneva: World Health Organization.
ANON. 1971*a Progress Statement – Back-siphonage Committee.* Department of the Environment. London: H.M.S.O.
ANON. 1971*b International Standards for Drinking Water.* Geneva: World Health Organization.
ANON. 1971*c Artificial Groundwater Recharge.* Medmenham: Water Research Association.
ANON. 1972 *Cost of River Water Treatment.* Trent Research Programme, Vol. 5. Reading Water Resources Board.
ANON. 1973 *Forty-Fifth Report on Results of Bacteriological, Chemical and Biological Examination of the London Waters for 1971–73.* London: Metropolitan Water Board.
ANON. 1975*a Council Directive Concerning the Quality Required of Surface Water Intended for Abstraction of Drinking Water in Member States, R/848/75(ENV 38).* Brussels: The European Economic Community.
ANON. 1975*b Interim Primary Drinking Water Standards.* United States Environmental Protection Agency. *Federal Register, Washington* **40**, 11990–11996.
APPLETON, B. 1973 Coming apart at the seals. *New Civil Engineer* 6 December, 6–9.
BAAS-BECKING, L. G. M. & WOOD, E. J. F. 1956 Biological processes in the estuarine environment. VIII. Iron bacteria as gradient organisms. *Proceedings Koninklijk Nederlandse akademie van wetenschappen* **59**, 398–407.
BAYS, L. R. 1975 Treated water. *Institute of Water Engineers and Scientists Symposium. Maintenance of Water Quality.* 9–11 September, University of Cambridge, pp. 85–98.
BAYS, L. R., BURMAN, N. P. & LEWIS, W. M. 1970 Taste and odours in water supplies in Great Britain: A survey of the present position and problems for the future. *Water Treatment and Examination* **19**, 136–160.
BEAN, P. G. & EVERTON, J. R. 1969 Observations on the taxonomy of chromogenic bacteria isolated from cannery environments. *Journal of Applied Bacteriology* **32**, 51–59.
BERNHARDT, H. 1967 Aeration of Wahnbach reservoir without changing the temperature profile. *Journal of the American Water Works Association* **63**, 943–963.
BOON, A. & LISTER, A. R. 1975 Formation of sulphide in rising main sewers and its prevention by use of oxygen. *Progress in Water Technology* 7, 289–330.
BOOTH, G. H. & TILLER, A. K. 1960 Polarization studies of mild steel in cultures of sulphate-reducing bacteria. Part 1. *Transactions of the Faraday Society* **56**, 1689–1696.
BOOTH, G. H. & TILLER, A. K. 1962*a* Polarization studies of mild steel in cultures of sulphate-reducing bacteria. Part II. *Transactions of the Faraday Society* **58**, 110–115.
BOOTH, G. H. & TILLER, A. K. 1962*b* Polarization studies of mild steel in cultures of sulphate-reducing bacteria. Part III. *Transactions of the Faraday Society* **58**, 2510–2516.
BOOTH, G. H., ELFORD, L. & WAKERLEY, D. S. 1968 Polarization of mild steel in the presence of hydrogen bacteria and methane bacteria. *British Corrosion Journal* **3**, 68–69.
BOUCHER, P. L. 1967 Microstraining and ozonization of water and waste-water. *Proceedings of the Twenty-second Industrial Waste Conference, Purdue University Engineering Bulletin, External Series*, pp. 771–787.
BRIGHTON, W. D. 1958 Pollution of chalk boreholes by filamentous organisms. *Proceedings of the Society for Water Treatment and Examination* 7, 144–156.
BUNKER, H. J. 1940 Microbiological anaerobic corrosion. *Journal of the Society of Chemical Industry* **59**, 412–414.
BURMAN, N. P. 1961 Some observations on coli-aerogenes bacteria and streptococci in water. *Journal of Applied Bacteriology* **24**, 368–376.
BURMAN, N. P. 1965 Taste and odours due to stagnation and local warming in long lengths of piping. *Proceedings of the Society for Water Treatment and Examination* **14**, 125–131.

BURMAN, N. P. & COLBOURNE, J. S. 1976 The effect of plumbing materials on water quality. *Journal of the Institute of Plumbing* **3**, 12–13.

CLARK, F. M., SCOTT, R. M. & BONE, E. 1967 Heterotrophic iron precipitating bacteria. *Journal of the American Water Works Association* **59**, 1036–1042.

COLLINGWOOD, R. W. 1966 Occurrence, significance and control of organisms in distribution systems. *Journal of the British Waterworks Association* **48**, 541–553.

COLLINGWOOD, R. W. 1970 Removal of algae and animals. *Water Treatment in the Seventies Symposium. Society for Water Treatment and Examination and the Water Research Association.* 5–7 January, University of Reading, pp. 100–112.

CRAUN, G. F. & MCCABE, L. J. 1973 Review of causes of water-borne disease outbreaks. *Journal of the American Water Works Association* **65**, 74–84.

CROSS, D. F., BENCHIMOL, A. & DIMOND, E. G. 1966 The fawcet aerator – a source of pseudomonas infection. *New England Journal of Medicine* **274**, 1430–1431.

DRAKE, C. H. 1965 Occurrence of *Siderocapsa treubii* in certain waters of the Niederheim. *Gewässer und Abwässer* **39/40**, 41–63.

DUECKER, W. W., ESTEP, J. W., MAYBERRY, G. M. & SCHWAB, J. W. 1948 Studies of properties of sulfur jointing compounds. *Journal of the American Water Works Association* **40**, 715–728.

DUTKA, B. J. 1973 Coliforms are an inadequate index of water quality. *Journal of Environmental Health* **36**, 39–46.

EVANS, M. R. & OWENS, J. D. 1972 Factors affecting the concentration of faecal bacteria in land drainage water. *Journal of General Microbiology* **71**, 477–485.

EVISON, L. M. & JAMES, A. 1975 *Bifidobacterium* as an indicator of faecal pollution in water. *Progress in Water Technology* 7, 57–66.

FENNEL, H., JAMES, D. B. & MORRIS, J. 1974 Pollution of a storage reservoir by roosting gulls. *Water Treatment and Examination* **23**, 5–24.

FEDERICK, L. R. & STARKEY, R. L. 1948 Bacterial oxidation of sulfur in pipe sealing mixtures. *Journal of the American Water Works Association* **40**, 729–736.

FRY, J. C. & STAPLES, D. G. 1976 Distribution of *Bdellovibrio bacteriovorus* in sewage works river water and sediments. *Applied and Environmental Microbiology* **31**, 469–474.

GELDREICH, E. E. 1966 *Sanitary Significance of Faecal Coliforms in the Environment.* Water Pollution Control Research Series, Publication WP – 20–3. Washington: United States Department of the Interior.

GELDREICH, E. E. 1970 Applying bacteriological parameters to recreational water quality. *Journal of the American Water Works Association* **62**, 113–120.

GELDREICH, E. E., KENNER, B. A. & KABLER, P. W. 1964 Occurrence of coliforms, faecal coliforms and streptococci on vegetation and insects. *Applied Microbiology* **12**, 63–69.

GELDREICH, E. E., NASH, H. D., REASONER, D. J. & TAYLOR, R. H. 1972 The necessity of controlling bacterial populations in potable water; community water supply. *Journal of the American Water Works Association* **64**, 596–602.

GERBER, N. N. & LECHEVALIER, H. A. 1965 Geosmin, an earthy smelling substance isolated from actinomycetes. *Applied Microbiology* **13**, 935–938.

GILFILLAN, D. J. 1971 *Report on the Investigation of Back-siphonage Risks in Domestic Properties, Winter 1970/71.* TP 82. Medmenham: Water Research Association.

HALL, E. S. 1973 Quantitative estimation of disinfection interferences. *Water Treatment Examination* **22**, 153–171.

HOLDEN, W. S. 1970 *Water Treatment and Examination* London: Churchill.

HORNBY, F. P. 1954 Letter to Editor. *Proceedings of the Society for Water Treatment and Examination* **3**, 98–99.

HORVATH, J. & SOLTI, M. 1959 Beitrag zum Mechanismus der anaeroben microbiologischem Korrosion der Metalls im Boden. *Werkstoffe und Korrosion* **10**, 624–630.

HOUGHTON, G. U. 1970 Slow sand filtration and biological processes. *Water Treatment in the Seventies Symposium. Society for Water Treatment and Examination and the Water Research Association.* 5–7 January, University of Reading, pp. 79–99.

HUNTER, C. A. & ENSIGN, P. R. 1947 An epidemic of diarrhoea in a new-born nursery caused by *Pseudomonas aeruginosa. American Journal of Public Health* **37**, 1166–1169.

HUTCHINSON, M. 1974*a* Microbiological aspects of groundwater pollution. In *Groundwater Pollution in Europe.* Port Washington, N.Y.: Water Information Center Inc.

HUTCHINSON, M. 1974*b* WRA Medlube: An aid to mains disinfection. *Water Treatment and Examination* **23**, 174–189.

HUTCHINSON, M., JOHNSTONE, K. I. & WHITE, D. 1966 Taxonomy of the acidophilic thiobacilli. *Journal of General Microbiology* **44**, 373–381.

HUTCHINSON, M., RIDGWAY, J. W. & CROSS, T. 1975 Biodeterioration of rubbers in contact with water, sewage and soil. In *Microbial Aspects of the Deterioration of Materials*, eds Lovelock, D. W. & Gilbert, R. J. Society for Applied Bacteriology, Technical Series No. 9. London & New York: Academic Press.

JEBB, W. H. H. 1965 Effects of cross-connections and back-siphonage. *Proceedings of the Society for Water Treatment and Examination* **14**, 102–105.

JOHNSON, J. D. & OVERBY, R. 1970 Bromine and bromamine disinfection chemistry. *Proceedings of the National Specialty Conference on Disinfection. 8–10 July, University of Massachusets,* pp. 37–60.

LEEFLANG, K. 1963 Microbiologic degradation of rubber. *Journal of the American Water Works Association* **55**, 1523–1535.

LEWIS, W. M. 1966 Odours and tastes in water derived from the River Severn. *Proceedings of the Society for Water Treatment and Examination* **15**, 50–74.

LITSKY, W., MALLMANN, W. L. & FIFIELD, C. W. 1955 Comparison of the most probable numbers of *Escherichia coli* and enterococci in river waters. *American Journal of Public Health* **45**, 1049–1053.

MACRAE, I. C. & EDWARDS, J. F. 1972 Adsorption of colloidal iron by bacteria. *Applied Microbiology* **24**, 819–823.

MILLINER, R., BOWLES, D. A. & BRETT, R. W. 1972 Biological pretreatment at Tewkesbury. *Water Treatment and Examination* **21**, 318–326.

ORCHARD, V. A. & GOODFELLOW, M. 1974 The selective isolation of *Nocardia* from soil using antibiotics. *Journal of General Microbiology* **85**, 160–162.

PALIN, A. 1950 A study of the chloro-derivatives of ammonia and related compounds with special reference to their formation in the chlorination of natural and polluted waters. Parts I, II and III. *Water and Water Engineering* **54**, 151–159, 189–200, 248–256.

PARKER, C. D. 1945 The corrosion of concrete. *Australian Journal of Experimental Biology and Medical Science* **23**, 81–98.

PARKER, S. S. 1972 Biological pretreatment at Strensham. *Water Treatment and Examination* **21**, 315–317.

PIET, G. J., ZOETEMAN, B. C. J. & KRAAYEVELD, A. J. A. 1972 Earthy smelling substances in surface waters of the Netherlands. *Water Treatment and Examination* **21**, 281–286.

POSTGATE, J. R. 1963 The microbiology of corrosion. In *Corrosion,* Vol. 1, ed. Shreir, L. L. London: Newnes.

POYNTER, S. F. B. & MEAD, G. C. 1964 Volatile organic liquids and slime production. *Journal of Applied Bacteriology* **27**, 182–195.

POYNTER, S. F. B. & STEVENS, J. K. 1975 The effects of storage on the bacteria of hygienic significance. In: *The Effects of Storage on Water Quality*. Medmenham: Water Research Centre.

PRATT, A. C. R. 1965 Symposium on Consumer Complaints – The tap washer. *Water Treatment and Examination* **14**, 135–141.

PUGH, N. J. 1949 The treatment of doubtful waters for public supplies. *Journal of the Institute of Water Engineers* **3**, 123–166.

ROBECK, G. C., CLARKE, N. A. & DOSTAL, K. A. 1962 Effectiveness of water treatment processes in virus removal. *Journal of the American Water Works Association* **54**, 1275–1292.

ROMERO, J. C. 1970 The movement of bacteria and viruses through porous media. *Ground Water* **8**, 37–48.

SHORT, C. S. 1973 *Removal of Ammonia from River Water; a Pilot-scale Investigation of Biological Filtration, Sedimentation and Air Stripping, together with Cost Estimates.* TP 101. Medmenham: Water Research Centre.

SILVEY, J. K. G. 1953 Newer concepts of tastes and odour in surface water supplies. Part 1. *Water and Sewage Works* **100**, 426–429.

SILVEY, J. K. G. 1954 Newer concepts of tastes and odour in surface water supplies. Part II Isolation and identification of noxious actinomycetes in raw water and distribution systems. *Water and Sewage Works* **101**, 208–211.

SMALLS, I. C. & GREAVES, G. F. 1968 A survey of animals in distribution systems. *Water Treatment and Examination* **17**, 150–181.

TAYLOR, E. W. 1960 The pollution of surface and underground waters. *Journal of the British Waterworks Association* **42**, 582–603.

THOFERN, E. & SPEH, K. 1974 Untersuchungen zur Verkeimung von Trinkwasser. Die Bedeutung des Kondeswassers in Trinkwasserspeichern. *GWF – wasser Abwasser* **115**, 538–541.

TYLER, P. A. & MARSHALL, K. C. 1967*a* Microbial oxidation of manganese in hydro-electric pipelines. *Antonie van Leeuwenhoek* **33**, 171–183.

TYLER, P. A. & MARSHALL, K. C. 1967*b* Form and function in manganese-oxidizing bacteria. *Archiv für Mikrobiologie* **56**, 344–353.

VICTOREEN, H. T. 1969 Soil bacteria and color problem in distribution systems. *Journal of the American Water Works Association* **61**, 429–431.

VICTOREEN, H. T. 1974 Control of water quality in water transmission and distribution mains. *Journal of the American Water Works Association* **66**, 369–370.

WILLIAMS, D. N. 1974 An infestation by a parthenogenetic chironomid. *Water Treatment and Examination* **23**, 215–231.

WILLIAMS, B. M. & RICHARDS, D. W. 1976 Salmonella infection in the herring gull (*Larus argentatus*). *Veterinary Record* **98**, 51.

WOLFE, R. S. 1964 Iron and manganese bacteria. In *Principles and Applications in Aquatic Microbiology*. New York & Chichester: John Wiley & Sons.

WOLZOGEN-KUHR, C. A. H. & VAN DER VLUGT, L. S. 1934 The graphitization of cast iron as an electro-biological process in anaerobic soil. *Water, den Haag* **18**, 147–165.

WORMWELL, F. 1973 *Corrosion of Metals Research: Corrosion of Mild Steel by Sulphur.* London: H.M.S.O.

The Identification, Cultivation and Control of Iron Bacteria in Ground Water

D. R. CULLIMORE

Regina Water Research Institute, University of Regina, Regina, Saskatchewan, Canada

AND

ANNETTE E. MCCANN

Canadian Water Resources Engineering Corporation, Saskatchewan, Canada

CONTENTS

1. Introduction

(a) *General introduction*

IRON BACTERIA may be defined as "that group of aerobic bacteria which appear to utilize the oxidation of ferrous and/or manganous ions as an essential component in their metabolic functioning". The resultant production of ferric and/or manganic salts (usually the hydroxide) within the cell or cell coatings gives the bacteria their typical brown colouration. Studies of pyrite deposits laid down 300 million years ago have shown them to contain a range of fossil bacteria including representatives of two major groups of iron bacteria, *Gallionella* and *Sphaerotilus* (Schopf *et al.* 1965).

Fossil iron bacteria were first reported by Ehrenberg (1836); while analysing ochre microscopically he mistook them for diatoms but named them *Gallionella ferruginea.* Other genera of iron bacteria were first described from living specimens, such as *Leptothrix ochracea* which is mainly responsible for the formation of bog ore (Kutzing 1843), and *Crenothrix polyspora*, discovered and described by Cohn (1870) while studying the brown flocculent precipitates which had rendered well waters undrinkable. In addition to a very detailed description, Cohn stained the bacteria using the Prussian blue reaction in which potassium ferricyanide with hydrochloric acid stained with an intensity varying with the concentration of ferric compounds present. Cohn's studies also highlighted the nuisance value of iron bacteria.

Throughout history, there are reports of water from wells, rivers and creeks being stained blood red or brown and becoming undrinkable presumably due to the growths of iron bacteria. In Berlin in 1877, the whole water-main system had to be replaced because it was clogged with iron bacteria growing within the system. Typical symptoms of iron bacterial growths in water supplies are: (a) discolouration of the waters (yellow to rust-red or brown); (b) reduction in flow rates through the system caused by coatings of iron bacteria inside the pipes; (c) development of thick red or brown coatings on the sides of reservoirs, tanks and cisterns; sometimes sloughing off to form either fluffy specks in the water or gelatinous clumps of red to brown filamentous growths; (d) rapid clogging of filter screens; (e) heavy surface and sedimented growths of a red or brown colour, sometimes irridescent (ochre), in water.

Frequently, the heavy growths of iron bacteria form a substrate for other bacteria which may degrade these materials anaerobically to form acidic products and hydrogen sulphide. These in turn can cause taste, odour and corrosion problems. Controlling the growth of iron bacteria has always posed a problem due to the heavy deposits of ferric and/or manganic salts around the cell and the cell coatings themselves, forming a natural barrier to any bactericidal agent. Stott (1973) summarized the difficulty of controlling iron bacteria by the statement "Iron bacteria are tenacious and continue to grow even after the severest kind of treatment . . . and if relief is to be had it is likely to be temporary."

It was the nuisance nature of iron bacteria in water supplies that focussed attention upon their activities and led Winogradsky (1888) to postulate a biological grouping of the organisms characterized by their relation to iron. He concluded from his cultural studies that the iron bacteria were chemoautotrophic and derived energy from the transformation of iron from the ferrous to the ferric state. Winogradsky's postulate of obligate chemoautotrophy was shown to be incorrect at least for some iron bacteria since they grew well heterotrophically (Molisch 1892). Two schools of thought generated around these two opposing postulates of chemoautotrophic and heterotrophic nutritional pathways for the iron bacteria (Pringsheim 1952). Mulder (1974) reviewed the literature with particular emphasis on the chemolithotrophy of the sheath-forming iron bacteria. Confusion in determining their true nutritional status has been made more intense because these organisms grow only at pH values above 6, a pH range in which ferrous ions will oxidize rapidly by purely chemical reaction. Mulder contends that it is therefore very difficult to prove that the energy released from such a reaction can be utilized by the bacteria. He postulates that "it is highly probable that in many cases the so-called biological iron oxidation by these bacteria is confined to absorption of chemically oxidised iron by the sheaths or slime layer surrounding the sheaths." Mulder (1964) does, however, demonstrate that the organisms of the *Leptothrix* group are able to convert manganous ions readily to manganic oxide over the pH range of 6.0–7.5. There is evidence that this conversion is due to the presence of proteinaceous substances promoting manganese oxidation on the outside of the sheaths.

Many attempts have been made to differentiate the iron bacteria into groups based upon differences in nutritional requirements. Stott (1973) divided them into three groups, viz., group 1, those that precipitate ferric hydroxide from solutions of ferric bicarbonate, using the carbon dioxide set free and the available energy of the reaction for their life processes; group 2, those that do not require ferrous bicarbonate for their vital processes but which cause the deposition of ferric hydroxide when either inorganic or organic salts are present; group 3, those that attack iron salts of organic acids, utilizing heterotrophically the organic acid radical while eventually converting the basic salt to ferric hydroxide. In practice, the iron bacteria are almost always differentiated by their morphological characteristics.

(b) *The classification of iron bacteria*

(i) *Based on* Bergey's Manual of Determinative Bacteriology

Considerable changes have occurred in the classification of the iron bacteria over the last decade. In the 7th edition of *Bergey's Manual of Determinative Bacteriology* (Breed *et al.* 1957), the iron bacteria were listed in the Caulobacteriaceae, Siderocapsaceae, Chlamydobacteriaceae and Crenotrichaceae, while in the 8th edition (Buchanan & Gibbons 1974), they were incorporated in 15 genera, listed in the following parts: part 2, (gliding bacteria) *Toxothrix*; part 3, (sheathed bacteria), *Sphaerotilus, Leptothrix, Lieskeella, Crenothrix* and *Clonothrix*; part 4, (budding and/or appendaged bacteria), *Pedomicrobium,*

Gallionella, *Metallogenium* and *Kusnezovia* and part 12, (Gram negative chemolithotrophic bacteria), *Thiobacillus* (one species only, *Thiobacillus ferrooxidans*), *Siderocapsa, Naumanniella, Ochrobium* and *Siderococcus*.

Diagrammatic presentations of these genera are shown in Fig. 1, their dominant habitats in Table 1, and a proposed dichotomous key for their identification in Table 2.

(ii) *Other genera reported to include iron bacteria*

Although the 8th edition of *Bergey's Manual of Determinative Bacteriology* considerably reduces the confusion in the classification of iron bacteria, species of iron bacteria have been reported in other bacterial genera.

Mann & Quastel (1946) showed that the manganese oxide present in soil was mainly the result of biological oxidations. Van Veen's (1973) review showed in studies *in vitro* that several different types of fungi and some bacterial genera, (*Cryptococcus, Pseudomonas* and *Hyphomicrobium*) were all capable of oxidizing manganese independently of hydrocarboxylic compounds. *Bacillus* spores and the chlamydospores of several fungi could become impregnated with manganic oxides after prolonged incubation in agar media enriched with $MnSO_4$ or $MnCO_3$. *Arthrobacter* (strain 216) and *Coniothyrium fuckelii* were both found to oxidize manganese salts the most rapidly. Schweisfurth (1973) studied a range of manganese-oxidizing strains of *Pseudomonas* and proposed a species group, *Ps. manganoxydans*. Clark *et al.* (1967) briefly reviewed the definition of the term iron bacteria and considered that it included all organisms capable of precipitating iron biologically. Using synthetic media containing ferric ammonium citrate, they found that isolates of *Aerobacter aerogenes, Serratia indica* and *Bacillus pumulis* could all precipitate iron mainly through the utilization of the citrate. Ivarson & Heringa (1972) in a study of the micro-organism oxidizing manganese in manganese deposits in soil, isolated some organisms resembling *Hyphomicrobium* and also a fungus belonging to the genus *Cephalosporium*, both of which could actively oxidize manganese, but failed to detect any of the noted genera of soil iron bacteria. In similar studies of Illinois waters, *A. aerogenes* was most commonly isolated. All bacteria possessing the iron precipitating characteristic had two features in common, viz., the ability to utilize citrate and the possession of capsular material. Clearly, in the presence of citrate and iron or manganese organic molecules, the degradation in total or in part of the organic fraction could lead to a biological precipitation of the iron or manganese. This is a pattern of precipitation different from that associated with the traditional grouping of iron bacteria.

(c) *Studies on the major genera of iron bacteria*

(i) Gallionella

Although described first by Ehrenberg (1836), a very detailed morphological study of *Gallionella* was reported by Cholodny (1924). Isolation techniques and

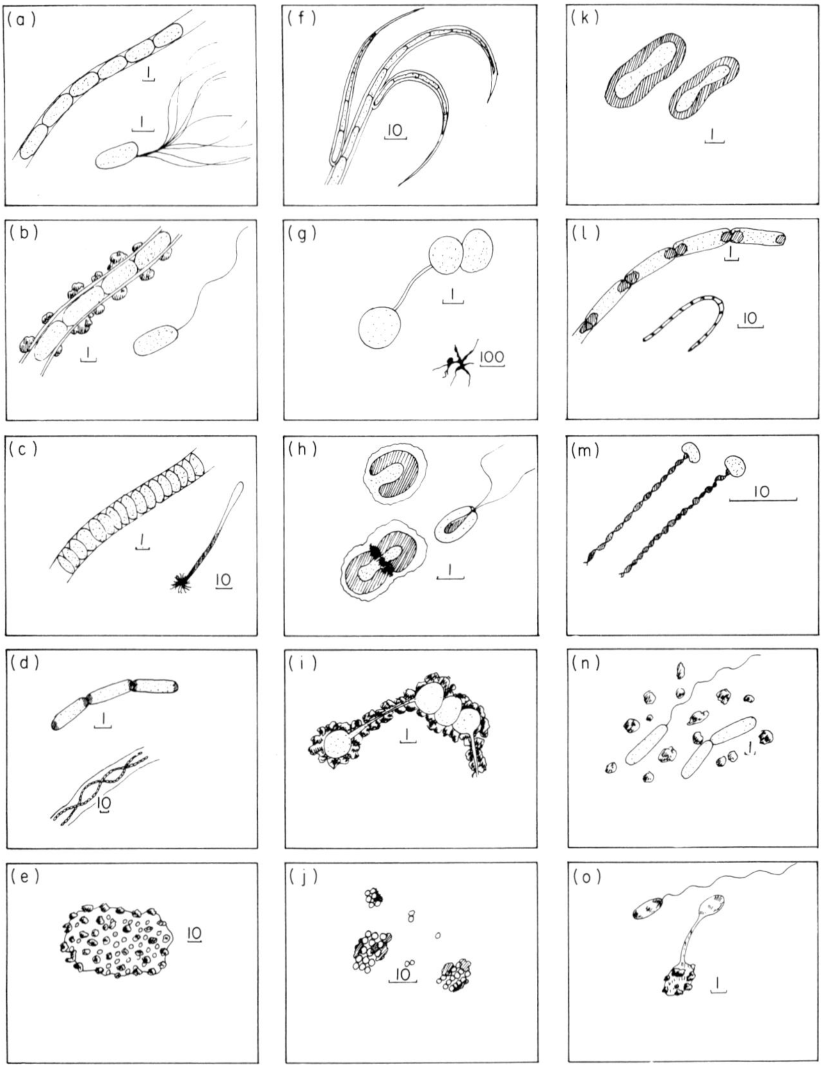

Fig. 1. The principal morphological features of iron bacteria. (a) *Sphaerotilus*, upper: shows cells in sheath; lower: single cell. (b) *Leptothrix*, upper: shows cells in encrusted sheath; lower: single cell. (c) *Crenothrix*, upper: shows cells in sheath; lower: complete filament. (d) *Lieskeella*, upper: cells stained with methylene blue; lower: spirally wound filaments in capsule. (e) *Siderocapsa*, section of cells embedded in common capsule encrusted with iron and manganic compounds. (f) *Clonothrix*, apical tips of filaments. (g) *Metallogenium*, upper: free cells; lower: microcolony. (h) *Ochrobium*, upper: single cell; lower left: paired cell; lower right: motile cell. (i) *Kusnezovia*, cells and interconnecting filaments heavily encrusted with manganic deposits and oxalic acid. (j) *Siderococcus*, cell arrangements showing encrustations of micro colonies with ferric hydroxides and oxides. (k) *Naumanniella*, single cells. (l) *Toxothrix*, upper: cells in filaments; lower: U-shaped trichome. (m) *Gallionella*, single cells. (n) *Thiobacillus ferrooxidans*, cells among precipitated iron. (o) *Pedomicrobium*, upper: swarmer cell; lower: stained preparation, mother cell encrusted with deposits giving off a hypha with bud at tip. The scale number is in μm. *Thiobacillus* refers only to the species, *T. ferrooxidans*.

Table 1

Range of habitats in which iron bacteria occur and their ability to oxidize reduced iron and manganese salts

Genus	Habitat						Oxidizes		
	Soil	Fresh water	Bogs	Mud sediments in lakes and rivers	Well water and piped systems	Acidic mine drainings	Fe only	Mn only	Fe and/or Mn
Toxothrix	–	+	+	±	–	–	+	–	–
Sphaerotilus	–	+	–	+	+	–	–	–	+
Leptothrix	–	+	–	+	+	+	–	–	+
Lieskeella	–	+	–	+	–	–	+	–	–
Crenothrix	–	+	–	–	+	–	–	–	+
Clonothrix	–	+	–	–	+	–	–	–	+
Pedomicrobium	+	+	–	–	–	–	–	–	+
Gallionella	+	+	+	+	+	–	+	–	–
Metallogenium	+	+	–	+	–	–	–	–	+
Kusnezovia	–	–	–	+	–	–	–	+	±
Thiobacillus ferrooxidans	+	+	+	+	+	+	+	–	–
Siderocapsa	–	+	–	–	+	–	–	–	+
Naumanniella	+	+	–	–	+	–	–	–	+
Ochrobium	–	+	–	–	+	–	+	–	–
Siderococcus	–	–	–	+	–	–	+	–	–

Table 2

Dichotomous key to the genera of iron bacteria

1.	Cells reproduce by budding, never possess tapering filaments	2
	Cells do not reproduce by budding unless filaments are tapered	4
2.	Cells possess cellular extensions resembling hyphae	*Pedomicrobium*
	Cells do not possess cellular extensions resembling hyphae	3
3.	Do not possess rigid cell walls, buds are atypical 'elementary bodies'	*Metallogenium*
	Possess rigid cell walls, buds resemble vegetative cells	*Kusnezovia*
	Possess rigid cell walls, cells become pear-shaped prior to budding	*Siderococcus*
4.	Cells occur in chains or filaments enclosed in a sheath and may also be present as single cells or swarmers	5
	Cells never enclosed in a sheath	8
5.	Cells occur in chains when within a sheath	6
	Cells occur in filaments when within a sheath	7
6.	Motile cells possess a bundle of subpolar flagella, sheath not encrusted with ferric or manganic oxides	*Sphaerotilus*
	Motile cells possess a single polar flagellum, sheaths tend to be encrusted with ferric or manganic oxides	*Leptothrix*
7.	Tapering filament enclosed in distinct sheath	*Clonothrix*
	Filament not tapering and may exhibit false branching, sheath thin and indistinct	*Crenothrix*
8.	Cells possess a torus or marginal thickenings, resembling a diatom	9
	Cells do not possess torus or marginal thickenings	10
9.	Torus resembles horseshoe, cells often in pairs	*Ochrobium*
	Torus does not resemble horseshoe, never in pairs	*Naumanniella*
10.	Cells occur in long filaments, frequently U shaped	*Toxothrix*
	Cells never in long filaments	11
11.	Cells possess a long spirally twisted stalk arising from centre of cell	*Gallionella*
	Cells do not possess stalks	12
12.	Cells embedded in common capsule	*Siderocapsa*
	Cells not embedded in common capsule	13
13.	Cells derive energy from oxidation of reduced sulphur compounds, cells never in spiral arrangement	*Thiobacillus ferrooxidans*
	Cells do not derive energy from oxidation of reduced sulphur compounds, cells occur in double sprial arrangement	*Lieskeella*

some of the difficulties of obtaining pure cultures were outlined by Nunley & Krieg (1968), who found that pure cultures of *Gallionella ferruginea* could be obtained by incubating for 1–2 days in Wolfe's medium containing 0.5% of formalin to prevent contaminant heterotrophic growth. Enrichment of *Gallionella* cultures was reviewed by Balashova (1967) and excellent growth was obtained in an atmosphere of 6% O_2, 59% N_2 and 35% CO_2. The ultrastructure (Balashova & Cherni 1970; Hanert 1970), and the form of iron in the filaments (Mardanyan & Balashova 1971), have been subjected to intensive investigation. These studies revealed that *Gallionella* was a mycoplasmodial organism with a stalk consisting of helically coiled uniquely mineralized oxidized fibres in a structure quite different from that of the common brown compounds of oxidized iron. The nature of the cell led Balashova (1969) to suggest that a relationship could be drawn between *Gallionella, Metallogenium*, and *Mycoplasma* since all three bacterial genera lacked a cell wall. Extensive physiological studies of *G. ferruginea* were undertaken by Hanert (1968), who found that it grew well under low oxygen concentrations (0.1–0.2 $mg.l^{-1}$ of O_2), while higher levels (>2.75 $mg.l^{-1}$ of O_2) were inhibitory. It was also able to fix $^{14}CO_2$ in measurable quantities, and to undertake autoxidation of ferrous iron, stimulated by carbon dioxide in a medium containing ferrous sulphide.

(ii) Sphaerotilus/Leptothrix *group*

The *Sphaerotilus/Leptothrix* group of sheathed bacteria has been extensively reviewed by Dondero (1975), and earlier by Phaup (1968). A further role of *Sphaerotilus* as a component of 'sewage fungus' was discussed by Curtis (1969), and its ultrastructure described by Bissett & Brown (1969). These bacteria (in particular, *Sphaerotilus natans*) are a major component of 'sewage fungus', defined by Curtis (1969) as "massive growth of slimy cotton-wool-like plumes (white, grey or brown) which can rapidly colonise surfaces." Typically it is found in rivers below sources of organic pollution. Dondero (1975) indicated that members of this group could also grow in pipes and, along with deposits of iron and manganese, reduce the rate of flow of water through the system. However, most reported occurrences are of surface water infestations, the slimy growths interfering with the development of benthic animals and consequently of fish (Avery 1970).

(iii) Toxothrix

Krul *et al.* (1970) in a review and study of *Toxothrix* found it to be present in a large number of the iron springs in and around Lansing, East Michigan. They used a partially submerged microscopic technique and also grew the organisms in the laboratory. It was noted that the trichomes would disintegrate rapidly during laboratory observation, a feature of the genus which may have restricted its observation in routine screening studies. Hasselbarth & Ludemann (1967) observed *Toxothrix* in well waters in districts of Germany.

(iv) Thiobacillus ferrooxidans

Thiobacillus ferrooxidans is a unique species obtaining energy simultaneously from the oxidation of inorganic sulphur compounds and iron (Temple & Colmer 1951), a feature which led to its implication as a causative factor in corrosion problems involving acidic mine waters. The difficulties in culturing and enumerating *T. ferrooxidans* were overcome by Tuovinen & Kelly (1973) using a medium with a pH of 1.3 and $FeSO_4 \cdot 7H_2O$, $(NH_4)_2SO_4$, $MgSO_4 \cdot 7H_2O$ and H_2SO_4 as the sources of sulphur and iron. Enumeration was performed using selected non-inhibitory membrane filters (Tuovinen *et al.* 1971*a*) and agar base. The mechanisms of sulphur and iron oxidation have been studied (Temple & Colmer 1951; Vestal & Lundgren 1971). High tolerances to some metals (Zn, Ni, Cu, Co, Mn and Al) have been reported (Tuovinen *et al.* 1971*b*); the organism was resistant to these at concentrations of more than 10 $g.l^{-1}$. Other metals, for example Ag, Te, As and Se were more toxic, and could not be tolerated at concentrations above 50–100 $mg.l^{-1}$. Increased tolerance was noted during oxidation of sulphur and iron. This may be a significant ecological feature favouring the growth of *T. ferrooxidans* in heavy metal solutions toxic to other potential microbial competitors.

2. Problems Caused by Iron Bacteria

(a) *General survey*

The occurrence of iron in geological strata is already well established (Lepp 1975), but little attention has been paid to the role of iron bacteria in the strata or aquafers, mainly because of the difficulty of monitoring the organisms. The biological cycling of manganese and iron in water has been reviewed (Mulder 1972) and the chemical factors affecting the equilibria and availability of iron and manganese have been discussed (Hem 1972). For soils, minerological aspects of the biological iron and manganese cycles have been given by Iwasa (1970), all of which clearly illustrates a diverse potential habitat for the iron bacteria. It is not therefore surprising that problems have been reported in well and mining operations, piped water and sewage systems and in open bodies of water.

Iron bacteria have caused problems in water supplies since the dawn of civilization and there are many references in history to 'red' water, undrinkable water covered in slime, and plugged wells.

In wells, the major problems are (a) growths plugging the screens; (b) coating of the piped systems, impellars and motors, thereby reducing flow rates; (c) reduced potability of the water and finally (d) total plugging of the well. Hasselbarth & Ludemann (1972) reported encrustations of wells which caused rapid decreases in yields particularly at times of maximum demand. The encrustations were considered by Hasselbarth & Ludemann (1972) to be caused by: (a) iron and manganese bacteria, various species of which exist in the soil and could presumably enter a well during the initial boring operations or by seepage into the aquafer feeding the well; (b) sulphate-reducing bacteria which reduce sulphates to sulphides to meet their respiratory needs (the sulphides when excreted react with the iron to form iron sulphide deposits); (c) corrosion of metallic tubing and extension pipes.

Mogg (1972) elaborated on the remedial aspects of encrustations in wells, and recommended that the reduction or elimination of this problem could be achieved by providing an increase in screen open årea in newly designed wells, reduced draw down in wells through lower pumping from the screened area of the well by the use of packers and vacuum seals, and periodic chemical treatment (sulphamic or hydrochloric acids) to reduce microbial numbers. Mogg (1972) considered the dominant iron bacterial genera in well encrustations to be *Gallionella, Crenothrix* and *Leptothrix* and postulated that cells of iron bacteria may be transported easily from one well to another on well repair tools and equipment. He cites an example of a North Carolina well which became rapidly infected after undergoing minor repairs. Grainge & Lund (1969) indicated that "iron bacteria cause serious uncontrolled fouling in a high proportion of the wells throughout the world" and further indicated that "there is a proliferation of recommended controls, many of which we have found to be ineffective. . . ." There is considerable evidence to support this (see Section 4).

Rao (1970) reported that in the Howrah district of India, iron bacteria occurred widely in the wells and water supply system, causing serious clogging or corrosion of the pipes. The iron content of the wells was between 0.2 and 0.8 $mg.l^{-1}$ and the dominant genus was *Clonothrix.* Other genera also frequently encountered were *Crenothrix, Leptothrix* and *Siderocapsa.* The occurrence of iron bacteria in water supplies was described (*Anon.* 1939) in which the major problems were stated to be the limitation of hydraulic efficiency owing to the presence of growths up to 1.5 cm thick on the inside of pipes, impaired water quality (taste, odour and colour), and potential plugging of filters and pipes. Some of these aspects were further described by Omerod (1974). Considerable problems have been generated in mines and mine drainage which is rich in dissolved reduced sulphur compounds and iron. In the presence of air, *T. ferrooxidans*, in conjunction with other sulphide-oxidizing bacteria, generates a very acidic product due to the release of sulphuric acid as a terminal metabolic product. The factors involved have been extensively studied (Lorenz & Stephan 1967; Baker & Wilshire 1970), and in particular the rate at which it occurs (Singer & Stumm 1970). Methods for reducing the acidity in drainage water were proposed by Walsh & Mitchell (1972) and included (a) the partial neutralization of the water to a pH value above 4.3 to inhibit stalked bacteria and *T. ferrooxidans*, (b) the use of surface active agents to detach stalks (holdfasts) at the site of pyrite degradation and (c) the introduction of heterotrophic bacteria capable directly or indirectly of parasitizing upon the stalked bacteria. Most treatments currently employed utilize (a).

(b) *Global distribution of iron bacteria*

Apart from frequent statements indicating that the problems involving iron bacteria in ground water are worldwide, there has been no attempt to determine the true extent of the problem. Cullimore & McCann (1974) carried out a survey of 150 countries by correspondence with the respective Departments of Environment or equivalent bodies. The global distribution of iron bacterial problems in ground water was drafted on the basis of the governmental replies (Fig. 2). Some countries elaborated extensively on the nature of the problems (summarized in Table 3).

From this survey based on responsive countries only, it is clear that iron bacteria occur on all continents (except Antartica for which no data is available) and other countries including China, Germany, Yugoslavia and Holland presumably also experience problems since papers on the subject have originated from them. Indeed, it seems likely that iron bacteria are present in the ground water and surface waters throughout the temperate and tropical zones of the world.

(c) *Local problems with iron bacteria in Saskatchewan*

The southern half of Saskatchewan is prairie, with an annual rainfall of 20–40 cm, and is fed by one major river system and several large aquafers running from the northern half of the province which is a part of the Canadian Shield. In the south, the dominant industry is agriculture (grain crops). The low rainfall restricts the source of water to spring run-off and well supplies. Problems with iron bacterial growth in wells occur across the southern part of the province (Fig. 3) to a varying degree (*Anon.* 1972). W. A. Meneley (pers. comm.) has calculated that the approximate annual cost of remedial measures to the provincial economy is four to six million dollars. The principal effects of iron bacterial growths were listed (*Anon* 1972) as being the corrosion of water pumps, pressure tanks, galvanized pipes and fittings; the clogging of metal and plastic pipes; the reduction of water flow and water pressure and the coating of the resin beds of water softeners with slime, reducing efficiency and imparting unpleasant tastes and odours to the water. Dominant iron bacteria listed were *Gallionella* and *Sphaerotilus.* The authors, in a survey of well water samples from across the southern half of Saskatchewan, found that over 90% of the wells contained iron bacteria (see Section 3c) and that the dominant genera were, in

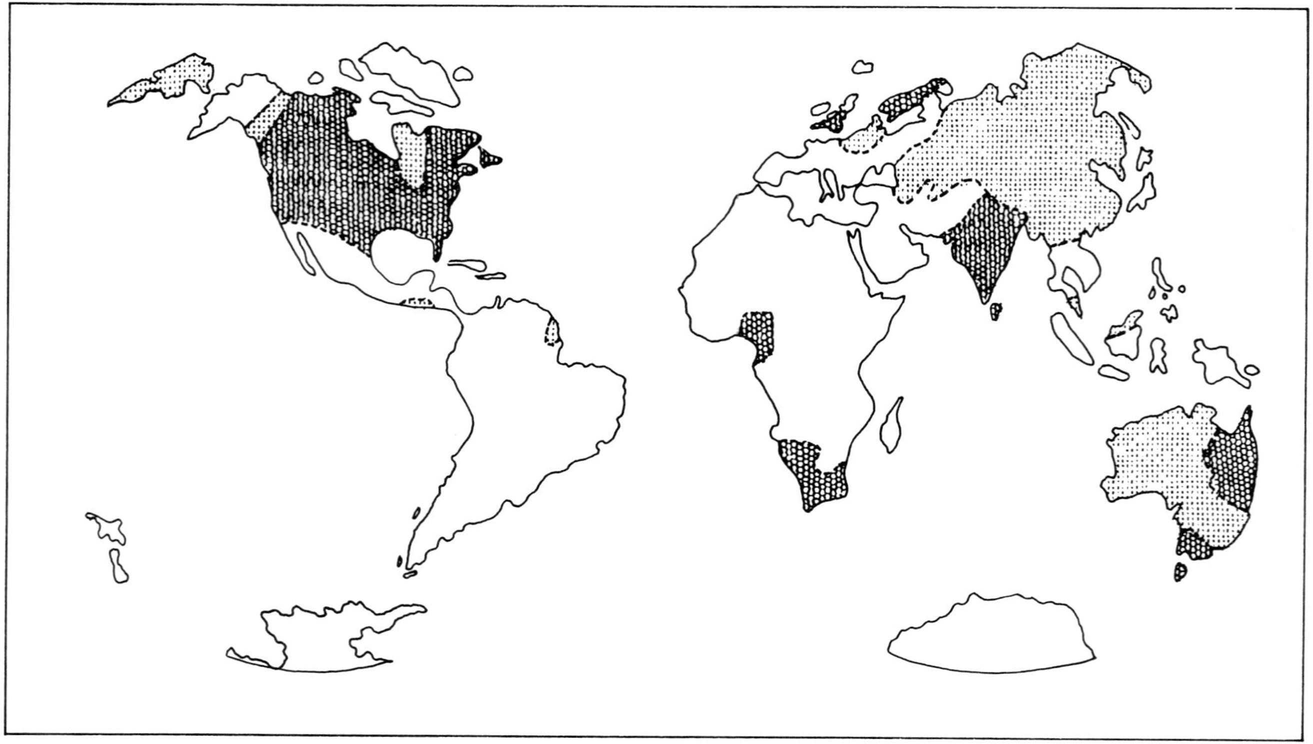

Fig. 2. Global distribution of iron bacteria problems acknowledged by government agencies. The shaded areas represent districts where governments have indicated problems from iron bacteria. Documented information and locally severe occurrences are shown in the dark shade. The light shade indicates that these problems are suspected to be caused by iron bacteria. Unshaded areas represent the countries which failed to respond to the survey and about which there is no specific mention in the literature concerning the distribution of the bacteria or which reported ignorance of the occurrence of iron bacteria. The map does not therefore give a complete distribution.

decreasing order of percentage occurrence: *Crenothrix, Leptothrix, Gallionella, Sphaerotilus* and occasionally *T. ferrooxidans* and *Siderocapsa.* Control methods used include treatment with hypochlorite ('shock' chlorination); sulphamic or sulphuric acids and a number of proprietary products, but there is a high incidence in the re-occurrence of problems after all types of treatment. Some attempts are made to enforce higher hygienic standards during construction and maintenance operations in an effort to prevent cross-contamination of wells.

3. The Growth and Enumeration of Iron Bacteria

(a) *Factors influencing the growth of iron bacteria*

The impact of iron bacterial growths on the quantity and quality of water supplies has focussed some attention on the factors controlling their growth. Each chemical and physical factor will be discussed separately.

(i) *Iron*

As suggested by the name iron bacteria, iron can perform a key role in controlling the growth of the organisms. Wolfe (1960) isolated a strain of *Clonothrix* growing in waters devoid of manganese and containing only 0.02 $mg.l^{-1}$ of iron, but in general, growths occur only at substantially higher concentrations of iron. Hasselbarth & Ludemann (1972) in their review of the large iron and manganese bacteria (*Gallionella, Leptothrix*) found that in static water conditions growth occurred at between 1.6 and 12 $mg.l^{-1}$ iron, and was prevented at 14 $mg.l^{-1}$ iron. In flowing water, such as a pumping well, encrustrations of iron bacteria could be expected if the iron concentration exceeded 0.2–0.5 $mg.l^{-1}$ because of the continuous flow of nutrients. Similar concentrations of iron ($mg.l^{-1}$) are reported elsewhere in the literature, for example, 0.2–0.8 (Rao 1970); 0.3 (Luthy 1964) and 1 (Mogg 1972). Starkey (1945) and Stephenson (1950), calculated the reaction dynamics assuming that the oxidation of ferrous ions generated the sole source of energy for the synthesis of cell material. The ratios calculated were 500 : 1 and 448 : 1 (for iron to cell material) respectively, clearly indicating that very large deposits and encrustration of iron bacteria occur relative to cell mass.

(ii) *Manganese*

This may fulfil the same role as iron in the metabolism of iron bacteria, but has been ignored to some extent in earlier well monitoring programmes. Luthy (1964) considered 0.05 $mg.l^{-1}$ of manganese to be undesirable since it would cause staining but stated that the problem became more severe at manganous ion concentrations > 0.15 $mg.l^{-1}$. Studies carried out by the authors appeared to indicate that for most strains of iron bacteria, the uptake of iron and manganese

Table 3

Information received from various governmental agencies concerning the occurrence of iron bacterial infestations

Country	Area	Problem specified	Causative agent
Australia	Victoria and Queensland	Irrigation jets plugged, plugged pipes	*Gallionella* dominant in ground water (Queensland)
Canada	Most provinces	Extensive growth in wells	*Gallionella, Crenothrix, Leptothrix, Clonothrix, Sphaerotilus* and *Siderocapsa* (rare)
El Salvador	Apopa and Soyapango	Clogging well screens	Iron bacteria, not specified
Guyana	–	Water supplies	Not specified
India	Widespread	Reduction in flow rates and potability	*Clonothrix* predominant in Calcutta area
Malaya	–	Contamination of wells and irrigation water for rice culture	Not specified
Nigeria	Bomore	Plugging of screens	Not specified
Norway	Widespread	Growths in systems of hydroelectric power plants	*Gallionella*

Singapore	City	Deterioration in water supply	Not specified
South Africa	Widespread	Pipe scaling	Not specified
Sri Lanka	Widespread	Plugging through water supply systems	*Crenothrix*
Sweden	Widespread	Discoloured water, plugged pipes	*Gallionella*, *Crenothrix*, and *Leptothrix*
United Kingdom	–	Discolouration of water supplies	*Gallionella* (less organic matter present) or *Leptothrix*/*Crenothrix* (organic matter relatively high)
United States of America	Alabama	5% wells discoloured, 62% wells contain iron bacteria	*Gallionella* and *Sphaerotilus*
	Indiana, New England, New Jersey, Ohio	Problems with iron bacteria particularly common	Not specified
	California, Colorado, Indiana, Kansas, Kentucky, Louisiana, Missouri, New York, Ohio, South Carolina, Vermont, West Virginia	969 communities surveyed and 10% of water examined exceeded 0.3 $mg.l^{-1}$ iron and could support iron bacterial infestations	Not specified
Union of the Socialist States of Russia	Widespread	Not specified	Not specified

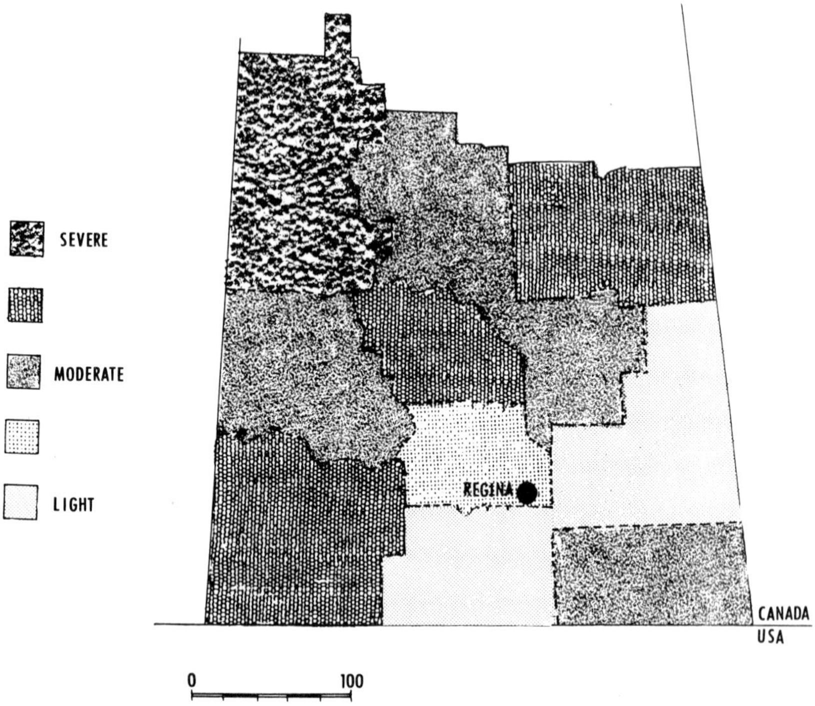

Fig. 3. Intensity of iron bacterial contamination of wells in the agricultural region of Saskatchewan. Severity of the contamination problem is given in a sliding shade scale displayed to the left of the map. The scale is in miles. The clear zone to the north is outside the agricultural region of Saskatchewan.

did not interact with each other in any fixed ratio. In bioassays of freshly isolated iron bacteria (mainly *Crenothrix* sp.), most displayed a preferential growth in media containing higher concentrations of ferrous ions than in media containing the equivalent molar concentrations of manganous ions. However, for one strain of *Crenothrix* subjected to intensive uptake studies, it was noted that: (a) the rate of reduction of dissolved manganese was affected by the concentration of iron present; (b) when iron was in excess of manganese ($>5:1$) then the rate of utilization of manganese was accelerated; (c) when the total concentration of iron and manganese was low (<10 mg.l^{-1} mixed iron and manganese), then the excess of iron necessary to stimulate manganese uptake was more extreme ($>100:1$). In these bioassays, few strains grew more efficiently on manganese than on iron. Clearly, the alternative use of manganous ions by iron bacteria requires more study, and the name iron bacteria may be a poor one since it focusses attention on that element rather than on manganese and perhaps some other metallic elements.

(iii) *pH*

Iron bacteria, except *T. ferrooxidans*, generally grow well over the pH range of 5.4 to 7.2 (Hasselbarth & Ludemann 1972). The pH of individual culture media are within this range: Prévot's medium, pH 6.0; iron bacteria medium, pH 6.5; *Leptothrix* medium, pH 5.8–6.8; Lieske's medium, pH 6.6; van Niel's medium, pH 6–7 (Rodina 1965). Under more alkaline conditions (that is, pH of 7.5–9.0), ferrous and manganous ions tend to oxidize rapidly by normal physicochemical processes and become less available as a potential energy source.

(iv) *Oxygen*

All iron bacteria are either aerobic or microaerophilic, and massive growths of iron bacteria have been reported in wells containing less than 5 $mg.l^{-1}$ O_2 (Hasselbarth & Ludemann 1972). Growth may be suppressed under saturated oxygen conditions. In addition, Weart & Margrave (1957) noted that the growth of some species of iron bacteria was restricted to a defined oxidation–reduction range.

(v) *Temperature*

There has been no direct study of the growth range of iron bacteria directly isolated from wells but some ranges are listed in *Bergey's Manual of Determinative Bacteriology* (Buchanan & Gibbons 1974) and are presented in Table 4. Studies on the growth of iron bacteria in the wells of southern Saskatchewan would suggest that all the strains are obligate or facultative psychrophiles, since the water temperature varies between 3 and 14°C. Although only limited thermal gradient growth studies have been undertaken, evidence from field studies suggests that temperature elevation in well water may trigger off iron bacterial growths. For example, in one refinery well (#6, Consumers' Co-operative Refinery, Regina), significant increases in iron bacterial numbers followed a rise in the water temperature above 5.5–5.8°C. When the temperature returned to a value below this level the bacterial numbers declined.

(vi) *Carbon*

Most wells contain sufficient dissolved carbon dioxide/bicarbonate (Hasselbarth & Ludemann 1972) to meet the growth requirements for the chemoautotrophic iron bacteria (*Gallionella*) so that carbon is not likely to be restrictive for this genus. The other genera requiring organic carbon may be restricted by the availability (qualitative and quantitative of these compounds). In Rodina's (1965) listed media, organic carbon is provided in a variety of forms – citrate, acetate, glucose, peptone, asparagine and infusates of various leaves (for example, lettuce and willow). Comparative studies indicate that the most universal source of carbon is citrate (as ferric ammonium citrate). It is

Table 4
Growth temperature ranges for iron bacteria

Genus	Growth temperature range (°C)		
	minimal	optimal	maximal
Sphaerotilus	15	25–30	37
Leptothrix	10	20–25	35
**Crenothrix*	6	26–28	34
Thiobacillus ferrooxidans	?	15–20	25

* Cullimore & McCann (unpublished).

difficult to predict a threshold concentration below which iron bacterial growth would be restricted or totally inhibited since the total available organic carbon would be a function of flow rate as well as concentration.

The range of cultural procedures which have been used for the iron bacteria are shown in Table 5.

(b) *Cultural differentiation of iron bacteria*

Although there has been a considerable level of activity in determining the types and physiological grouping of the traditional iron bacteria, they remain a poorly defined group (Macrae & Edwards 1972). This stems from the fact that many bacteria can become encrusted with iron precipitated from a sol, as demonstrated by Macrae & Edwards (1972) for *K. pneumoniae, Micrococcus, E. coli, Corynebacterium pseudodiphtheriticum, Ps. fluorescens, Mycobacterium phlei* and *Caulobacter.* Macrae *et al.* (1973) cast further doubt on the delineation of iron bacteria by reporting that isolates of *Pseudomonas, Alcaligenes* and *Moraxella* were able to utilize organic iron complexes (as ferric ammonium citrate and in some cases, as ferric malonate and/or ferric galate). In all cases iron was precipitated, often as a heavy red precipitate. Clearly, the definition of iron bacteria needs resolving if an accurate determination is to be conducted into the relative importance of iron bacteria in well waters.

Of primary concern must be the name iron bacteria which in itself does not accurately define their function since many bacteria in this group can alternatively use manganese or even use manganese exclusively. Furthermore, there is no evidence which would exclude the use of other metallic elements from performing the same function, albeit in a minor role. Therefore, the group could be more precisely defined by the term metallo-oxidizing bacteria (MOB). This would eliminate from the group the heterotrophic bacteria capable of utilizing the organic component of iron or manganese organic complexes leading to the precipitation of iron or manganese (i.e. the metallo-precipitating

Table 5

Cultural procedures for iron bacteria

Method	Specific notes	References
Thiobacillus ferrooxidans		
9K medium	9000 $mg.l^{-1}$ Fe, pH 3.3 developed by	Silverman & Lundgren (1959)
9K medium	used by	Margalith *et al.* (1966), Remsen & Lundgren (1966), Korczynski *et al.* (1967), Wang *et al.* (1970), Howard & Lundgren (1970), Vestal & Lundgren (1971), Lapteva *et al.* (1971), Niemela & Tuovinen (1972)
9K medium	Added 0.5M solution of $FeSO_4.7H_2O$	Vestal & Lundgren (1971)
9K medium	Omitted chloride salts	Howard & Lundgren (1970)
Medium for use in acid bituminous coal mine effluents	$FeSO_4.7H_2O$, 2000 $mg.l^{-1}$; $MgSO_4.7H_2O$,0.1%; $(NH_4)_2SO_4$, 0.05%, pH adjusted using H_2SO_4 to 2–2.5	Temple & Colmer (1951)
Medium for coal mine effluents	$(NH_4)_2SO_4$, 0.015%; KCl, 0.0005%; $MgSO_4.7H_2O$, 0.05%; KH_2PO_4, 0.005%; $Ca(NO_3)_2$, 0.001%. $FeSO_4.7H_2O$ added to final strength of 0.1%	Leathen *et al.* (1956)
Solid agar medium	2% agar added to basal medium	Temple & Colmer (1951)
Solid agar medium	Enumeration using agar media fails due to inhibition of organisms	Bryner & Jameson (1958), Unz & Lundgren (1961)
Silica gel as gelling agent in media	Variable success	Leathen *et al.* (1951), Leathen *et al.* (1956). Bryner & Jameson (1958), Beck (1960), Lapteva *et al.* (1971)
Enumeration of viable organisms	K_2HPO_4, 0.08%; $MgSO_4.7H_2O$, 0.08%; $(NH_4)_2SO_4$, 0.08%; $FeSO_4$, $7H_2O$, 6.66% pH, 1.55. Prewashed membrane filters. Agarose at 0.3–0.5% (w/v)	Tuovinen & Kelly (1973)

Continued

Table 5 – *Continued*

Method	Specific notes	References
Incorporation of organic compounds in medium	Reported T. *ferrooxidans* inhibited by specific compounds	Kelly (1971), Tuovinen *et al.* (1971b), Usami & Sugitani (1971)
Incorporation of organic compounds in medium	Only some strains inhibited	Shafia & Wilkinson (1969), Tabita & Lundgren (1971)
Medium containing glucose, basal salts, *p*-amino benzoic acid	Transition from chemolithotrophy to organotrophy	Shafia *et al.* (1972)
Sphaerotilus/Leptothrix		
Cultured	Autotrophically on thiamin, biotin, cyanocobalamin-supplemented medium	Ali & Stokes (1971)
Methionine synthesis	Cyanocobalamin shown to be essential	Johnson & Stokes (1965)
Growth on manganous ions	*Sphaerotilus discophorus* could exclusively utilize manganous rather than ferrous ions	Ali & Stokes (1971)
Heterotrophic media	Recommended for group	Mulder & Van Veen (1963), Petitprez *et al.* (1969), Stokes & Powers (1965 and 1967). Bisset & Brown (1969)
Carbon substrate utilization by strains of *Sphaerotilus*	Wide range of sugars could be metabolized	Lackey & Wattie (1940), Stokes (1954), Scheuring & Hohnl (1956)
Carbon substrate utilization by strains of *Sphaerotilus*	Fructose utilized	Hohnl (1955). Mulder & Van Veen (1963)
	Lactose utilized by some strains	Lackey & Wattie (1940), Scheuring & Hohnl (1956), Razumov (1961)
	Lactose not utilized	Stokes (1954)
	Xylose and ribose utilized	Curtis (1969)
	Glucose suppressed the oxidation of other organic compounds	Stokes & Powers (1967)

	Organic alcohols could be utilized but not methanol	Stokes (1954)
	Glycerol utilization confirmed	Hohnl (1955), Mulder & Van Veen (1963)
	Succinic, fumaric, butyric, lactic, pyruvic and acetic acids utilized	Stokes (1954)
	Additional acids including malic, propionic, citric, gluconic, malonic and tartaric all utilized	Scheuring & Hohnl (1956)
Simple medium	Gelatin supplemented with mineral salts	Pringsheim (1949) Zikes (1915)
Amino acid utilization	0.1 $mg.l^{-1}$ methionine stimulates uptake	Wilson (1960)
	Asparagine, aspartic acid, glutamine and glutamic acid could serve as carbon and nitrogen sources only	Scheuring & Hohnl (1956)
	Methionine, threonine, tyrosine, glycine and cystine could serve as nitrogen sources only	Scheuring & Hohnl (1956)
	Disparity in results postulated to be due to different isomers or concentrations	Harrison & Heukelekian (1958), Hohnl (1955)
Inorganic nitrogen utilization	Nitrate utilized	Linde (1913)
	Grew with ammonium or nitrate ions	Cataldi (1939), Lackey & Wattie (1940)
	Ammonium ion utilized only when carbon source was sucrose, glycerol or succinate	Stokes (1954)
	Nitrate proved superior source of nitrogen	Hohnl (1955), Razumov (1961), Phaup (1968)
	Cyanocobalamin (B_{12}) stimulated uptake of NH_4, NO_2 and NO_3	Dias & Heukelekian (1967), Mulder & Van Veen (1962), Okrend & Dondero (1964)
	B_{12} stimulation of uptake could be duplicated using higher concentrations of methionine	Wuhrmann & Koestler (1950)
	Thiamine and biotin also essential for uptake in some strains	Stokes & Johnson (1965) Johnson & Stokes (1965)

Continued

Table 5 – *Continued*

Method	Specific notes	References
Mineral salt requirements	Basic medium suggested	Lackey & Wattie (1940)
	$FeCl_3$ could cause inhibition	Johnson & Stokes (1966)
	Filament formation calcium dependent	Dias & Dondero (1967)
	Strontium could not be used as a substitute for calcium dependency	Dias *et al.* (1968)
	Phosphate inhibitory at	
	0.15 M	Hohnl (1955)
	0.05 M	Gaudy & Wolfe (1961)
	Sphaerotilus dominant component of 'sewage fungus' at 150 $\mu g.l^{-1}$ phosphorus	Wuhrmann *et al.* (1966) Ornerod *et al.* (1966)
Basic medium	Enrichment medium using ferrous sulphide at pH 6.0	Kucera & Wolfe (1957)
	Modified above medium to monitor pH changes with indicators	Wolfe (1960)
	Eliminated contaminants by 0.5% formalin pretreatment for 2 days	Nunley & Krieg (1968)
Preservation of viable cultures	15% glycerol at −80°C	Nunley & Kreig (1968)
Optimization of growth	Maximal at 0.1–0.2 $mg.l^{-1}$ O_2	Hanert (1968)
	1% CO_2 enrichment	Van Iterson (1958)
Recommended medium	Lieske's medium containing 3% (w/v) iron filings or flat plates	Rodina (1965)

non-oxidizing bacteria, MPNB). While the MOB group would oxidize specifically either manganous or ferrous compounds, with encrustation and/or precipitation of the metal as the oxides or hydroxides, the MPNB would cause only precipitation of the metal, with encrustation only when iron is present in a colloidal form (Table 6). In most instances, however, the MOB group would be encrusted with the metallic oxides or hydroxides while the MPNB group would cause precipitation. Confirmation of the types of bacteria present in any primary isolation/enumeration would therefore have to include at least microscopic examination for the types of bacteria present. Many of the media recommended for isolating MOB will allow the growth of the MPNB. Rodina (1965) has listed a

Table 6

Genera comprising the metallo-oxidizing bacteria and the metallo-precipitating non-oxidizing bacteria

Group	Genus	Autotrophic	Heterotrophic	Non-colloidal metal encrusted on cell	Metal precipitated only
MOB	*Gallionella*	+	–	+	–
MOB	*Toxothrix*	–	+	∓	±
MOB	*Sphaerotilus*	–	+	–	+
MOB	*Leptothrix*	–	+	+	–
MOB	*Lieskeella*	–	+	+	–
MOB	*Crenothrix*	–	+	+	–
MOB	*Clonothrix*	–	+	+	–
MOB	*Pedomicrobium*	–	+	+	–
MOB	*Metallogenium*	–	+	+	–
MOB	*Kusnezovia*	–	+	+	–
MOB	*Thiobacillus*	–	+	±	∓
MOB	*Siderocapsa*	–	+	+	–
MOB	*Naumanniella*	–	+	+	–
MOB	*Ochrobium*	–	+	+	–
MOB	*Siderococcus*	–	+	±	∓
*MPNB/MOB	*Pseudomonas*	–	+	∓	+
*MPNB/MOB	*Hyphomicrobium*	–	+	∓	+
*MPNB/MOB	*Arthrobacter*	–	+	∓	+
MPNB	*Aerobacter*	–	+	–	+
MPNB	*Serratia*	–	+	–	+
MPNB	*Bacillus*	–	+	(∓)	+
MPNB	*Klebsiella*	–	+	–	+
MPNB	*Alcaligenes*	–	+	–	+
MPNB	*Moraxella*	–	+	–	+
MPNB	*Corynebacterium*	–	+	–	+
MPNB	*Caulobacter*	–	+	–	+
MPNB	*Mycobacterium*	–	+	–	+
MPNB	*Escherichia*	–	+	–	+

* Individual strains capable of oxidizing ferrous and/or manganous inorganic salts have been reported.

range of media suitable for culturing iron bacteria (that is, MOB). Cullimore & McCann (1975), using a range of these media at pH values of 6.0 and 7.4 to culture MOB from 15 water samples taken from Saskatchewan wells, found by means of a microscopic examination of pellicular growths of the liquid media that MOB had grown to differing extents in all media (Table 7). Genera recovered included *Crenothrix, Leptothrix, Sphaerotilus, Gallionella* and *Siderocapsa*, with *Gallionella* and *Crenothrix* being the most commonly isolated. The Winogradsky and the iron peptone media gave the best recovery rates. *Gallionella* could not, however, be reliably subcultured on these media but *Crenothrix, Leptothrix* and *Sphaerotilus* grew well.

(c) *Qualitative examination of iron bacteria*

Many techniques have been developed for the rapid screening of water samples for the detection of iron bacteria. Direct microscopic examination, the original preferred technique (Biswas 1937; *Anon.* 1939), is still widely used (Barbic & Bracilovic 1974). The most successful stain in the author's experience is that of Meyers (1958) in which the cells stain red and the iron deposits blue. The technique consists of: (a) separating the cells by centrifugation (if water is to be examined); (b) smearing the centrifuged pellet onto a slide (if a culture is to be examined then direct smear of the pellicle may be made); (c) air-drying slide; (d) placing it in methanol for 15 min; (e) heating to boiling point a 1 : 1 mixture of 2% potassium ferricyanide and 5% acetic acid (both solutions in distilled water); (f) immersing slide in boiling mixture for 2 min; (g) washing it gently with

Table 7

Percentage occurrence of different genera of metallo-oxidising bacteria on twelve media using well water samples

Medium*	pH	*Crenothrix*	*Gallionella*	*Leptothrix*	*Sphaerotilus*	*Siderocapsa*
Winogradsky	6	60	27	0	7	0
Winogradsky	7.4	80	33	7	0	0
Prévot	6	40	20	13	13	0
Prévot	7.4	27	13	7	0	0
Iron peptone	6	87	20	13	0	0
Iron peptone	7.4	87	13	7	7	0
Leptothrix	6	33	13	7	0	7
Leptothrix	7.4	27	20	0	7	7
Ferric ammonium citrate	6	20	13	0	7	0
Ferric ammonium citrate	7.4	73	27	0	0	0
Lieske's	6	20	13	13	7	0
Lieske's	7.4	20	13	0	0	0

* Formulae as listed in Rodina (1965).

distilled water after cooling; (h) staining for 5 to 10 min with 2% aqueous safranin; (i) rinse, dry and examine. Alternatively, wet mounts of water samples rich in iron bacteria can be efficiently examined using phase-contrast microscopy (V. G. Collins, pers. comm.). For samples low in bacterial numbers, Leuschow & Mackenthun (1962) used filtration through a 0.45 μm pore size membrane filter, followed by drying at 100°C and saturating the filter with immersion oil having the same refractive index as the filter material. The filter is examined under an oil immersion objective and the types of iron bacteria observed and enumerated. Leuschow & Mackenthun (1962) applied the technique to wells in Wisconsin and found that, using 100 ml samples, 55% were positive for iron bacteria. Dominant bacteria were *Gallionella* and *Leptothrix* and the highest counts were $>10^7$ cells.ml^{-1} in reddish and turbid waters.

Several very simple cultural techniques have been developed to indicate presence or absence of iron bacteria in water samples. These include the following.

1. Water is placed in a wide-necked sample bottle and left overnight. The appearance of flakes resembling cotton wool indicates the presence of iron bacteria, and is confirmed by microscopic examination of the flakes (Rodina 1965).
2. Water is placed in an aquarium jar together with sediment. A cork with several cover glasses inserted vertically into its lower side is floated on the water. The appearance of rust spots and/or cotton-like accumulates above the precipitated sediment indicates the presence of iron bacteria, and examination of the cover glasses (air dried and stained) after 24 h incubation will reveal the types of any iron bacteria which have become attached to the glass (Cholodny 1953).
3. The sample is placed in a conical flask to which a chemically cleaned soft steel washer is added. An extruded plastic rod is now placed vertically in the water. After two days, a translucent filamentous growth occurs at and below the water line on the rod and develops a brown tint indicative of the presence of iron bacteria (Grainge & Lund 1969). The authors recommended the use of this technique to ascertain the potential effectiveness of control programmes.
4. Cullimore & McCann (1975) have developed a three-day field test for detection of both the MOB and MPNB in water, based on a modification of Winogradsky's medium; with nutrient levels established at the minimal concentration of each element to achieve optimal growth assuming an adequate supply of all other nutrients, using a strain of *Crenothrix* (Table 8). The test is performed in a 25 ml capacity screw-capped tube containing 0.75 ml of concentrated medium (X20) evaporated to dryness at 65°C under aseptic conditions. The water sample is placed directly into the tube whereupon the medium is rehydrated and returns to its normal concentration, and incubated in the dark at room temperature (22 ± 3°C) for

Table 8

Critical minimal concentrations for maintenance of optimal growth of a strain of Crenothrix

Element	Concentration ($mg.l^{-1}$)	
	Original medium*	Modified medium
Fe	1000	600
N	750	450
C	2500	1500
Na	140	0
K	220	220
P	90	90
Mg	50	50
S	67	67
Ca	90	0
Cl	16	0

pH adjusted to 7.4.
* Winogradsky's medium.

three days. In Saskatchewan, this period has been found to be sufficiently long for growth to occur. In general, the iron bacteria develop a thick pellicle or flaky deposit on the surface of the medium which itself becomes yellow or brown. If no bacteria are present then the medium will slowly auto-oxidize to a green colour. Several alternative reactions can occur in this test (Table 9). Reaction patterns A to D all indicate the presence of iron bacteria but no clear categorization of the genera can thereby be ascertained. Reaction patterns D and E are both the result of the presence of hydrogen sulphide-producing micro-organisms since the black deposits are iron sulphides. In pattern E, no iron bacteria have been recovered and the test indicates the presence of sulphide-producing micro-organisms only. The test

Table 9

*Alternative reactions occurring in the field test**

Reaction type	Visible changes in medium						
	Colour of medium			Brown pellicle or plug	Brown flakes on surface	White effervescence	Black deposits
	colourless	yellow	brown				
A	–	∓	±	+	∓	–	–
B	–	∓	±	–	+	–	–
C	–	+	–	–	–	–	–
D	∓	∓	±	+	∓	–	+
E	+	–	–	–	–	+	+

* Cullimore & McCann (1975).

therefore indicates not only the presence of iron bacteria but also of the corrosive hydrogen sulphide producers. Tests on ground waters in Saskatchewan have been 95% positive for the presence of iron bacteria and microscopic examination of the surface growths has revealed the dominant types to be *Crenothrix, Leptothrix, Sphaerotilus* and *Gallionella.*

(d) *Quantitative examination of iron bacteria*

Modifications of several of the qualitative procedures can be used to achieve quantitative results by the application of a serial dilution technique prior to the test procedure, or by the use of the membrane filtration method of Leuschow & Mackenthun (1962). Another MF technique has been developed by Cullimore & McCann (1975) in which the water sample (100 ml, 10 ml equiv.; 0.1 ml equiv.) is membrane filtered using a 0.22 μm pore size filter, and subsequently cultured on the modified Winogradsky's medium containing 2% agar, and incubated at 28°C for 3 days. The iron bacteria grow as brown colonies, sometimes irridescent, with irregular edges, such colonies being counted as iron bacteria. A typical pattern of iron bacterial populations in well waters north east of Pilot Butte (east of Regina) is given in Fig. 4 and Table 10. This clearly shows that the

Table 10

Distribution of iron bacteria in wells north east of Pilot Butte, Saskatchewan

Well site (see Fig. 4)	Iron bacterial numbers (organisms.ml^{-1})	
	Before run-off*	After run-off
1	88	13
2	25	>3000
3	85	1
4	50	336
5	100	>3000
6	300	27
7	>3000	1
8	30	376
9	10	>3000
10	>3000	>3000
11	300	400
12	200	–
13	300	>3000
14	0	446
15	50	70

* Run-off is the period of snow melt which occurs each spring.

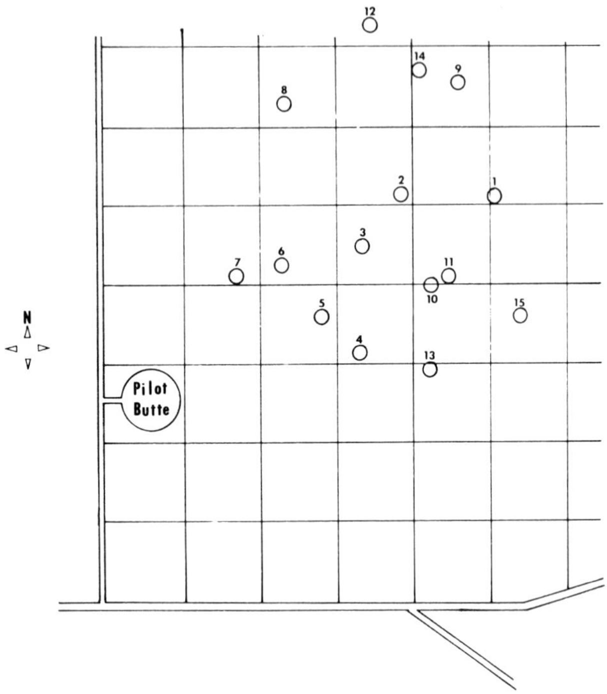

Fig. 4. Distribution of wells tested for iron bacteria north east of Pilot Butte, Saskatchewan. The grid pattern divides the land into square mile sections.

numbers present are subject to considerable variation depending upon factors such as the water temperature, level of pumping activity and quality of the water.

4. Control of Iron Bacteria in Ground Water Supplies

(a) *Chemical control of iron bacteria*

Many chemical treatments have been suggested for the control of iron bacteria in ground water supplies including bactericidal compounds, halogens and halogenated compounds, organic and inorganic acids, copper and copper salts. Many of the individual treatments recommended are summarized in Table 11; most frequently used are calcium or sodium hypochlorite, hydrochloric acid, sulphamic acid and some proprietary preparations. Cullimore & McCann (1975), using an automatically recording densitometer, examined the ability of some of these compounds to inhibit the growth of strains of *Crenothrix* and *Gallionella* in Winogradsky's medium (pH 7.4). The effectiveness of each compound

TABLE 11

Recommended chemical treatments for the control of iron bacteria in ground waters

Treatment*	Site	Reference
110 lb oxalic acid, 50 lb sulphamic acid, 50 lb wetting agent, 0.25 lb inhibitor	Wells	Grainge & Lund (1969)
Elimination of dissolved CO_2 by elevation of pH to above 8.3	Wells	Ellis (1932)
Residual chlorine, 0.2 mg.l^{-1}	Lab tests	Grainge & Lund (1969)
Hydrogen peroxide, 100 mg.l^{-1} and phosphate inhibitor	Lab tests	Grainge & Lund (1969)
Hypochlorite, 0.438%	Wells	Machmeier (1971)
Residual chlorine, 50–100 mg.l^{-1} for 2 h	Wells	Machmeier (1971)
Shock chlorination with 5.25% hypochlorite	Wells	Machmeier (1971)
Hydrochloric acid (muriatic acid), 14–21%	Screened wells	Schafer (1974)
Sulphamic acid, 7.5–10% (several hours contact time)	Screened wells	Schafer (1974)
Hydroxyacetic acid, 4.7–7% (contact time related to pH of water)	Screened wells	Schafer (1974)
Chlorine gas to give 500 mg.l^{-1}	Wells	Schafer (1974)
LBA (Liquid Antibacterial Acid, USA Patent 3085929), 5% (treat for 36 h)	Wells	Luthy (1964)
Recycling of hypochlorite solutions	Water supplies	Rao (1970)
Hydrochloric acid treatment followed by 300 mg.l^{-1} chlorine, 18 h contact	Wells	Mogg (1972)
Calcium hypochlorite, 715 mg.l^{-1}	Wells	Schafer (1974)
Lithium hypochlorite, 0.14%	Wells	Schafer (1974)
Sodium hypochlorite, 0.14%	Wells	Schafer (1974)
Chlorine dioxide gas (limited use)	Wells	Schafer (1974)
Potassium permanganate, 0.1–0.2%	Wells	Schafer (1974)
Continuous chlorination	Wells	Woods (1973)
Acrolein, 0.1–30 mg.l^{-1} (restricted use)	Water systems	Woods (1973)

*All concentrations mentioned refer to final concentrations in ground water.

differed very significantly with the number of cells present in the 15 ml culture (Table 12). Javex (a commercial preparation of sodium hypochlorite) prevented growth at 250 mg.l^{-1} for up to 100 cells.15 ml^{-1} while 1% was necessary to prevent the growth of 7×10^5 to 1×10^6 cells.15 ml^{-1}. Clearly, the selection of concentration for the chemical control must reflect the number of cells present

Table 12

Extrapolated effective control concentration ($mg.l^{-1}$) for five disinfectants against a range of cell concentrations of iron bacteria

Disinfectant	Cell concentration of iron bacteria (cells.15 ml^{-1})				
	+30 to 100	300 to 1000	+3000 to 10,000	70,000 to 100,000	+700,000 to 1,000,000
Javex*	250	750	850	1,000	10,000
HTH*	2000	5000	7500	10,000	100,000
IP (iodine polymer)	10	20	30	40	NC
$CuSO_4$	500	1000	5000	50,000	10,000
LBA	50,000	50,000	50,000	50,000	ND
$KMnO_4$	10–20	100	250	500	5000

All data for pH of 7.4.
+ Extrapolated data.
NC, no control since the solubility of IP in water is very low; ND, not determinable from data.
* Both based on hypochlorite concentration.

within the treated system. The most effective of the tested compounds was an iodine polymer synthesized by Levine, Chemistry Department, University of Regina, the structural formula of which (Fig. 5) is similar to that of crystal violet. Field trials have yet to be conducted on the polymer since little is known of its potential environmental effect. Potassium permanganate was also highly

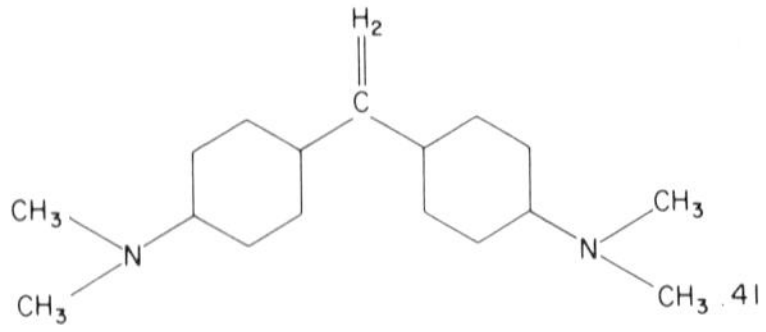

Fig. 5. Molecular structure of iodine polymer.

effective but is perhaps potentially dangerous since it contains manganese which could become a substrate for iron bacterial regrowths.

In Saskatchewan, recommended chemical treatments are widely applied but there is a history of recolonization of the treated wells over even short periods of time; this has led to some doubt as to the effectiveness of control by chemical means. A summary of some of the data for limited field studies on the efficiency of recommended treatments applied to wells in Saskatchewan is shown in Fig. 6. It will be seen that inadequate control practices frequently lead to a post-treatment surge in the iron bacterial populations after a few days. The mechanisms which could influence this may be postulated to be as follows.

1. The chemical control agent may have a differential effect on the iron bacteria growing in clumps or as a slime coating. Inhibition may first occur in the outermost cells; penetration of the inner (protected) cells might be more a function of the ability of the agent to become transported through the cell and the copious slime coatings than upon its toxic potential. Thus, as a result of poor penetration, very high concentrations of an agent (for example,

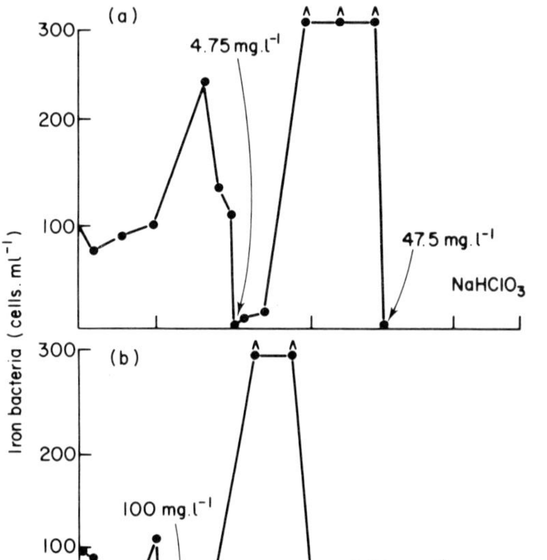

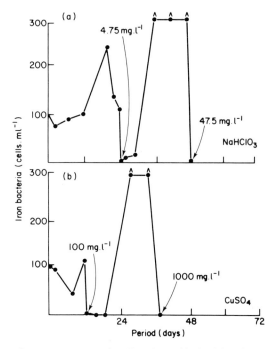

Fig. 6. Effect of two treatments of sodium hypochlorite (a) and copper sulphate (b) on the iron bacterial population in a shallow bore well in Saskatchewan. Arrows represent termination of treatment.

sodium hypochlorite) could be employed without achieving adequate control.

2. Iron bacteria may be growing extensively outside of the treatment zone, and be pulled back into the treatment zone upon the resumption of pumping.
3. The control agent may become neutralized by dead organic material and non-target bacteria, thus reducing its effectiveness.
4. Water temperature may be a critical factor affecting the metabolic activity of the iron bacteria. Well water temperatures in Saskatchewan vary between 4 and 12°C and there is some evidence (unpublished) that the iron bacteria are by and large facultative psychrophiles with growth initiation between 5 and 9°C. In one industrial well which was closely monitored, the population of iron bacteria increased rapidly when the well water temperature rose between 5.4 and 6.0°C. Laboratory studies on a strain of *Crenothrix* indicated that its minimum growth temperature was 7.5°C. The optimum growth temperature was found to be 26 to 29°C, and inhibition occurred at 34°C. Further studies are under way to determine the temperature growth range of a number of iron bacteria. The temperature, by influencing the rate of metabolic activity, would affect the rate at which chemical control agents would be taken up and transported into the cell (when an active transport system is necessary).

These factors undoubtedly contribute to the sometimes ineffective chemical treatment of wells and have led to the search for alternative systems for achieving control.

(b) *Physical control of iron bacteria*

Comparatively little attention has been paid to the physical control of iron bacteria using ultrasonics, heat, cathodic protection and u.v. irradiation. Mogg (1972) discussed some of the physical factors involved in the construction of a well which could influence subsequent growths of iron bacteria, and suggested that metal or plastic parts should be resistant to hydrochloric or sulphamic acids and recommended (in decreasing order of resistance and suitability) most plastics and fibreglass, stainless steel types 304 and 316, silicon manganese bronze, silicon red brass, armco iron, low carbon steel and concrete. He disputed the beliefs that non-conductors such as plastics, fibreglass, concrete and transite would not be subject to encrustations by iron bacteria. Mogg also recommended that problems could be reduced by provision of a greater screen open area in the well; by reduction of draw-down through more efficient design and lower pumping rates; by provision of regular chemical treatment at recommended rates regardless of the appearance of symptoms; and by keeping oxygen away from the screened areas with vacuum seals and packers. Mogg (1972) also commented that "there seem to be more cases of iron bacteria today, and we believe that iron bacterial spores (cells) can be transported from one well to another." He illustrated this by citing a trouble-free well which had been in service for a number of years and became plagued with iron bacterial problems after the pump had been removed for minor repair. Indeed, it is now recommended practice in many parts of the world to disinfect all tools, equipment and drilling materials prior to drilling a new well, in order to reduce the risk of contamination by iron bacteria from another site. In some instances, a new well is treated immediately with hydrochloric or sulphamic acid.

The direct physical treatment of well water has received little attention, but an initial laboratory-based study was undertaken by Cullimore & McCann (1975). Since sonication (at between 500 and 900 kHz) is a well-documented method of disintegrating microbial cells (Carpenter 1972), it was thought that sonication of wells contaminated with iron bacteria might control the growth of the organisms. Several experiments were conducted on cultures of *Crenothrix* and *Gallionella* but no significant reductions in cell numbers were observed even after 60 min of exposure to a tissue sonicator operating at maximum power (800 kHz). This abnormal survival capability is at least in part due to the extensive mucoid and gelatinous coatings of the cell which serve to dampen the effects of the sonication. Sonication is therefore an unsatisfactory method for the control of iron bacterial growth. Similarly, no real control of iron bacteria was achieved by providing cathodic protection.

On the other hand, studies on the heat sensitivity of 22 cultures of iron bacteria (Fig. 7) revealed that all were fairly sensitive to temperatures close to the pasteurization range used for milk (Table 13). Laboratory trials in a simulated well (a 3.1 m length of 10 cm plastic casing) gave complete elimination of all iron bacteria (3000 cells.ml^{-1}) in a 2 m head of water within 20 min of reaching holding temperatures of 52, 55, 60 or 70°C. This indicates

Table 13

Time–temperature combinations necessary to kill 100% of the cells in 9 ml amounts of 22 cultures of iron bacteria

Holding time (min)	Temperature (°C)
100	56
50	57.5
30	58
20	59
10	59.5
5	62.5
3	64
2	65
1	66
0.1	67.5

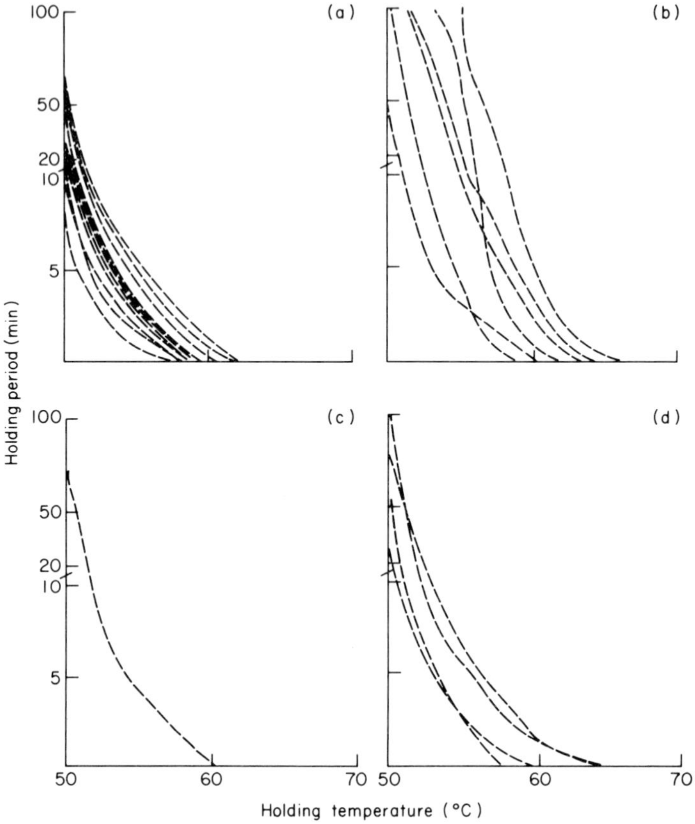

Fig. 7. Influence of holding time (min) and temperature on the survival of cells of 22 strains of iron bacteria. (a) Crenothrix, (b) Sphaerotilus, (c) Gallionella, (d) CG. The individual graphs indicate time–temperature combinations just sufficient to kill all cells.

that pasteurization should eliminate completely all iron bacteria from the system by the end of the holding temperature phase. The initial field trials were conducted on Well #24, Imperial Oil Refinery, Regina, using high pressure steam injected down the well as the heat source for pasteurization. The temperature was raised to 65°C and held for 40 min before being lowered by pumping out the treated water. This water was turbid and dark brown in colour. Two pasteurization treatments (Fig. 8) had the composite effect of elevating the hourly flow rate from 3000 to 7000 imperial gallons.h^{-1}. Previous to this treatment, the well had been subjected to a number of chemical treatments (HTH, A9 bactericide and sulphamic acid), some of which gave increased flow rates which were however of short duration as the iron bacterial level in the well built-up again. This also occurred after the two steam pasteurizations but did not

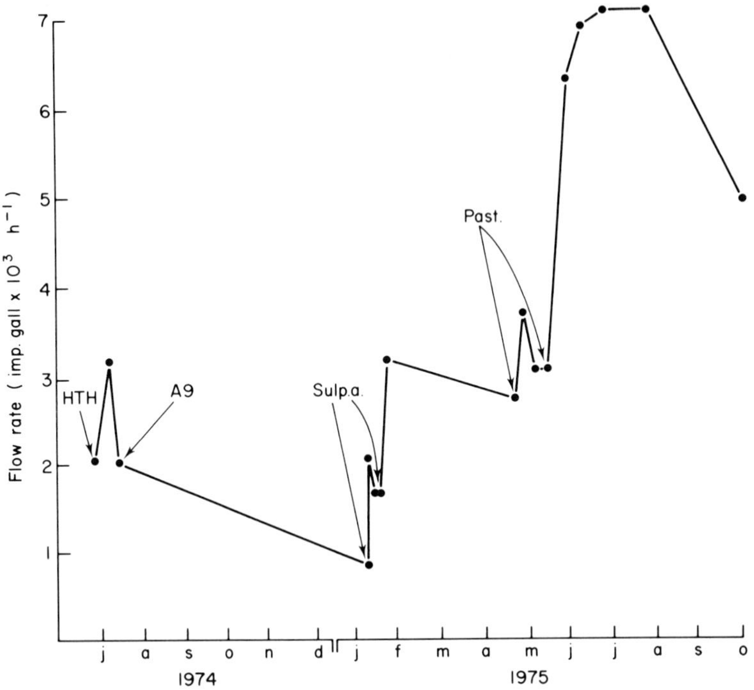

Fig. 8. Influence of treatments to eliminate iron bacteria on the flow rate of Well #24, Imperial Oil Refineries, Regina. HTH, commercial preparation of sodium hypochlorite; A9, disinfectant; Sulp. a., sulphamic acid; Past., pasteurized by steam to 65°C for 40 min. Arrows represent termination of treatment. Iron bacterial growths will reduce flow rates. Flow rate is in imp gall x $10^3 h^{-1}$.

reach a level which would affect refinery operations. No further studies were conducted at this well since the refinery was closed down and operations moved to Edmonton, Alberta.

During the spring of 1975, a pilot pasteurization system was installed in Well #6, Consumers Co-operative Refineries Ltd (Fig. 9). Despite several technical difficulties in providing the steam supply, three pasteurization runs were performed and the influence of these on the iron bacterial population was recorded (Fig. 9). In each case, steam was applied gradually until the well water and pump-out water registered between 60 and 70°C. The system was left for 40 min and then the contents were pumped out. This water was dark brown and contained very high numbers of iron bacteria. As pump-out continued the water became clear. In Well #6, the pump-out water after pasteurization continued to contain iron bacteria at populations of between 30 and 300 cells ml^{-1}. If the well had been pasteurized effectively (and all indications are that it was), then the recurrence of a low level population would indicate that iron bacteria were

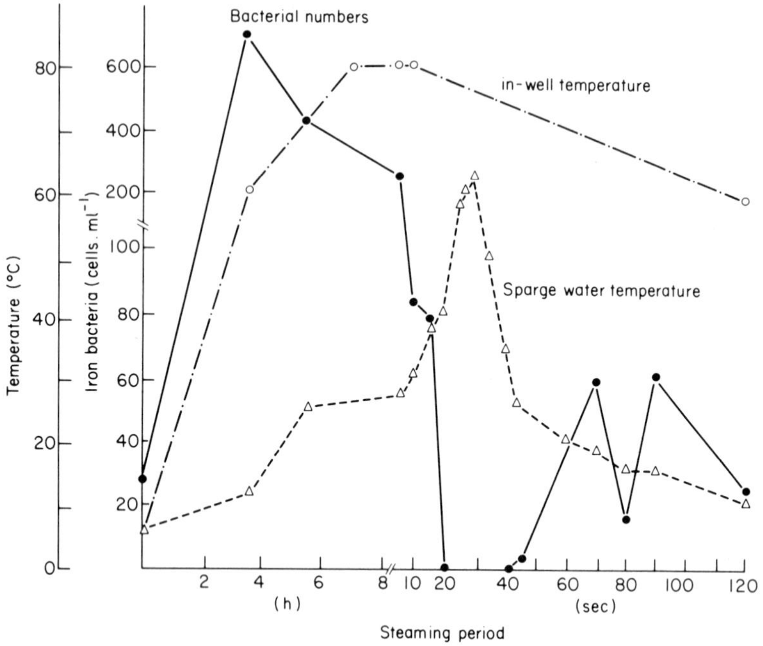

Fig. 9. Impact of steam pasteurization of Well #6, Consumers Co-operative Refineries Ltd, Regina. Bacterial numbers peaked during the steaming period of 8 h due to the slime breaking up on screen and internal walls and fixtures prior to temperature becoming lethal. Recurrence of iron bacteria after treatment due to bacteria being drawn into well through screen after pasteurization.

growing or surviving in the sand and aquafer outside of the screen and were being pulled into the well with the pumping activity. From this it can be deduced that a secondary build-up in iron bacterial numbers will occur in the well on the screen, casings and pumping equipment, so that maintenance of high flow rates would be dependent upon regular pasteurization depending on the rapidity of the secondary build-up. Further trials are now being planned to develop an automatic well pasteurization unit and also to investigate its potential for the treatment of farm wells.

(c) *General discussion*

From all this, it is clear that the iron bacteria are in general very resistant to chemical methods of control, perhaps due to the protective slime layers and other coatings which surround the cells, together with the tendency for the cells to clump and/or form thick layers. Furthermore, these coatings are often heavily impregnated with ferric and manganic oxides and hydroxide deposits which could restrict the diffusion of the chemical agents and perhaps enter into some direct chemical reaction with them. Clearly the lack of success of chemical treatments is in part due to insufficient concentration and/or contact time to allow total penetration of the iron bacteria in the wells, or is "too little, too late". To overcome this, more attention needs to be paid to the accurate categorization of wells to determine the necessary level of treatment if chemical methods are to be used. One possible scheme is given below.

Category A. No iron bacteria present; iron bacteria test negative.

Category B. Iron bacteria present in low numbers but no surface colonial growth present; iron bacteria test positive. Iron bacteria numbers: 1 to 300 cells.ml^{-1}. (In isolated cases, there could be as many as 5000 cells.ml^{-1}.)

Category C. Iron bacteria present in moderate numbers with some surface colonial growth present; iron bacteria test positive. Iron bacteria numbers: 300–5000 cells.ml^{-1}.

Category D. Iron bacteria present in high numbers with extensive surface colonial growth but no 'plugging', iron bacteria test positive (may also show sulphide reduction, i.e. a black precipitate). Iron bacteria numbers: 5000 to 50,000 cells.ml^{-1}.

Category E. Iron bacteria present in excessive numbers with such extensive surface colonial growth that 'plugging' occurs; iron bacteria test positive (frequently accompanied by heavy sulphide reduction, i.e. a black precipitate). Iron bacteria numbers: 50,000–10,000,000 cells.ml^{-1}.

Note that screens could become plugged in any of categories B to E but that the rate of plugging would increase dramatically with the severity of the problem.

Three category groups may be established using the simple field test system developed by Cullimore & McCann (1975). These are: category A wells, no reaction; category B and C, reaction type A, B or C (see Table 9); category D and E, reaction type D, E or occasionally A, B or C.

In Saskatchewan, the vast majority of wells (80% of those tested) fit into the middle group (category B, 60%; category C, 20%). Less than 5% of the wells are in category A and the remaining 15% of those tested are in category D or E, many of which have been abandoned. A possible relationship between chemical treatment and the effective control of iron bacteria may be achieved, at least in theory, for category B, C and D wells (Table 14), but not for category E wells, since the iron bacterial coatings would probably be too thick to allow complete penetration of the chemical even over a prolonged period. The frequency of treatment would depend upon the rate at which the quality of the well water degenerated. In some instances, very rapid growth of iron bacteria has been

Table 14

Potential treatment time–chemical concentration for the control of iron bacteria in wells in categories B to D

Well category		Chemical treatment		
		Sodium hypochlorite	Sulphamic acid	Hydrochloric acid
B	Conc (%)	0.5	7.5	14
	Time (h)	6	12	6
C	Conc (%)	1	10	18
	Time (h)	24	24	12
D	Conc (%)	10	10	21
	Time (h)	48	48	24

observed to appear on screens in a matter of hours (Oliver 1975). Steam pasteurization offers a more rapid (total operation time $1\frac{1}{2}$ to 6 h) and complete destruction of the iron bacteria. It is therefore ideally suited to the treatment of high capacity wells such as those employed in industry and by water authorities, where it can be installed as a permanent facility. For small wells, treatment by steam pasteurization can be achieved using portable steam generators. In conclusion, some reduction in the level of iron bacterial infections could be achieved by steam pasteurization of: (a) all newly installed wells; (b) all wells undergoing major repairs; (c) all category C, D, and E wells, on a routine basis until the wells return to a stable category B or A stage whereupon chemical treatment should suffice.

Treatments (a) and (b) would reduce the risk of a well unaffected by iron bacteria from becoming infected with contaminants from the soil, drilling equipment, or repair and replacement materials. Treatment (c) would reduce the population of iron bacteria dramatically within the well and keep the population down to manageable levels by subsequent regular treatments.

5. Acknowledgements

The authors wish to express their sincerest gratitude to the following agencies for their sponsorship of sections of the study discussed in this chapter: Family Farm Improvement Branch, Department of Agriculture, Provincial Government of Saskatchewan (Section 4), Saskatchewan Research Council (Section 3), National Research Council of Canada (monitoring equipment), and the Consumers Co-operative Refineries Ltd, Regina (steam pasteurization studies of Well #6). Also the following individuals for their invaluable aid: Mrs Frances Bright, Mrs Jean Hudey (manuscript preparation); Mr Stephen Yang, Miss Jane Anweiler, Mr Michael Riley, Mr Lewis Cocks, Miss Denise Bean (technical assistance); Dr W. Meneley, Mr F. Snell, Mr W. Franko, Mr W. Protz, and Mr R. Torrie (helpful advice and enthusiastic support) and finally Dr E. Tward for his invaluable aid in developing the systems for pasteurization.

6. References

ALI, H. S. & STOKES, J. L. 1971 Stimulation of heterotrophic and autotrophic growth of *Sphaerotilus discophorus* by manganous ions. *Antonie van Leeuwenhoek* **37**, 519–528.

ANON. 1939 Iron bacteria in water supplies. *The Johnson Drillers' Journal* **11**, 1–5.

ANON. 1972 *Iron Bacteria in Rural Water Supplies.* Technical Tips, Water and Sewage 1. Family Farm Improvement Branch, Saskatchewan Department of Agriculture, Regina, Canada.

AVERY, E. L. 1970 Effects of domestic sewage on aquatic insects and salmonids of the East Gallatin River, Montana. *Water Research* **4**, 165–177.

BAKER, R. A. & WILSHIRE, A. G. 1970 Microbial factors in mine drainage formation. *U.S. Water Pollution Control Series* 1410 DKN. July 1970.

BALASHOVA, V. V. 1969 The relationship of *Gallionella* to *Mycoplasma. Doklady Akademii Nauk SSSR* **184**, 1429–1434.

BALASHOVA, V. V. & CHERNI, N. E. 1970 Ultrastructure of *Gallionella filamenta. Mikrobiologiya* **39**, 348–351.

BARBIC, F. F. & BRACILOVIC, D. M. 1974 Iron and manganese bacteria in Ranney Wells. *Water Research* **8**, 895–898.

BASASHOVA, B. B. 1967 Enriched culture of *Gallionella filamenta. Mikrobiologiya* **36**, 646–650.

BECK, J. V. 1960 A ferrous-ion oxidising bacterium. I. Isolation and some general physiological characteristics. *Journal of Bacteriology* **79**, 502–509.

BISSET, K. A. & BROWN, D. 1969 Some electron microscope observations on the morphology of *Sphaerotilus natans. Giornale di Microbiologia* **17**, 97–99.

BISWAS, K. 1937 Studies on iron bacteria. *Biologia Generalis*, Band XIII: 421–435.

BREED, R. S., MURRAY, E. G. D. & SMITH, N. R. (eds) 1957 *Bergey's Manual of Determinative Bacteriology, 7th Edn.* Baltimore: Williams & Wilkins Co.

BRYNER, L. C. & JAMESON, A. 1958 Microorganisms in leaching sulphide materials. *Applied Microbiology* **6**, 281–287.

BUCHANAN, R. E. & GIBBONS, N. E. (eds) 1974 *Bergey's Manual of Determinative Bacteriology, 8th Edn.* Baltimore: Williams & Wilkins Co.

CARPENTER, P. L. 1972 *Microbiology*, 3rd Edn. Toronto: W. B. Saunders Co.

CATALDI, M. S. 1939 Estudio fisiologico systematico de algunas Chlamydobacterials. Thesis, University of Buenos Aires.

CHOLODNY, N. G. 1924 Zur Morphologie der Eisenbakterien *Gallionella* und *Spirophyllum. Deutsch Botanische Gesellschaft. Berlin Berichte* **42**, 35–44.

CHOLODNY, N. G. 1953 *Zhelezobakterii.* Moscow: Publishing House of Russia, Academy of Sciences.

CLARK, F. M., SCOTT, R. M. & BONE, E. 1967 Heterotrophic iron-precipitating bacteria. *Journal of the American Water Works Association* **59**, 1036–1042.

COHN, F. 1870 *Beitrage zur Biologie der Pflanzen* **1**, 108.

CULLIMORE, D. R. & MCCANN, A. E. 1974 *The Global Distribution of Iron Bacteria in Water.* Unpublished report #9, Regina Water Research Institute, University of Regina, Canada.

CULLIMORE, D. R. & MCCANN, A. E. 1975 *The Control of Iron Bacteria in Water.* Final report #10, Regina Water Research Institute, University of Regina, Canada.

CURTIS, E. J. C. 1969 Sewage fungus: its nature and effects. *Water Research* **3**, 289–311.

DIAS, F. F. & DONDERO, N. C. 1967 Calcium nutrition of *Sphaerotilus. Bacteriological Proceedings* GISO.

DIAS, F. F. & HEUKELEKIAN, H. 1967 Utilization of inorganic nitrogen compounds by *Sphaerotilus natans* growing in a continuous flow apparatus. *Applied Microbiology* **16**, 1191–1199.

DIAS, F. F., OKREND, H. & DONDERO, N. C. 1968 Calcium nutrition of *Sphaerotilus* growing in continuous flow apparatus. *Applied Microbiology* **16**, 1364–1369.

DONDERO, N. C. 1975 The *Sphaerotilus-Leptothrix* group. *Annual Review of Microbiology* **29**, 407–428.

EHRENBERG, C. G. 1836 Vorlaufige Mitteilungen uber das wirkliche Vorkommen fossiler Infusorien und ihre grosse Verbreitung. *Poggendorf's Annalen* **38**, 213–227.

ELLIS, D. 1932 *Iron Bacteria.* New York: Fredrick A. Stokes & Co.

GAUDY, E. & WOLFE, R. S. 1961 Factors affecting the growth of *Sphaerotilus natans. Applied Microbiology* **9**, 580–584.

GRAINGE, J. W. & LUND, E. 1969 Quick culturing and control of iron bacteria. *Journal of the American Water Works Association* **61**, 242–245.

HANERT, H. 1968 Untersuchungen zur Isolierung, Schoffwechselphysiologie und Morphologie von *Gallionella ferruginea* Ehrenberg. *Archiv für Mikrobiologie* **60**, 348–376.

HANERT, H. 1970 Struktur und Wachstum von *Gallionella ferruginea* Ehrenberg am naturalichen Standort in den ersten 6 Std der Entwicklung. *Archiv für Mikrobiologie* **75**, 10–24.

HARRISON, M. E. & HEUKELEKIAN, H. 1958 Slime infection – literature review. *Sewage and Industrial Wastes* **30**, 1278–1302.

HASSELBARTH, U. & LUDEMANN, D. 1967 Die biologische Verockerung von Brunnen durch Massenentwicklung von Eisen – und Manganbakterien. *Bohrtechnik – Brunnenbaukohrleitungsbau* **10**, 11–20.

HASSELBARTH, U. & LUDEMANN, D. 1972 Biological incrustation of wells due to mass development of iron and manganese bacteria. *Water Treatment and Examination* **21**, 20–29.

HEM, J. D. 1972 Chemical factors that influence the availability of iron and manganese in aqueous systems. *Journal of the Geological Society of America* **83**, 443–450.

HOHNL, G. 1955 Ernahrungs und stoffwechselphysiologische Untersuchungen an *Sphaerotilus natans. Archiv für Mikrobiologie* **23**, 207–250.

HOWARD, A. & LUNDGREN, D. G. 1970 Inorganic pyrophosphatase from *Ferrobacillus ferrooxidans (Thiobacillus ferrooxidans) Canadian Journal of Biochemistry* **48**, 1302–1307.

IVARSON, K. C. & HERINGA, P. K. 1972 Oxidation of manganese by microorganisms in manganese deposits in a Newfoundland soil. *Canadian Journal of Soil Science* **52**, 401–416.

IWASA, Y. 1970 Minerological studies of iron minerals in soils. *Bulletin of the National Institute of Agricultural Sciences Series B* **15**, 187–236.

JOHNSON, A. H. & STOKES, J. L. 1965 Effect of amino acids on growth of *Sphaerotilus discophorus. Antonie van Leeuwenhoek* **31**, 165–174.

JOHNSON, A. H. & STOKES, J. L. 1966 Manganese oxidation by *Sphaerotilus discophorus. Journal of Bacteriology* **91**, 1543–1547.

KELLY, D. P. 1971 Autotrophy: concepts of lithotrophic bacteria and their organic metabolism. *Annual Review of Microbiology* **25**, 177–210.

KORCZYNSKI, M. S., AGATE, A. D. & LUNDGREN, D. G. 1967 Phospholipids from the chemoautotroph *Ferrobacillus ferrooxidans. Biochemical and Biophysical Research Communications* **29**, 457–462.

KRUL, J. M., HIRSCH, P. & STALEY, J. T. 1970 *Toxothrix trichogenes* (Chol.) Beger and Bringmann: the organism and its biology. *Antonie van Leeuwenhoek* **36**, 409–420.

KUCERA, S. & WOLFE, R. S. 1957 A selective enrichment method for *Gallionella ferruginea. Journal of Bacteriology* **74**, 344–349.

KUTZING, F. T. 1843 *Phycologia Generalis.*

LACKEY, J. B. & WATTIE, E. 1940 The biology of *Sphaerotilus natans* Kutzing in relation to bulking of activated sludge. *Public Health Report, Washington* **55**, 975–987.

LAPTEVA, A. M., FRUICHKOV, V. A. & GOLOMZIK, A. I. 1971 Application of gel plates impregnated with the medium 9K for quantitative control and isolation of *Thiobacillus ferrooxidans*. *Mikrobiologiya* **40**, 572–574.

LEATHEN, W. W., MCINTYRE, L. D. & BRAILEY, S. A. 1951 A medium for the study of the bacterial oxidation of ferrous iron. *Science, New York* **114**, 280–281.

LEATHEN, W. W., KIRSEL, N. A. & BRAILEY, S. A. 1956 *Ferrobacillus ferrooxidans*: a chemosynthetic autotrophic bacterium. *Journal of Bacteriology* **72**, 700–704.

LEPP, H. 1975 *Geochemistry of Iron.* Stroudsberg, Penn: Dowden, Hutchinson and Ross, Inc.

LEUSCHOW, L. A. & MACKENTHUN, K. M. 1962 Detection and enumeration of iron bacteria in municipal water supplies. *Journal of the American Water Works Association* **54**, 751–756.

LINDE, P. 1913 Zur Kenntnis von *Cladothrix dichotoma* Cohn. *Zentralblatt für Bakteriologie, Parasitenkunde, Infektionskrankheiten und Hygiene, Abteilung II* **39**, 369–394.

LORENZ, W. D. & STEPHAN, R. W. 1967 Factors that affect the formation of coal mine drainage pollution in Appalachia. *The Incidence and Formation of Mine Drainage Pollution 18.* U.S. Bureau of Mines Report, Washington, D.C., attachment C to Appendix C.

LUTHY, R. G. 1964 New concept for iron bacteria control in water wells. *Water Well Journal* **24**, 29–30.

MACHMEIER, R. E. 1971 *Chlorination of Private Water Supplies.* Agriculture Extension Service, University of Minnesota, Bulletin M-156.

MACRAE, I. C. & EDWARDS, J. F. 1972 Absorption of colloidal iron by bacteria. *Applied Microbiology* **24**, 819–823.

MACRAE, I. C., EDWARDS, J. F. & DAVIS, N. 1973 Utilisation of iron gallate and other organic iron complexes by bacteria from water supplies. *Applied Microbiology* **25**, 991–995.

MANN, P. J. O. & QUASTEL, J. H. 1946 Manganese metabolism in soils. *Nature, London* **158**, 154–156.

MARGALITH, P., SILVER, M. & LUNDGREN, D. G. 1966 Sulfur oxidation by the iron bacterium *Ferrobacillus ferrooxidans. Journal of Bacteriology* **92**, 1706–1709.

MARDANYAN, S. S. & BALASHOVA, V. V. 1971 State of iron in the filaments of *Gallionella. Mikrobiologiya* **40**, 121–123.

MEYERS, G. E. 1958 Staining iron bacteria. *Staining Techniques* **33**, 283–285.

MOGG, J. L. 1972 Practical corrosion and incrustation guide lines for water wells. *Ground Water* **10**, 6–11.

MOLISCH, H. 1892 *Die Pflanze in ihren Beziehungen zum Eisen.* Jena.

MULDER, E. G. 1964 Iron bacteria, particularly those of the *Sphaerotilus – Leptothrix* group, and industrial problems. *Journal of Applied Bacteriology* **29**, 44–71.

MULDER, E. G. 1972 Le cycle biologique tellurique et aquatique du fer et du manganese. *Revues Ecologique et Biologique du Sol.* T **IX 3**, 321–348.

MULDER, E. G. 1974 Genus *Leptothrix* Kutzing 1843, 198. In *Bergey's Manual of Determinative Bacteriology*, 8th Edn. eds Buchanan, R. E. & Gibbons, N. E. Baltimore: Williams & Wilkins Co.

MULDER, E. G. & VAN VEEN, W. L. 1962 The *Sphaerotilus – Leptothrix* group. *Antonie van Leeuwenhoek* **28**, 236–237.

MULDER, E. G. & VAN VEEN, W. L. 1963 Investigations on the *Sphaerotilus – Leptothrix* group. *Antonie van Leeuwenhoek* **29**, 121–153.

NIEMELA, S. I. & TUOVINEN, O. H. 1972 Acidophilic *Thiobacilli* in the river Sirppujoki. *Journal of General Microbiology* **73**, 23–29.

NUNLEY, J. W. & KRIEG, N. R. 1968 Isolation of *Gallionella ferruginea* by use of formalin. *Canadian Journal of Microbiology* **14**, 385.

OKREND, H. & DONDERO, N. C. 1964 Requirements of *Sphaerotilus* for cyanocobalamin. *Journal of Bacteriology* **87**, 286–292.

OLIVER, D. 1975 *Report on a Television Survey of a Borehole at Shipmeadow Belonging to the East Anglian Water Company.* Geophysical Services Report, March 1975. Water Research Centre, Medmenham Laboratory, England.

OMEROD, K. 1974 Problemer med slam og dyn i distribusjonsnett for vann. *Norsk Institutt for Vannforsking. Temarapport* **2**, 57.

ORNEROD, J. G., GRYNNE, B. & ORNEROD, K. S. 1966 Chemical and physical factors involved in the heterotrophic growth response to organic pollution. *Verh. int. Vereintheor. Angew. Limnol* **16**, 906–910.

PETITPREZ, M., PETITPREZ, A., LECLERC, H. & VIVIER, E. 1969 Quelques aspects structuraux de *Sphaerotilus natans. Annales de L'Institut Pasteur de Lille* **20**, 103–114.

PHAUP, J. D. 1968 The biology of *Sphaerotilus* species. *Water Research* **2**, 597–614.

PRINGSHEIM, E. G. 1949 The filamentous bacteria *Sphaerotilus, Leptothrix, Cladothrix*, and their relation to iron and manganese. *Philosophical Transactions of the Royal Society of London B* **233**, 453–462.

PRINGSHEIM, E. G. 1952 Iron organisms. *Endeavour* **11**, 208–214.

RAO, C. S. G. 1970 Occurrence of iron bacteria in the tube well water supply of Howrah. *Environmental Health* **12**, 273–280.

RAZUMOV, A. S. 1961 Microbial indicators of organic pollution of waters by industrial effluents. *Mikrobiologiya* **30**, 764–768.

REMSEN, C. & LUNDGREN, D. G. 1966 Electron microscopy of the cell envelope of *Ferrobacillus ferrooxidans* prepared by freeze-etching and chemical fixation techniques. *Journal of Bacteriology* **92**, 1765–1771.

RODINA, A. G. 1965 *Methods in Aquatic Microbiology* (Revised translation 1972), eds Colwell, R. R. & Zambruski, M. S. Baltimore: University Park Press.

SCHAFER, D. C. 1974 The right chemicals are able to restore or increase well yield. *The Johnson Drillers' Journal* Jan.-Feb, 1–4, 13.

SCHEURING, L. & HOHNL, G. 1956 *Sphaerotilus natans* seine Ökologie und Physiologie. *Schriften des Vereins der Zellstoff – und Papier-chemiker und – Ingenieure* **26**, 1–151.

SCHOPF, E. G., EHLERS, E. G., STILES, D. V. & BIRLE, J. D. 1965 Fossil iron bacteria preserved in pyrite. *Proceedings of the American Philosophical Society* **109**, 288–308.

SCHWEISFURTH, R. 1973 Manganoxydierende Bakterien. I. Isolierung und Bestimmung einiger Stamme von Manganbakterien. *Zeitschrift für Allgemeine Mikrobiologie* **13**, 341–347.

SHAFIA, F. & WILKINSON, R. F., Jr. 1969 Growth of *Ferrobacillus ferrooxidans* on organic matter. *Journal of Bacteriology* **97**, 256–260.

SHAFIA, F., BRINSON, K. B., HEINZMAN, M. W. & BRADY, J. M. 1972 Transition of chemolithotroph *Ferrobacillus ferrooxidans* to obligate organotrophy and metabolic capabilities of glucose grown cells. *Journal of Bacteriology* **111**, 56–65.

SILVERMAN, M. P. & LUNDGREN, D. G. 1959 Studies on the chemoautotrophic iron bacterium *Ferrobacillus ferrooxidans.* I. An improved medium and a harvesting procedure for securing high cell yields. *Journal of Bacteriology* **77**, 642–647.

SINGER, P. C. & STUMM, W. 1970 Acidic mine drainage: the rate determining step. *Science, New York* **167**, 1121–1123.

STARKEY, R. L. 1945 Precipitation of ferric hydrate by iron bacteria. *Science* **102**, 523–533.

STEPHENSON, M. 1950 *Bacterial Metabolism.* New York: Longmann, Green & Co.

STOKES, J. L. 1954 Studies on the filamentous sheathed iron bacterium *Sphaerotilus natans. Journal of Bacteriology* **67**, 278–291.

STOKES, J. L. & JOHNSON, A. H. 1965 Growth factor requirements of two strains of *Sphaerotilus discophorus. Antonie van Leeuwenhoek* **31**, 175–180.

STOKES, J. L. & POWERS, M. T. 1965 Formation of rough and smooth strains of *Sphaerotilus discophorus. Antonie van Leeuwenhoek* **31**, 157–164.

STOKES, J. L. & POWERS, M. T. 1967 Stimulation of poly-β-hydroxybutyrate oxidation in *Sphaerotilus discophorus* by manganese and magnesium. *Archiv für Mikrobiologie* **59**, 295–301.

STOTT, G. A. 1973 The tenacious iron bacteria. *The Johnson Drillers' Journal* July-August, 4–7.

TABITA, R. & LUNDGREN, D. G. 1971 Utilisation of glucose and the effect of organic compounds on the chemolithotroph *Thiobacillus ferrooxidans*. *Journal of Bacteriology* **108**, 328–333.

TEMPLE, K. L. & COLMER, A. R. 1951 The autotrophic oxidation of iron by a new bacterium *Thiobacillus ferrooxidans*. *Journal of Bacteriology* **62**, 605–611.

TUOVINEN, O. H. & KELLY, D. P. 1973 Studies on the growth of *Thiobacillus ferrooxidans*. I. Use of membrane filters and ferrous iron agar to determine viable numbers, and comparison with $^{14}CO_2$-fixation and iron oxidation as measures of growth. *Archiv für Mikrobiologie* **88**, 285–298.

TUOVINEN, O. H., NIEMELA, S. I. & GYLLENBERG, H. G. 1971*a* Toxicity of membrane filters to *Thiobacillus ferrooxidans*. *Zeitschrift für Allgemneine Mikrobiologie* **11**, 627–631.

TUOVINEN, O. H., NIEMELA, S. I. & GYLLENBERG, H. G. 1971*b* Tolerance of *Thiobacillus ferrooxidans* to some metals. *Antonie van Leeuwenhoek* **37**, 489–496.

UNZ, R. F. & LUNDGREN, D. G. 1961 A comparative nutritional study of three chemoautotrophic bacteria: *Ferrobacillus ferrooxidans, Thiobacillus ferrooxidans* and *Thiobacillus thiooxidans*. *Soil Science* **92**, 302–313.

USAMI, S. & SUGITANI, T. 1971 The effect of organic substances on the growth and iron oxidation activity of iron oxidizing bacteria. *Journal of Fermentation Technology* **49**, 587–591.

VAN ITERSON, W. 1958 *Gallionella ferruginea* Ehrenberg in a different light *Verhandelingen der Koninklijke Nederlandse Akademii van Wetenschappen AFD Naturkunde*, 2nd set **52**, 1.

VAN VEEN, W. L. 1973 Biological oxidation of manganese in soils. *Antonie van Leeuwenhoek* **39**, 657–662.

VESTAL, J. R. & LUNDGREN, D. G. 1971 The sulfite oxidase of *Thiobacillus ferrooxidans (Ferrobacillus ferrooxidans)*. *Canadian Journal of Biochemistry* **49**, 1125–1130.

WALSH, F. & MITCHELL, R. 1972 Biological control of acid mine pollution. *Journal of the Water Pollution Control Federation* **44**, 763–768.

WANG, W. S., KORCZYNSKI, M. S. & LUNDGREN, D. G. 1970 Cell envelope of an iron-oxidising bacterium: studies of lipopolysaccharide and peptidoglycan. *Journal of Bacteriology* **104**, 556–565.

WEART, J. G. & MARGRAVE, G. E. 1957 Oxidation-reduction potential measurements applied to iron removal. *Journal of the American Water Works Association* **59**, 1223.

WILSON, I. S. 1960 The treatment of chemical wastes. In *Waste Treatment*, ed. Isaac, P. C. C. Oxford: Pergamon Press.

WINOGRADSKY, S. 1888 Uber Eisenbakterien. *Botanische Zeitung* **46**, 261–270.

WOLFE, R. S. 1960 Microbial concentration of iron and manganese in water with low concentrations of these elements. *Journal of the American Water Works Association* **62**, 1335–1337.

WOODS, G. A. 1973 Bacteria: friends or foes? *Chemical Engineering* **80**, 81–84.

WUHRMANN, K. & KOESTLER, S. 1950 Uber den Vitaminbedarf des Abwassenbakteriums, *Sphaerotilus natans* kutz. *Verhandlungen der Schweizerischen Naturforschenden Gesellschaft Wissenschaftlicher Teil* **130**, 177–178.

WUHRMANN, K., RUCHTI, J. & EISENBERGER, E. 1966 Quantitative experiments on self-purification with pure organic compounds. *Proceedings of the 3rd International Conference on Water Pollution Research* **1**, 229–251.

Microbiology of Polluted Estuaries with Special Reference to the Bristol Channel

AVRIL E. ANSON* AND G. C. WARE

Department of Bacteriology, University of Bristol, Bristol, Avon, England

CONTENTS

1. Introduction

BY VIRTUE of their geography, estuaries combine the advantages of living on the open coast with those of living some distance inland. Since historic times the sheltered surroundings and ease of access both to the sea and inland via rivers has attracted trading communities to the shores of estuaries. Furthermore, the natural beauty of estuaries, the extensive wildlife and the proximity of the sea and rivers with all their recreational advantages, have made many estuaries into holiday and tourist centres. The larger communities have now developed into major ports, thriving cities or important holiday resorts all of which produce large quantities of industrial and domestic wastes.

The fast currents visible in estuaries led to the assumption in the past that wastes disposed of into them were rapidly washed out to sea by the tide and dispersed. This view, still prevalent today, was used to justify the dumping of untreated wastes of all kinds directly into estuaries. The majority of domestic wastes discharged into estuaries today still recieve no treatment whatsoever (*Anon*. 1973).

* Present address: Department of Biological Sciences, Hatherly Laboratories, Prince of Wales Road, University of Exeter, Exeter, Devon, England.

Industry has been further encouraged to settle along estuarine coastlines by the more lenient attitude of the law towards pollution of tidal waters than of inland rivers since tidal waters cannot be used for drinking purposes. The law permits free discharge of all industrial and municipal wastes into these uncontrolled waters so there is no financial incentive for the polluters to treat their wastes prior to discharge.

These factors have resulted in the extensive industrial and domestic pollution which now exists in the estuaries of this country (*Anon.* 1973).

2. Estuaries as Closed Systems

Until recently very little was known of the microbiology of polluted estuaries since most studies were concentrated on the chemistry or hydrology of these areas, with only a few contributing some microbiological information. (*Anon.* 1964, 1972*a, b*). However, these and other studies (Ware *et al.* 1972; Hamilton 1973) have in fact demonstrated the slowness of the net movement of water towards the open sea. Even in the highly energetic system of the Bristol Channel where the geography of the region gives rise to some exceptionally fast tidal currents, net water movement towards the open sea is very slow. Both salinity contour studies (Hamilton 1973) and computer simulations of water movements (Fig. 1; *Anon.* 1973/4) have demonstrated the presence of an anticlockwise gyre of surface water in the outer section of the Bristol Channel which results in a long retention time of polluted water within the estuary. Hamilton (1973) estimated a replacement time for the region of between 190 and 260 days. The situation in the Bristol Channel is further aggravated by the adsorption of pollutants on to sediment particles and the subsequent retention of the pollutants in the estuary for unknown periods (*Anon.* 1972*b*). It is perhaps wiser, therefore, to regard estuaries not as extensions of the open sea but rather as closed systems in which self-purification is of greater importance than was previously suspected.

3. Processes of Self-purification

The self-purification of an estuary from sewage wastes is brought about by two simultaneous processes: (a) biological oxidation by organisms, some of which, such as bacteria, may be indigenous to the sewage and which use up the nutrients in it; (b) the death of organisms, mainly bacteria, present in sewage. Death of sewage bacteria in sea water has been attributed to predation and other biological factors, competition for food, the bactericidal action of sunlight, the normal physicochemical properties of sea water and toxic agents (Pramer *et al.* 1963; Anson 1975).

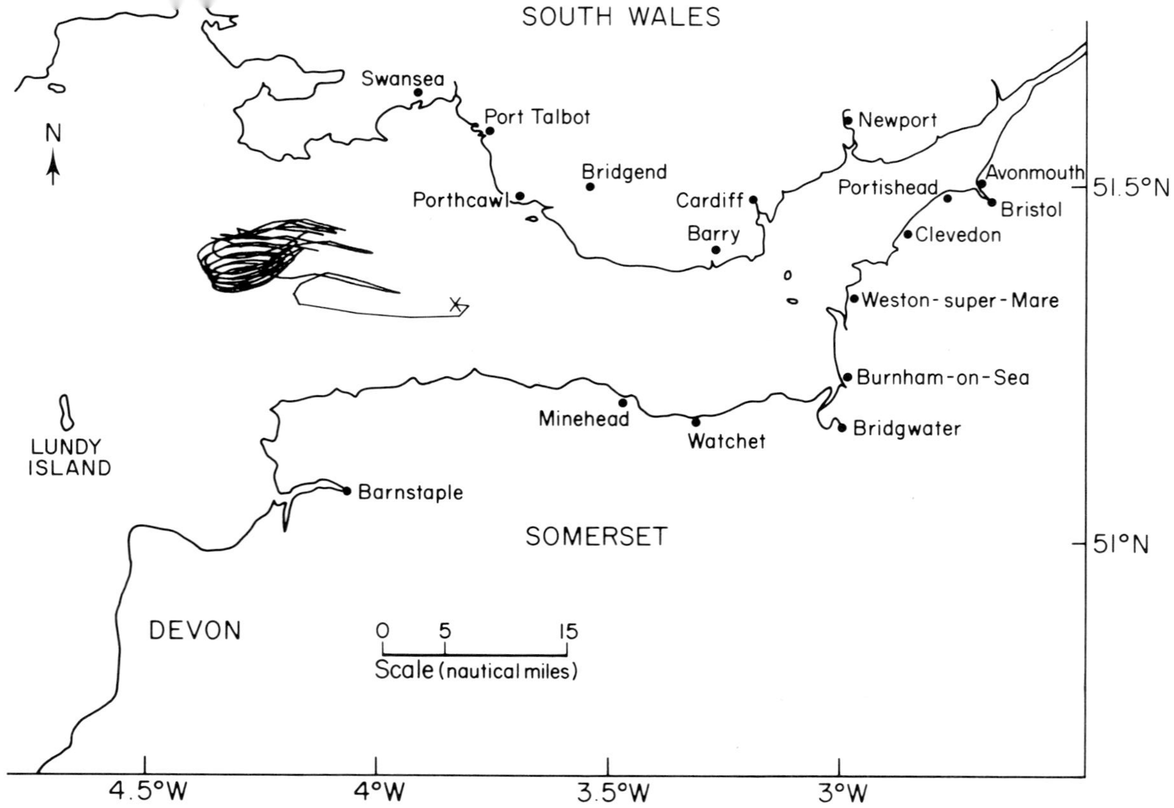

Fig. 1. Computer simulation of surface water movements in the Bristol Channel. The trace starts at the point marked X at high tide, and continues hour by hour for 15 tide cycles. The water moves in a clockwise ellipse each tide cycle, but the overall movement of water during the seven days of the trace demonstrators part of an anti-clockwise gyre.

4. Survey of Sewage Pollution in the Bristol Channel

(a) *Methods*

In order to study the distribution and self-purification of microbial pollutants in an estuary, we have carried out a bacteriological survey of the Bristol Channel. During two successive winters 18 sampling trips were made by boat and a total of 270 samples collected at any state of the tide and wherever practicable. Each sample was analysed in the laboratory for coliform bacteria which indicate the presence of recent sewage pollution, and for aerobic heterotrophic bacteria as a guide to the level of readily available organic matter in the water (Kriss *et al.* 1967). As the population of aerobic heterotrophs was found to be stable in numbers for long periods in estuarine water (Anson 1975), whereas coliform bacteria which are not adapted to the water environment died very rapidly (Anson 1975), the ratio of coliform to non-coliform bacteria was considered by us to be a guide to the relative age of the sewage pollution and was therefore calculated for each sample.

(b) *Production of contour maps*

The results from this survey are displayed as contour maps showing the distribution of the different types of bacteria (Figs 2, 3 and 4). The maps have been corrected by a computer programme, using standard navigation techniques, to one state of the tide, namely mean sea level on a falling tide. This occurs one-quarter of a tide cycle (*ca.* 3 h) after high tide, when the water level is midway between that of high and low waters. This enables an 'instantaneous picture' of the estuary to be produced even though the samples were collected at many states of the tide. Mean sea level was chosen as the standard time because it is the only state of the tide when the water level is independent of the lunar (tidal) cycle (Rantzen 1961).

Allowance has also been made during the preparation of these maps for the mortality of coliforms in the interim period between collecting the samples and examining them in the laboratory. In this respect Figs 2–4 are superior to the maps published previously (Anson & Ware 1974) on which no such correction was made. Furthermore, the earlier contours were drawn by hand whereas those reproduced here were drawn by computer and are therefore, presumably, more accurate.

Rather than drawing contours from the raw data, the bacterial counts were given a grade on a convenient scale and contours of the grades were produced by computer. Exponential scales were used to grade the coliform and aerobic counts, so that the difference between one grade and the next took on an increasing significance towards the upper end of the scales and this should be borne in mind when examining the maps (Figs 2 and 3).

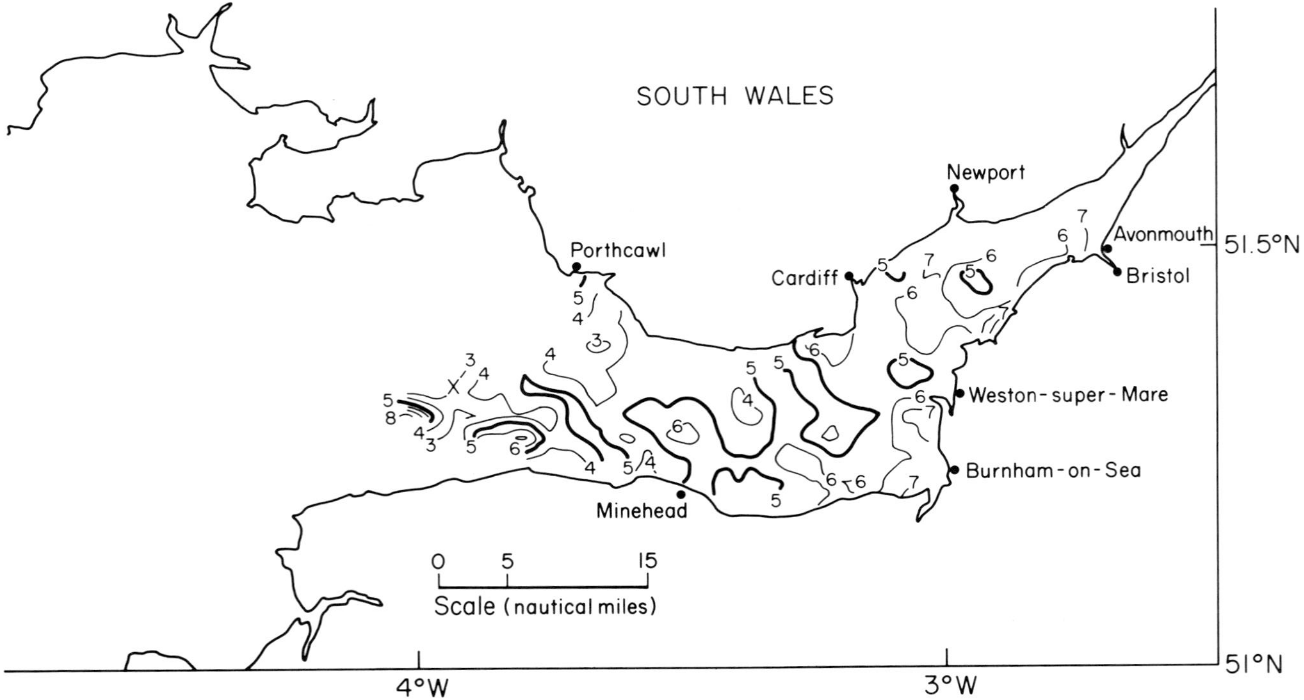

Fig. 2. Contour map showing the distribution of aerobic heterotrophic bacteria in the surface waters of the Bristol Channel. Distribution has been corrected for tidal drift to the standard time (see text). Bacterial counts have been graded on an exponential scale and the graded results contoured by computer. Key (grade, aerobic bacteria.ml^{-1}): 0, 0; 1, 1–4; 2, 5–18; 3, 19–65; 4, 66–222; 5, 223–742; 6, 743–2474; 7, 2475–8235; 8, 8236–27407; 9, 27408–290000. X marks the dumping ground for sludge from Bristol's sewage treatment works.

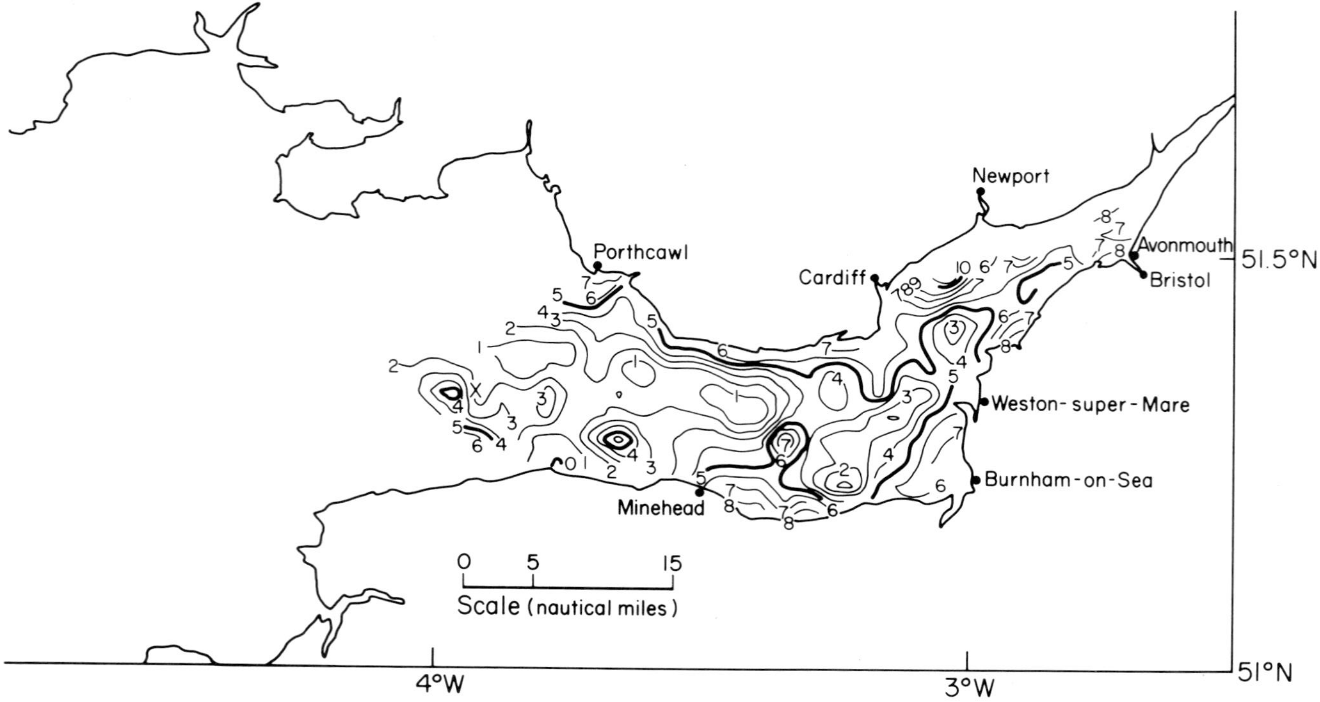

Fig. 3. Contour map showing the distribution of coliform bacteria in the surface waters of the Bristol Channel. Distribution has been corrected for tidal drift to the standard time (see text). Bacterial counts have been graded on an exponential scale and the graded results contoured by computer. Key (grade, coliforms.100 ml $^{-1}$): 0, 0; 1, 1–3; 2, 4–13; 3, 14–43; 4, 44–132; 5, 133–396; 6, 397–1181; 7, 1182–3512; 8, 3513–10435; 9, 10436–30997; 10, 30998–92068. X marks the dumping ground for sludge from Bristol's sewage treatment works.

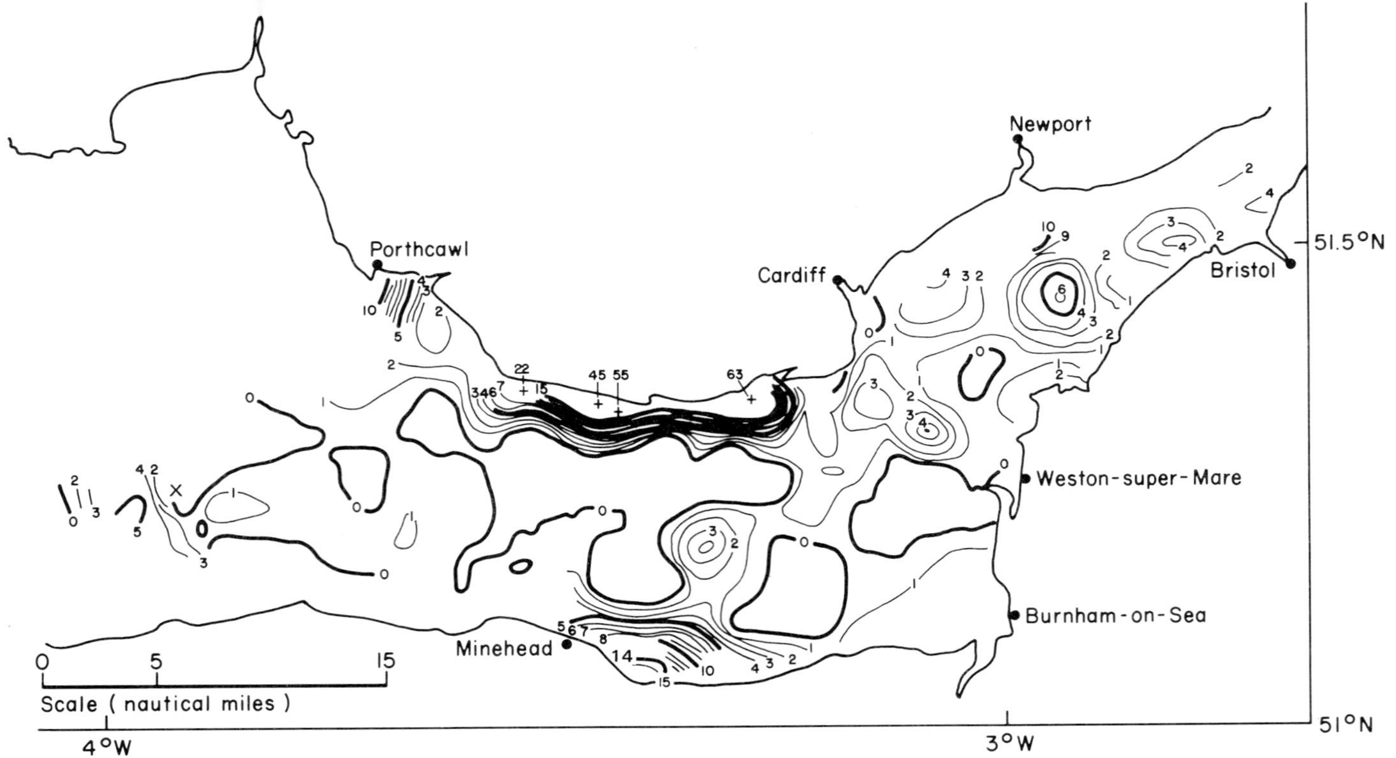

Fig. 4. Contour map showing the distribution of the ratios of coliform to non-coliform bacteria in the surface waters of the Bristol Channel. The distribution has been corrected for tidal drift to the standard time (see text). Ratios have been graded on a linear scale and the graded results contoured by computer. Key (grade, ratio of coliform to non-coliform bacteria expressed as a percentage): 0, 0–0.24%; 1, 0.25–0.74%; 2, 0.75–1.24%; 3, 1.25–1.74%; 4, 1.75–2.24%; 5, 2.25–2.74%; 6, 2.75–3.24%; 7, 3.25–3.74%; 8, 3.75–4.24%; 9, 4.25–4.74%; 10, 4.75–5.24%; etc. X marks the dumping ground for sewage sludge from Bristol's sewage treatment works.

A computer programme was chosen which could handle irregularly spaced data since our sample sites were not evenly spaced over the survey area. In order to convert the data into a form which could be contoured, a regular grid was superimposed on the study area and values obtained by computer for each grid intersection by extrapolation from the graded results. All the observations were used in the calculation of every grid value, each observation being given a weight dependant on its proximity to the grid intersection. The grid values were then used as data in a computer programme which plotted contour lines.

Contours were drawn at intervals of a single grade with every fifth grade starting from zero being presented in heavy ink. Since few points on the maps had grades higher than 15, these were not contoured but were marked on the map individually with a cross.

(c) *Results and discussion*

Maps of the distribution of bacteria in the estuary (Figs 2 and 3) showed that the levels of presumptive faecal pollution and organic matter (as measured by counts of aerobic heterotrophs) were disturbingly high. Counts of coliform bacteria (after allowing for mortality during transport) were as high as $4.7 \times 10^4.100\ ml^{-1}$ and counts of viable heterotrophs as high as $2.9 \times 10^5.ml^{-1}$.

The standards for recreational waters laid down by the European Economic Community (*Anon.* 1976) which have recently come in to force are 500 coliforms. 100 ml^{-1} (preferred level) and 10,000 coliforms. 100 ml^{-1} (mandatory level). In Fig. 3 the preferred level is equivalent to grade 6 and the mandatory level to grade 8. Much of the area around Avonmouth and along the English and Welsh coastlines has an unacceptably high level of sewage pollution by these standards.

Sewage pollution declined fairly steadily towards the open sea in the west with a few localized areas of relatively high counts especially near the area where digested sewage sludge from Bristol's sewage treatment works is dumped twice daily (Fig. 3). This sludge dumping was not found to be an important source of pollution, at least of the surface waters, probably because digestible sludge is low in readily available organic carbon and so is retractile to bacterial action. Its direct effect on the bacterial ecology of the region seemed to be restricted to a fairly localized area.

Sewage pollution was found to be more severe along the coastline than in midstream (Fig. 3) and in fact high coliform counts were obtained along the Welsh coastline for as far west as was surveyed. These were presumably caused by sewage discharges from Newport, Cardiff, Barry, Bridgend and Porthcawl. Along the English coastline, severe pollution was associated with sewage discharges from Avonmouth, Clevedon, Weston-super-Mare, Bridgwater and Minehead (Fig. 3).

The coliform map (Fig. 3) and the ratio map (Fig. 4) showed a similar pattern of contours particularly in the seaward section of the estuary. The close correlation in the western part of the estuary between recent pollution, as indicated by the ratio map, and high levels of sewage pollution, as shown by the coliform map, suggested that the estuary was undergoing rapid self-purification from coliform bacteria. In the eastern regions of the estuary, particularly around Avonmouth, high levels of sewage pollution were found which, judged by our rationale, were comparatively old, suggesting that self-purification in the upper stretches of the estuary was slower than on the seaward side. One possible explanation was that the region from Avonmouth westward to Burnham-on-Sea provided an estuarine environment as opposed to a marine environment, and, moreover, one that was rich in readily available nutrients. The brackish water of an estuary could well allow bacteria to survive better than in the open sea where they would have to tolerate an unsuitable osmotic pressure and compete with better-adapted indigenous microbes for nutrients to survive. Certainly, the osmotic pressures associated with brackish waters have been found by Anson (1975) to be very much more favourable to the survival of sewage bacteria than those associated with either fresh water or marine environments. A second explanation for the reduced rate of self-purification in the eastern section of the estuary could be that bacteria are protected from death in the upper stretches of the estuary by the presence of heavy loads of sediment in suspension in this region. These particles may protect the bacteria first by providing a substrate for the bacteria to adsorb to and grow upon and second by cutting down the amount of light which penetrates the water, thus reducing the lethal effect of sunlight on bacteria.

The distribution of bacteria in the Bristol Channel is further complicated by the large amounts of heavy metals released into the estuary, mainly in industrial discharges, which may have a significant effect on the self-purifying ability of the estuary (A. Sylvester, pers. comm.).

5. Pollution in Other Estuaries

The results reported in this study refer to the Bristol Channel, but they could apply equally to other polluted estuaries. Certainly, long retention times have been described for other estuaries including the Thames (*Anon.* 1964) which shows a net movement of water seawards of about one mile a day, and the Mersey Estuary which has a retention time of 1–3 weeks (Porter 1973). Under these circumstances, quite small discharges of wastes can build up to high concentrations of pollutants within the estuary.

6. The Economics of Pollution Abatement

The solution to estuarine pollution is simple – industrial and municipal wastes must receive satisfactory treatment before discharge. However, this solution, although simple, requires both time and money in order to build and maintain the necessary treatment plants. Porter (1973) estimated the cost of cleaning up the Mersey Estuary at one-hundred million pounds. It becomes necessary to ask what benefits would arise from such an improvement scheme for estuaries. The main benefit would be the restoration of commercial estuarine fisheries many of which have been damaged and some completely destroyed by pollution from industrial and municipal wastes. For example, Teeside water is lethal to fish because it contains toxic substances released in the industrial discharges (Porter 1973). Similarly, a severe decline has been noted in fish catches in the Mersey where pollution of the upper estuary occasionally becomes so severe that it turns anaerobic (Porter 1973). Not surprisingly, migratory fish cannot pass this barrier of polluted water, so they die. Even the less severely polluted estuaries of the Humber and Bristol Channel have shown declines in flat fish catches and the disappearance of migratory fish such as salmon (*Anon.* 1972*b*; Porter 1973). Many estuaries are nursery grounds of fish and loss of these grounds through pollution has wide-reaching effects on sea fishing.

It must be acknowledged, however, that the cost of cleaning up the estuaries would not be recovered by the restoration of commercial fisheries, and justification for such a programme must be mainly on the basis of aesthetic rather than economic values.

7. Conclusion

Where sea, land and river meet there exists, potentially, a varied and pleasant natural environment, not only for wildlife but also for man. Cities, ports, fishing villages and holiday resorts can make use of, while adding to, the assets of the estuary. But if pollution is not kept under control most of the assets of the estuary are lost. In the words of Porter (1973) "A severely polluted estuary serves only as a shipping lane and a sewer."

8. Acknowledgements

We gratefully acknowledge the help of Dr Thorpe and Linda Rogers at the Institute of Experimental Cartography, London, who produced the computer-drawn contour maps from our data. Permission for the use of this contouring programme was kindly granted by the Natural Environment Research Council who also financed one of us (A.E.A.). The extensive help of Yolande Arianayagam and Pauline Gosden in the survey is also gratefully acknowledged.

9. References

ANON. 1964 *Effects of Polluting Discharges on the Thames Estuary.* Report, Department of Scientific and Industrial Research. London: H.M.S.O.

ANON. 1972*a* *Out of Sight, Out of Mind. Report of a Working Party on the Disposal of Sludge in Liverpool Bay.* Report, Department of the Environment. London: H.M.S.O.

ANON. 1972*b* *The Severn Estuary and Bristol Channel. An Assessment of Present Knowledge.* Natural Environment Research Council Publications, Series C, No. 9.

ANON. 1973 *Report of a Survey of the Discharges of Foul Sewage to the Coastal Waters of England and Wales.* Report, Department of the Environment and Welsh Office. London: H.M.S.O.

ANON. 1973/4 *Annual Report of the Institute for Marine Environmental Research.* Natural Environment Research Council Publication.

ANON. 1976 Council directive of 8th December 1975 concerning the quality of bathing water. *Official Journal of the European Communities* **19**, 1–7 (No. **L**31, Feb 5th).

ANSON, A. E. 1975 *Distribution and Dynamic Balance of Certain Bacterial Populations in the Bristol Channel.* Ph.D. thesis, Department of Bacteriology, University of Bristol.

ANSON, A. E. & WARE, G. C. 1974 Survey of distribution of bacterial pollution in the Bristol Channel. *Journal of Applied Bacteriology* **37**, 657–661.

HAMILTON, P. 1973 The circulation of the Bristol Channel. *Geophysical Journal. Royal Astronomical Society* **32**, 409–422.

KRISS, A. E., MISHUSTINA, I. E., MITSKEVICH, N. & ZEMTSOVA, E. V. 1967 *Microbial Population of Oceans and Seas.* London: Edward Arnold.

PORTER, E. 1973 *Pollution in Four Industrialised Estuaries.* London: H.M.S.O.

PRAMER, D., CARLUCCI, A. F. & SCARPINO, P. V. 1963 The bactericidal action of sea water. In *Marine Microbiology*, ed. Oppenheimer, C. H. Springfield, Illinois: C. C. Thomas.

RANTZEN, M. J. 1961 *Little Ship Navigation*, 1st edn. London: Herbert Jenkins.

WARE, G. C., ANSON, A. E. & ARIANAYAGAM, Y. F. 1972 Bacterial pollution of the Bristol Channel. *Marine Pollution Bulletin* **3**, 88–90.

Coliphages in Sewage and the Marine Environment

P. A. AYRES

Ministry of Agriculture, Fisheries and Food,
Fisheries Laboratory, Burnham-on-Crouch,
Essex, England

CONTENTS

1. Introduction

ALMOST SIXTY YEARS ago d'Herelle introduced the term bacteriophage to describe the filterable agents which Twort (1915) had reported to have the ability to lyse bacterial cells. Initial research was directed towards the possibility of using these agents, or 'corpuscles' as they were then called, as bactericidal agents to combat infections in man and animals. The results were negative but led to the development of the idea that bacteriophage studies could be used as a model to study host–virus interactions. The results of subsequent work led eventually to such important discoveries in virology and microbial genetics that the bacteriophages have received more attention than any other group of viruses.

Bacterial and animal viruses have many similar physical, chemical and biological properties (Adams 1959), but the bacteriophages may be more easily and rapidly cultured, using relatively simple and economic techniques. Coliphages particularly have been widely favoured as virus models and include DNA-containing tailed phages of the T type and small spherical RNA phages such as MS2. The RNA phages have many similarities with enteroviruses such as polio and are frequently used as models, e.g. in disinfection studies, where the phage f2 is generally more resistant to chlorination than poliovirus, but is a particle of the same size and shape (Cramer *et al.* 1976).

The study of coliphages as a factor influencing the survival of coliform bacteria, including *E. coli*, has received some attention (Ware & Mellon 1956; Carlucci & Pramer 1960*b*) but much of the published research is concerned with its use as a model or potential indicator of enterovirus. This attention has been stimulated and perhaps prompted by the growing concern with viruses in sewage and the role of water and shellfish in the transmission of viral disease to man. Theoretically any human virus that is excreted in faeces may be transmitted through water contaminated by faecal matter. About 100 identifiable enteric viruses have been demonstrated in human faeces and more than 70 have been reported as occurring in water and sewage. Clarke *et al.* (1964) found that 39% of chlorinated, conventionally treated effluents contained viruses. It is ironic perhaps that virus transmission via water is best demonstrated by infectious hepatitis, an agent yet to be isolated in culture and characterized in the laboratory. Incidents of infectious hepatitis associated with contaminated water have been documented and there is the possibility of a yet undemonstrated viral agent(s) in epidemics of gastroenteritis and diarrhoeal disease (McDermott 1975).

In the marine environment the role of shellfish in the transmission of viral disease to man is again virtually limited to infectious hepatitis. The first reports appeared from Sweden (Roos 1956) and more recently from the USA (Dougherty & Altman 1962; Mason & McLean 1962; Mackowiak *et al.* 1976). The presence of enterovirus in shellfish has been demonstrated by Metcalf & Stiles (1968) who also showed that virus may be retained for long periods, particularly during winter. More recent evidence on the retention of viruses in shellfish (Mackowiak *et al.* 1976) emphasizes the inadequacy of existing bacteriological indices of faecal contamination for assessing risks of viral origin.

However, the risk to human health, whether existing or potential, stems from the basic fact that enteroviruses are present in sewage and may be transmitted to man via polluted water and shellfish. An understanding of the behaviour and fate of viruses in these materials is therefore fundamental to any assessment of the public health risks involved. The techniques for isolation, culture and identification of enterovirus from sewage and the marine environment have advanced dramatically in recent years but still remain expensive, time consuming and beyond the scope of many control agencies and research facilities. Coliphage offer some practical solutions to such difficulties and this paper attempts to review some of these applications and results of the use of coliphage both as a virus model and indicator in studies on sewage and the marine environment.

2. Coliphages in Sewage

Sewage is frequently used as a source of coliphage for research purposes, but although the occurrence of phage is recognized and utilized in this way, rarely have studies been directed towards determining the numbers and types of phage which may occur. Attention has been focussed on seeding sewage with coliphage as a model to study the efficiency of various treatment processes, including disinfection, on the reduction or removal of enteroviruses.

Limited work has been orientated towards an appraisal of coliphage as a factor involved in the observed reduction of coliforms during sewage treatment. Ware & Mellon (1956) showed that the ratio of phage-resistant to phage-sensitive organisms in sewage decreased from 1 to 1.3 in the crude influent to 1 to 4.5 in the effluent. Addition of a host *Escherichia coli* resulted in no significant change in the numbers of bacteriophage and the authors concluded that no evidence had been found to support the hypothesis that phage was responsible for the decline in the numbers of coliforms. Similar studies by Beckwith & Rose (1930) indicated that, although phages isolated from sewage were quite capable of lysing coliforms isolated from the same samples, it was unlikely that this occurred under natural conditions when *E. coli* and coliforms are not in a growth phase susceptible to phage attack. Calabro *et al.* (1972) later concluded that, although phages may play a role in the lysing of specific pathogens, they were unlikely to serve as biological control agents for coliforms.

My own interest has been to explore the potential of coliphage as an indicator of enterovirus in much the same way that *E. coli* is used to assess risks from bacterial pathogens in sewage or sewage polluted waters, and, where possible, to make a comparison between *E. coli* and coliphage. Any such assessment must logically commence with observations on sewage. Figures 1 and 2 show the results of estimates of coliphage and *E. coli* in samples of raw influent and final effluent sewage taken at weekly intervals for one year from an activated sludge type treatment plant. Coliphage were enumerated using the agar overlay technique of Adams (1959) with *E. coli* B (NCIB 9484) as host strain so that the phages counted were primarily of the T type and are not necessarily representative of the total coliphage. *Escherichia coli* were enumerated on MacConkey agar plates incubated for 24 h at 44°C.

The plotted data for coliphage (Fig. 1) show essentially random fluctuations within the range $3.5 \times 10^1 – 3 \times 10^4$ plaque forming units (PFU).ml^{-1} of raw influent and $>1.0–5 \times 10^3$ PFU.ml^{-1} of final effluent which could not be correlated with any seasonal factor such as flow. Similar results were also reported by Gilcreas & Kelly (1955) who demonstrated variation in the isolation of coxsackievirus, unrelated to numbers of coliphage. An overall reduction was observed between raw influent and final effluent samples but variation was considerable and random. Data for *E. coli* estimates on the same samples (Fig. 2) is included for comparison and, although counts are in the order of two orders of magnitude higher than estimates of coliphage, the random fluctuation in numbers and decrease from raw influent to final effluent exhibits the same overall trend. To examine in greater detail the sewage treatment process, a programme of weekly sampling at various stages of treatment was implemented. Estimates of coliphage and *E. coli* were made on each of the samples as before and the mean values of 52 weekly samples obtained. Estimates of *E. coli* and coliphage in raw influent were taken as 100% and used as a basis to compute the

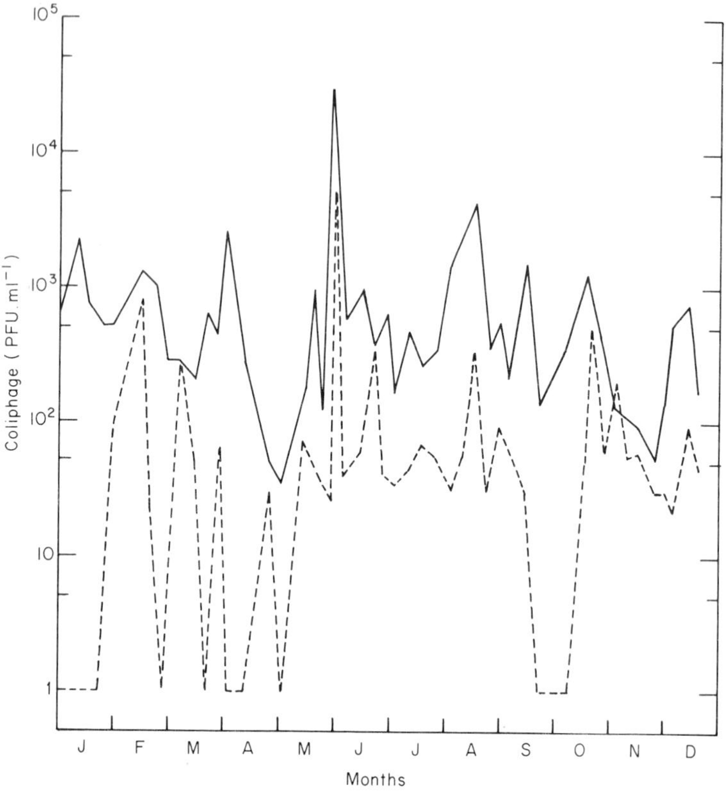

Fig. 1. Coliphage in raw influent and treated effluent sewage. ———, raw influent; -----, final effluent.

percentage increase or reduction in numbers at subsequent stages of treatment.

The results of this work, summarized in Fig. 3, indicate an initial increase in numbers during early stages of treatment. In the sewage treatment works sampled this is the only stage of the process where sewage is pumped, the remainder of the flow being by gravity beyond the grit traps and it is likely that mechanical disruption of particulate matter releases some of the trapped bacterial cells and phage into suspension. There is a marked reduction in both *E. coli* and coliphage as the result of activated sludge treatment, in good agreement with the findings of other workers. For comparative purposes the results of Bloom *et al.* (1959) have been plotted to illustrate that similar reductions have been observed with enteroviruses. Berg (1973) has reported that the activated sludge process is generally more effective at virus removal than other conventional biological treatments and various workers (Kelly & Sanderson 1959;

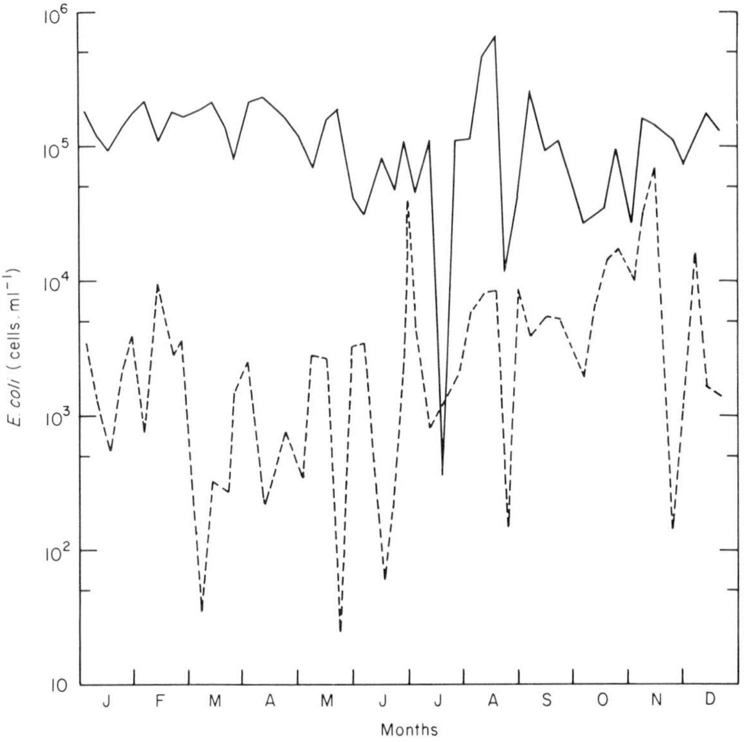

Fig. 2. *Escherichia coli* in raw influent and treated effluent sewage. ———, raw influent; -----, final effluent.

England *et al.* 1967; Safferman & Morris 1976) have shown that the virus reduction capacity of such a process can exceed 90%. Sproul *et al.* (1969) considered that virus inactivation in sewage treatment may be due to a number of mechanisms including: (a) loss of protective protein coat by enzymic action; (b) denaturation of the surface protein coat; (c) loss of structural integrity due to pH effects; (d) alteration of the nucleic acid core or surface protein by oxidants and inorganic or organic toxicants; (e) adsorption to various surfaces.

They considered that adsorption was the major factor involved in virus removal and certainly the available evidence seems to support this as the predominant mechanism (Kollins 1966; De Michelle 1974; Malina, Jr *et al.* 1974). The activated sludge process produces an adsorption area in terms of sludge volume in excess of that produced by conventional systems and this may well account for the greater efficiency in virus removal. Lund (1969) has queried the use of bacteriophage to model virus in such systems because the T phages, equipped with specific functional structures for the process of adsorption and

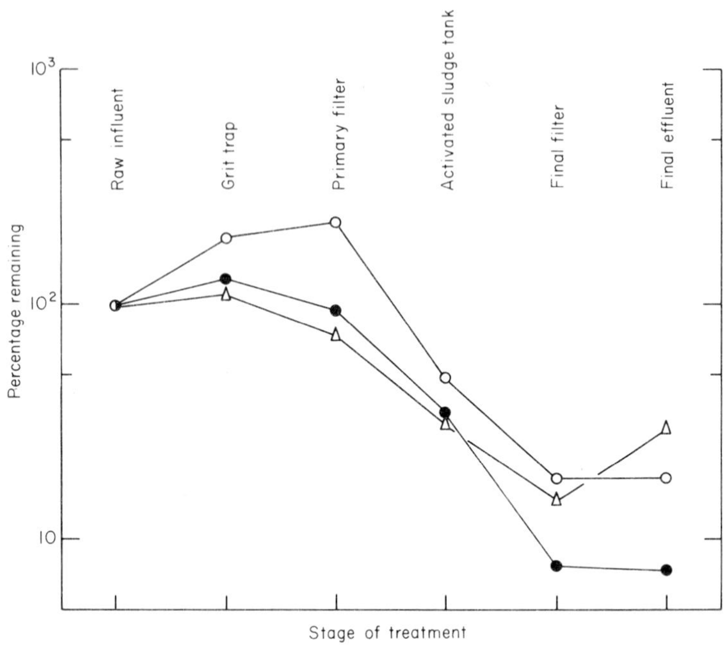

Fig. 3. Effect of sewage treatment on numbers of coliphage ($PFU.ml^{-1}$) and *E. coli* ($cells.ml^{-1}$). ○, *E. coli*),●, coliphage;△, enterovirus (after Bloom *et al.* 1959).

penetration into the host cell, may well attach to particulate matter in the same way. If this is so, it may be counterbalanced to some extent by the greater susceptibility of such phages to mechanical damage. Sorber *et al.* (1972), comparing the effects of reverse osmosis and ultrafiltration processes for sewage treatment, used both coliphage T_2 and Poliovirus I; they concluded that the different results obtained may have been due to tail damage of the T_2 phage particles. Poliovirus being a spherical particle with multiple receptor sites was potentially less likely to suffer random damage.

Table 1 gives the calculated ratios of *E. coli*/coliphage corresponding to the stages of treatment featured in Fig. 3 and some indication that the coliphage are less affected by treatment. Studies by the author on the fate of phage and *E. coli* in stored samples of primary effluent show that, while phage may persist for long periods (up to 20% remaining after 1 year), *E. coli* die rapidly, and this may account for the changing ratios in Table 1. From published figures by Clarke *et al.* (1964) and others it is apparent that coliphage in sewage may exceed enterovirus by a factor of 100 to 1000 times depending on the treatment process used. The coliform to enterovirus ratio, estimated by Clarke *et al.* (1964), was in

Table 1
Ratio of Escherichia coli *and coliphage numbers at various stages of sewage treatment*

Sampling site	Observed ratio* *E. coli* : Coliphage
Raw influent	512 : 1
Grit trap	570 : 1
Primary filter	392 : 1
Activated sludge tank	508 : 1
Final filter	229 : 1
Final effluent	152 : 1

* Mean of 52 samples at each site taken weekly for one year.

the order of 92,000:1 for sewage, decreasing to 50,000:1 in polluted surface water.

One of the few published studies on the types of coliphage occurring in sewage (Dhillon *et al.* 1970) concerns the distribution of coliphage in sewage taken at 11 sites representing urban and rural sewage in Hong Kong. Urban samples were slightly richer in phage than rural samples. Of 72 purified isolates studied, 50% were able to grow on three host *E. coli* strains ($K12F^+$, $K12F^-$ and B) and were represented at 10 of the 11 sites sampled. Only one strain appeared to be of a temperate nature and the authors concluded that virulent phages are far more widespread in nature as free virions than are the virions of temperate phages. It was suggested that temperate phages are present predominantly as prophage in host cells.

An interesting result of serological investigations reported in the same paper was that cylindrical, single-stranded DNA phages (e.g. AE2 and M13) produced turbid plaques in contrast to the clear plaques produced by the spherical, single-stranded RNA phages (e.g. MS2). Extensive distribution of male-specific RNA and DNA coliphages suggested that host cells must be quite widespread in nature but of 700 *E. coli* isolates from two sites none was lysed by phage MS2.

It is not within the scope of this paper to deal with applications of coliphage as a virus model in specific terms, as for example in studies of sewage treatment, but undoubtedly they have been usefully applied in disinfection studies (Cramer *et al.* 1976). My own studies have utilized the RNA phage MS2 because of its similarity to poliovirus and its spherical form which make it less prone to mechanical damage than the tailed phages, thus having good potential as a virus model.

3. Coliphages in Water

Coliphages have been shown to be present in sewage in numbers greater than corresponding levels of enteroviruses. Coliphages appear to be removed during sewage treatment at a rate and in a manner comparable to many enteroviruses but may be more resistant to disinfection procedures such as chlorination. Public health concern is, however, focussed on the fate of enteroviruses once released into the aquatic environment and it follows therefore that our studies should proceed with the object of understanding the fate of coliphages in water if we are to make a balanced assessment of their potential as virus models or indices of viral contamination.

As an initial experiment to study survival and to evaluate the duration of an experiment necessary to follow survival down to low levels, various types of water were seeded with MS2 phage. The infective dose of some viruses may be as low as 1 PFU and it was important to follow the fate of phage for as long as detectable numbers remained. Results of experiments with MS2 in distilled, deionized, tap water and sea water are shown in Fig. 4. The very rapid decline in number of PFU in distilled water was in marked contrast to their survival in deionized water although the pH values (6.8 and 6.4, respectively) were similar. Sea water and tap water (pH values of 8.2 and 7.9) gave similar results though low levels of phage persisted longer in tap water. In all treatments there is some indication of an initial agreement in survival rates linked to pH; i.e. although the overall survival in deionized water was greater than in other treatments, initial die-off was more rapid than with either sea or tap water at a higher pH suggesting that major pH effects are rapid. My work confirms that the optimum pH for MS2 phage is in the range pH 6 to 7 (for *E. coli* slightly higher at pH 7 to 8). At lower pH values die-off or inactivation was often so rapid that phage and *E. coli* could not be recovered even a few minutes after inoculation into test solutions. The overall results agree with published work but differ from those of Akin *et al.* (1976) working with Polio I virus, who reported greater survival in distilled water.

Coliphage released to the marine environment from a sewage discharge will often be subject to a range of salinities from fresh water to full sea water having a salt content of 3.5% (w/v). Carlucci & Pramer (1960*a*) reported that *E. coli* died more rapidly in the absence of salts than in sea water of 2.5% salinity but that above this value survival decreased. No significant difference was observed between survival in different sea water concentrations and solutions of NaCl of equivalent salinity. The results of our experiments on the effects of a range of salinity from 0 to 3.0% are shown in Fig. 5. Rapid mortality of *E. coli* was observed in fresh water and at the upper salinity range of 2.5 to 3.0% greater survival was observed between 0.5 and 1.5%.

Repeating the experiment with MS2 phage (Fig. 6) gave essentially similar

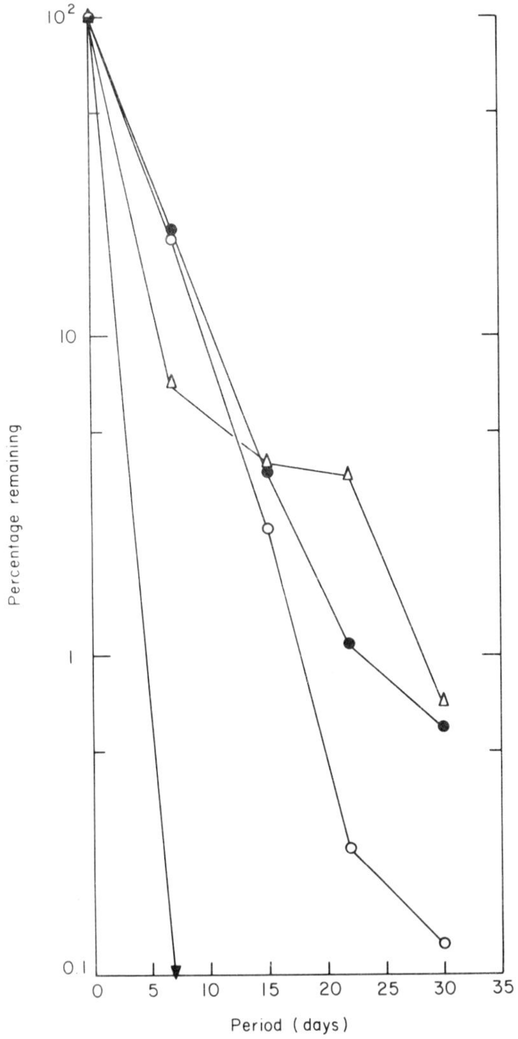

Fig. 4. Survival of coliphage MS2 in various waters. ○, Sea water, pH 8.2; ●, tap water, pH 7.9; ▼, distilled water, pH 6.8; △, deionized water, pH 6.4.

results with minimum survival at 0 and 3.0% but prolonged low level survival at 0.5 and 1.0%. Addition of host cells, *E. coli* NCIB 9481, reduced phage survival at all salinity levels possibly due to adsorption of phage onto distinguishing between the inactivation of phage and the binding of phage to

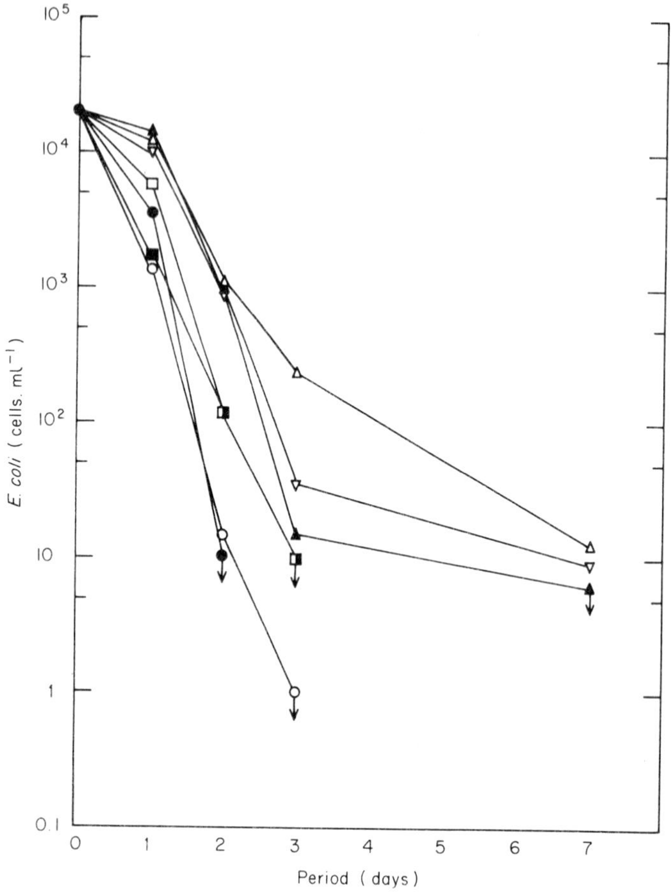

Fig. 5. The effect of salinity (parts per thousand) on the survival of *E. coli* in sea water. ●, 0; △, 5; ▲, 10; ▽, 15; □, 20; ■, 25; ○, 30.

particles (e.g. bacteria) which effectively removes them from suspension. Akin *et al.* (1976) looking at the loss of infectivity of poliovirus in sea water reported that maximum loss in artificial sea water occurred at 1.0% rather than 0.1 or 2.0%. Differences in the antiviral activity of natural and artificial sea water or NaCl solutions have been noted; Magnusson *et al.* (1966) suggest, therefore, that inactivation was of a biochemical nature but not directly caused by salinity of the water. A number of reports dealing with inactivation of both virus and bacteria in sea water have focussed on comparisons between sea water and NaCl solutions of equal salinity.

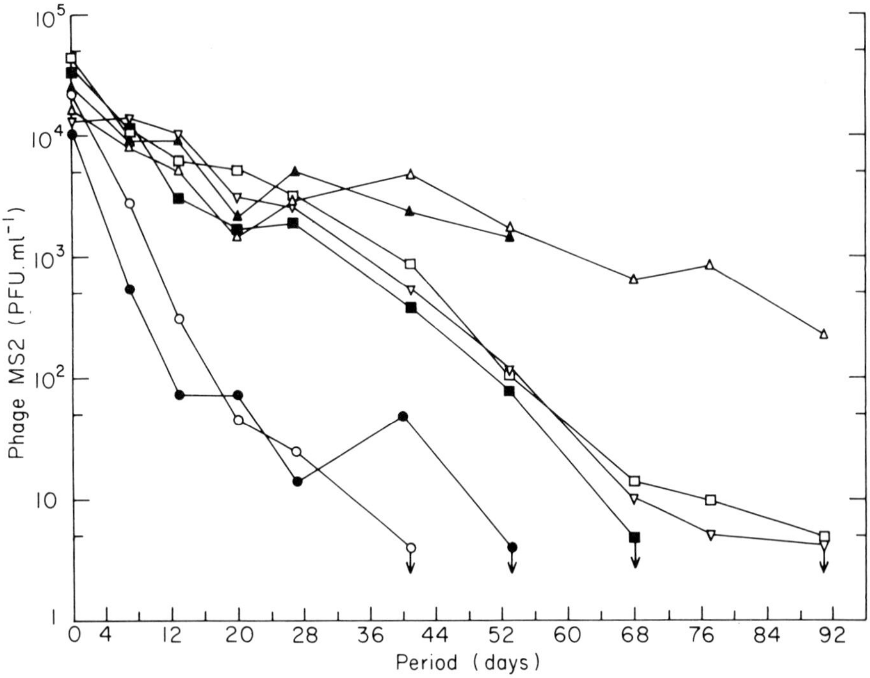

Fig. 6. The effect of salinity (parts per thousand) on the survival of coliphage MS2 in sea water. ●, 0; △, 5; ▲, 10; ▽, 15; □, 20; ■, 25; ○, 30.

On the basis of concentration, inorganic salts in sea water are potentially the most toxic substances in the sea and may affect bacterial survival by general osmotic effects or by specific ion toxicity. The latter factor may be important in virus survival also and experiments were therefore conducted with MS2 phage and *E. coli* in sea water from which several major component salts were absent (Figs 7 and 8). Solutions used included artificial sea water prepared according to a modified Wilder formula (Wood 1966) and several batches deficient in each of the major component salts, together with a NaCl solution of equivalent (3.0%) salinity. Survival of *E. coli* (Fig. 7) in artificial sea water and NaCl solution of the same salinity was very similar and exceeded only by survival in the treatment lacking calcium chloride. Considerable differences in the effects of the various chloride salts were observed, the solution lacking potassium chloride being associated with the poorest survival of all. In replicate experiments with MS2 phage (Fig. 8) improved survival occurred in NaCl and when any one of the four salts were omitted from artificial sea water; a pronounced decrease in

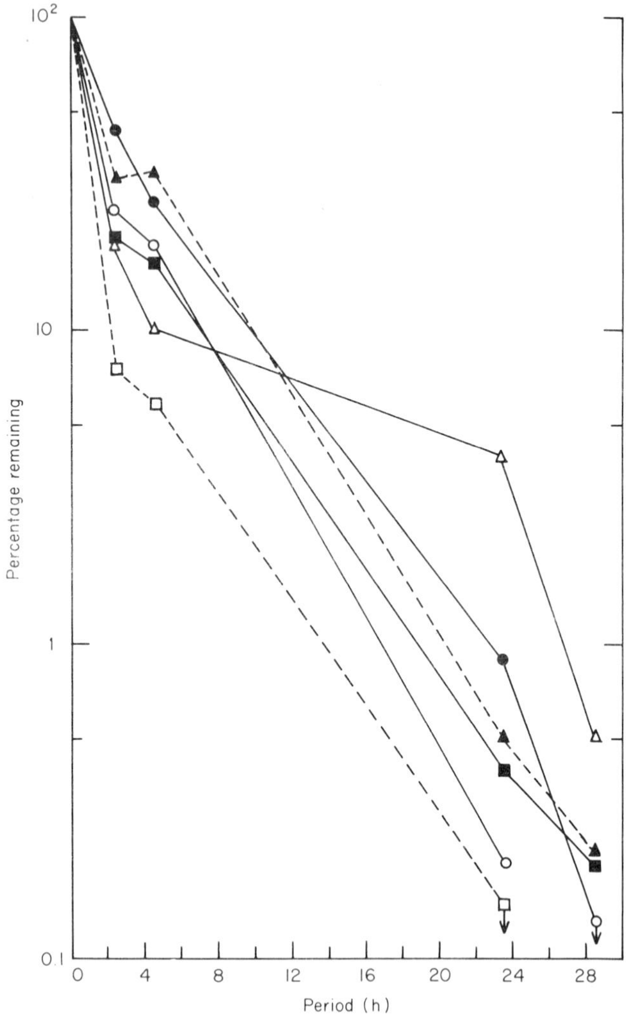

Fig. 7. The effect of major inorganic ions in sea water on the survival of *E. coli*. ●, Artificial sea water (ASW); ▲, NaCl only; ■, ASW less $MgSO_4$; ○, ASW less $MgCl_2$; △, ASW less $CaCl_2$; □, ASW less KCl.

survival occurred when magnesium sulphate was omitted. Kott *et al.* (1969) demonstrated that magnesium sulphate and NaCl exerted no significant influence on the survival of T phages when tested at the concentration normally found in sea water.

Consideration of the factors influencing survival of *E. coli* as outlined by Carlucci & Pramer (1960*a*) led us to study the effects of added nitrogen and

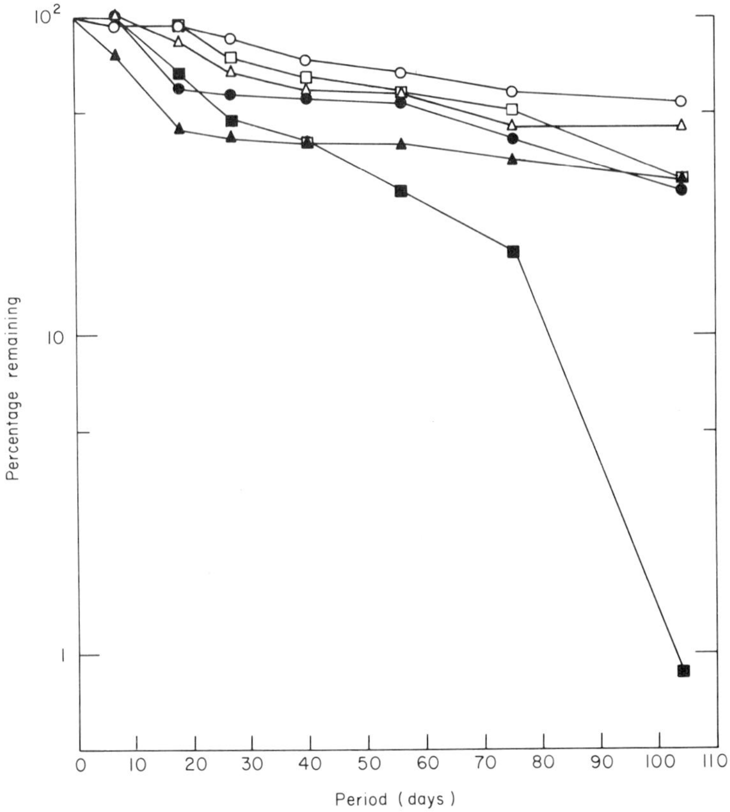

Fig. 8. The effect of major inorganic ions in sea water on the survival of coliphage MS2. ●, Artificial sea water (ASW); ▲, NaCl only; ■, ASW less $MgSO_4$; ○, ASW less $MgCl_2$; △, ASW less $CaCl_2$; □, ASW less KCl.

phosphate. We confirmed that, when using ammonium sulphate and ammonium phosphate, the beneficial effect of added nitrogen and phosphate was related to the concentrations used. Of particular interest, however, were parallel experiments with MS2 phage using concentrations of ammonium sulphate and phosphate from 0 to 100 $mg.l^{-1}$ in raw and autoclaved sea water. The results of adding ammonium sulphate (Fig. 9) demonstrate a striking difference between the two waters used regardless of the concentration of the salt added; i.e. superior survival in autoclaved sea water. When the experiments were repeated with ammonium phosphate (Fig. 10) the survival in raw sea water was similar to that observed in Fig. 9; i.e. it was unrelated to concentration of the salt added. With autoclaved sea water, however, survival decreased with increasing phosphate levels so that 100 mg.l^{-1} of phosphate gave results similar to those obtained with raw sea water.

Differences in survival patterns of bacteriophage, enterovirus and faecal bacteria have been demonstrated by many workers but the reasons proposed for such differences vary. It is apparent, however, that autoclaving removes or

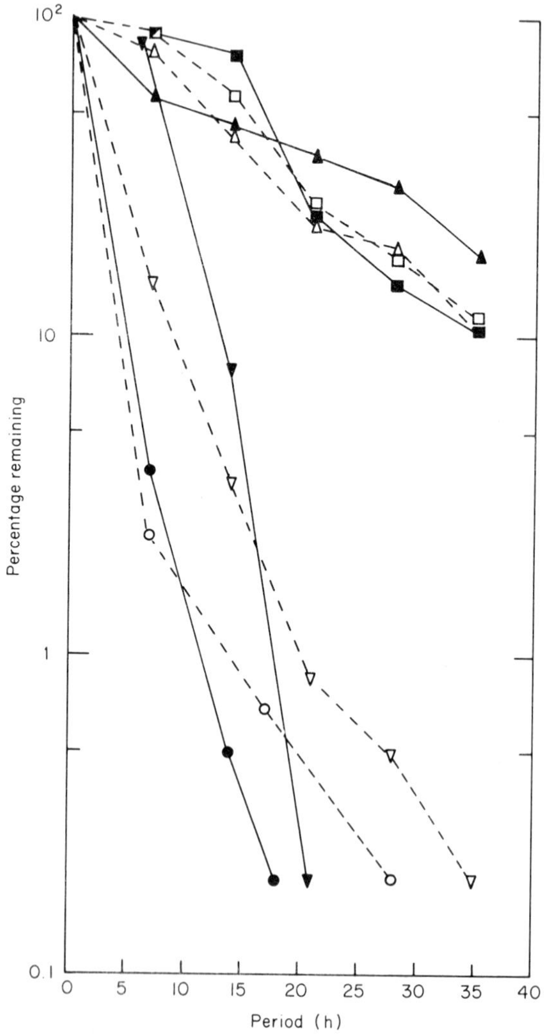

Fig. 9. The effect of added nitrogen (mg.l^{-1}) on the survival of coliphage MS2 in raw sea water and autoclaved sea water. Raw sea water: ○, 0 $(NH_4)_2SO_4$; ●, 1 $(NH_4)_2SO_4$; ▽, 10 $(NH_4)_2SO_4$; ▼, 100 $(NH_4)_2SO_4$. Autoclaved sea water: △, 0 $(NH_4)_2SO_4$; ▲, 1 $(NH_4)_2SO_4$; □, 10 $(NH_4)_2SO_4$; ■, 100 $(NH_4)_2SO_4$.

inactivates some component(s) of the antimicrobial capacity of sea water. The effects observed with added phosphate may be similar to those observed with added phosphate during sewage treatment. Sproul *et al.* (1969) showed that precipitation of 81 mg.l^{-1} of phosphate with calcium increased virus removal from sewage. Precipitation was observed to occur at high concentrations of

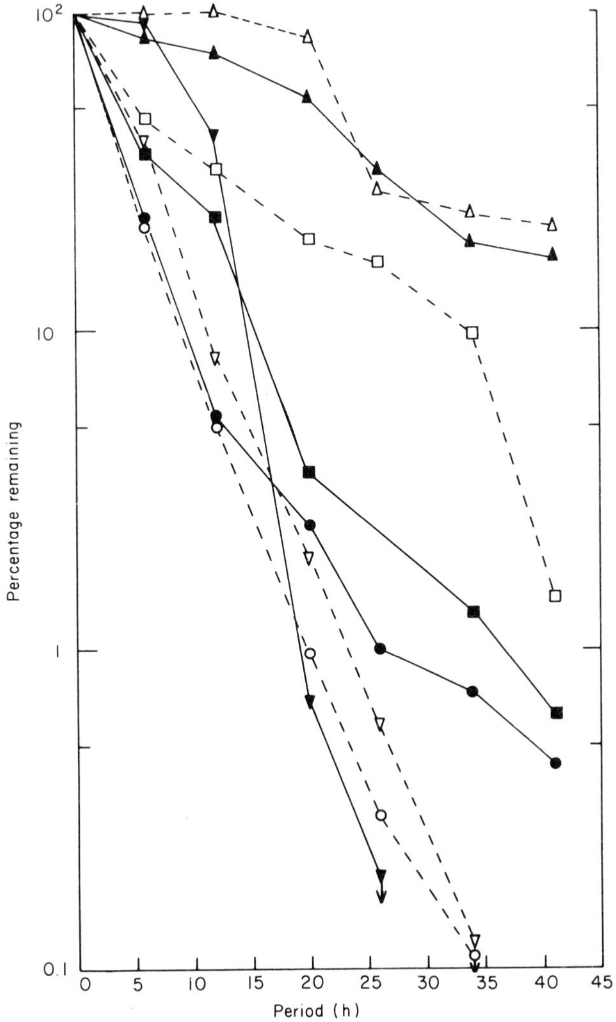

Fig. 10. The effect of added phosphate (mg.l^{-1}) on the survival of coliphage MS2 in raw and autoclaved sea water. Raw sea water: ○, 0 $(NH_4)_2HP)_4$; ●, 1 $(NH_4)_2HPO_4$; ▽, 10 $(NH_4)_2HPO_4$; ▼, 100 $(NH_4)_2HPO_4$. Autoclaved sea water: △, 0 $(NH_4)_2HPO_4$; ▲, 1 $(NH_4)_2HPO_4$; □, 10 $(NH_4)_2HPO_4$; ■, 100 $(NH_4)_2HPO_4$.

phosphate during my experiments and phage may have become adsorbed to these particles.

Temperature is another factor influencing survival of bacteria and viruses in sea water; and Vaughn & Metcalf (1975) have demonstrated extended survival of enterovirus in estuarine waters during winter months. They also showed that coliphage was considerably more persistent than coxsackievirus B when examined under similar conditions. Some of my data (Fig. 11) using MS2 phage

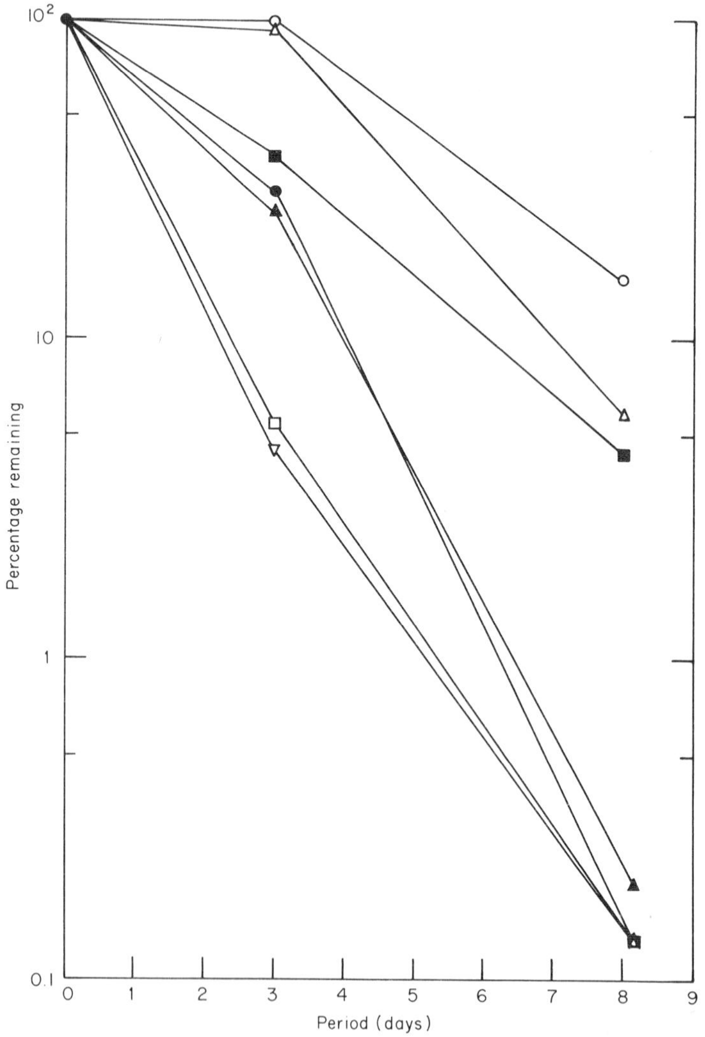

Fig. 11. The effect of temperature (°C) on the survival of coliphage MS2 in sea water. ○, 7.5; △, 15; ■, 20; ●, 25; ▲, 30; □, 37; ▽, outside (20°C).

indicated a definite temperature relationship, survival decreasing with increasing temperature. Of particlar interest was the experiment conducted in daylight at 20°C which exhibited greater inactivation than the laboratory experiment at 20°C under artificial light. Photodynamic inactivation of enteroviruses has been studied by Wallis & Melnick (1965) and although viruses are generally thought to be resistent, a degree of photosensitivity has been demonstrated under certain experimental conditions. Magnusson *et al.* (1966) failed to demonstrate a photodynamic inactivation of poliovirus in sea water though the phenomenon is well known with faecal bacteria and has been reported for phage T7 by Berry & Noton (1976). Our experiments with *E. coli* and MS2 seem to support a photo-inactivation theory (Fig. 12) but it would be difficult to postulate from this whether such inactivation occurs in the natural aquatic environment, and how to assess its importance.

In Section 2 dealing with sewage treatment mention was made of the fact that adsorption is probably one of the major factors in virus removal. Numerous studies on virus inactivation in sea water have produced much conflicting evidence as to the nature of the factors involved or their relative importance. Metcalf & Stiles (1967) suggested a correlation between virus survival and the normal biotic flora of natural waters. Matossian & Garabedian (1967) found that micro-organisms and particulate matter had no effect on the inactivation of poliovirus. In contrast Shuval *et al.* (1971), also using poliovirus, could not demonstrate antiviral activity in the absence of marine bacteria, and Gundersen *et al.* (1967) actually isolated a bacterium associated with antiviral activity. Mitchell & Jannasch (1969) using the phage X174 found that its survival was affected by 3 factors: (a) inactivation by micro-organisms; (b) inactivation by chemicals; (c) protective action of organic particulate matter.

Other workers have demonstrated that sewage exerts a protective effect on enteroviruses (Metcalf & Stiles 1967, Vaughn & Ryther 1974) but Berry & Noton (1976) have shown that inactivation of phage T7 is more rapid in polluted water. It is likely that in nature a combination of all of the factors (and probably others) listed by Mitchell & Jannasch (1969) occurs and that the contribution of each factor will vary from place to place with season and with the enterovirus under examination.

My own work with MS2 and wild type phages seemed to show that one overriding factor was the effect of autoclaving sea water. Thus, whichever factor was under investigation (salinity, pH, light or any factor) replicate treatments in raw sea water and autoclaved sea water always exhibited marked differences. Membrane filtration of sea water occasionally gave similar results but this was not consistent and seemed to have no seasonal pattern. Autoclaving sea water will obviously kill any micro-organisms but may also lead to various chemical changes and to the formation of precipitates including those of some heavy metals (Jones 1967).

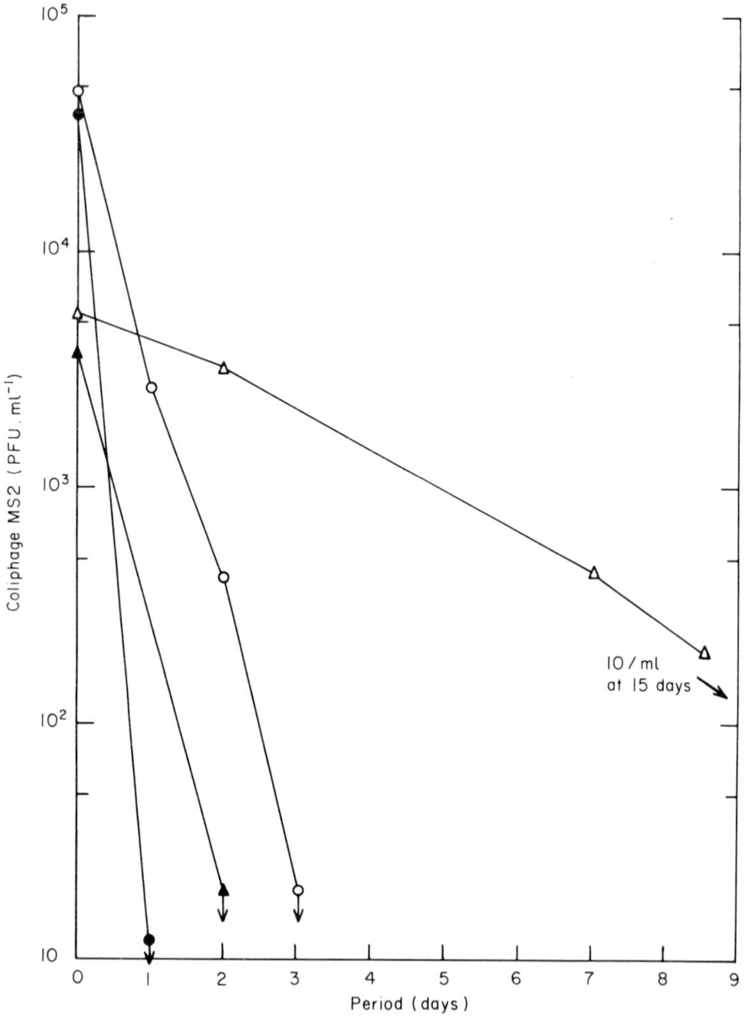

Fig. 12. The effect of sunlight on the survival of *E. coli* and coliphage MS2 in sea water. *E. coli*: ○, dark; ●, light; coliphage MS2; △, dark; ▲, light.

4. Coliphages in Sediments

Much of the work on coliphages in relation to sediments has been concerned with the use of phage as a model to study the fate of viruses in soils and has not been directly concerned with the marine environment. However, the results of this work on soils is equally applicable to marine sediments. Carlson *et al.* (1968)

reported that up to 99% adsorption of added phage T2 occurred on common clay materials such as kaolinite, montmorillonite and illite in concentrations of cations that occur in natural waters at pH 7. Virus adsorption to montmorillonite was reputed to occur fairly uniformly over the pH range 3.5 to 9.5 according to Schaub *et al.* (1974). Natural clays are normally a mixture of clay types and organic matter and virus adsorption to these may be lower (Gerba & Schaiberger 1975). Significant reduction in the inactivation of coliphage T7 in sea water plus 75 $\mu g.ml^{-1}$ of colloidal montmorillonite was reported by Bitton & Mitchell (1974). The authors concluded that protection was afforded by adsorption of viruses on to colloid surfaces but that this may not have rendered phage inactive. Cookson & North (1967) using activated carbon demonstrated that the adsorbed virus did not lose its infective ability and Carlson *et al.* (1968) also reported that adsorption was reversible and dependent on the concentration and type of cation present. In sea water the ionic strength is high and Roper & Marshalls' (1974) work, which showed how desorption could occur if the electrolyte concentration was diluted below a critical level, could explain in part the effects of salinity reviewed earlier. Clay minerals may also absorb lytic enzymes or antiviral toxins produced by antagonistic microflora, and such adsorption to sediment or particulate matter may contribute to some effects formerly attributed to other factors. Failure of virus-coated particulates to sediment has been postulated as a factor contributing to the mobility of virus in water (Metcalf *et al.* 1974). As a word of caution, considerable differences between the behaviour of coliphage and enteroviruses in soils have been reported (Lefler & Kott 1974, Gerba *et al.* 1975) and this may negate the use of coliphage as a model or indicator in such situations.

5. Coliphages in Shellfish

In the introduction to this paper mention was made of the role of shellfish in the transmission of infectious hepatitis and of the concern about the transmission of human enteroviruses. In view of the established and potential risks it is not surprising therefore that numerous investigations into the uptake and removal of viruses by shellfish have been reported (Hedström & Lycke 1964; Liu *et al.* 1966; Mitchell *et al.* 1966; Hoff & Becker 1968). In some studies the accumulation of viruses has not exceeded or even reached the concentration present in ambient waters but accumulation factors of up to 180 times have been reported for poliovirus (Hoff & Becker 1968). Studies using coliphage S-13 (Canzonier 1971) and the hard clam *Mercenaria mercenaria* have yielded accumulation factors >1500 times for some individual animals. The author attributed this apparently greater accumulation to the fact that he used low exposure levels of phage (1–8 $PFU.ml^{-1}$ sea water) compared with 40 to 1000 $PFU.ml^{-1}$ in other studies and to the high stability and recoverability of the phage compared with poliovirus. Laboratory studies by Vaughn & Metcalf (1975) showed that the accumulation of coliphage T7 was 5 to 30 times greater than that of coxsackievirus B-3. It has been stated or implied that viruses are eliminated from shellfish by direct physiological activity of the animals, but Canzonier (1971) suggested that the accumulation of low titres of virus over a long period could result in virus retention independent of physiological activity. Similar results were obtained by Seraichekas *et al.* (1968) working with poliovirus. The uptake of high initial titres of virus seems to result in rapid elimination in much the same way that faecal bacteria are eliminated, but prolonged exposure to low levels of virus may result in retention. Such observed loss of virus may be due solely to inactivation of virus in the animals. Canzonier (1971) also compared the accumulation of a large *Staphylococcus aureus* phage and the coliphage S-13 and obtained very different results. Obviously the nature and persistence of different viral particles in shellfish needs to be evaluated before any clear statement on accumulation and elimination can be attempted.

In the field situation, Vaughn & Metcalf (1975) examined the practicality of using a coliphage indicator system for human enterovirus in polluted waters. The study, which included shellfish, showed that while coliphage could be isolated from oysters throughout the period March to October, enteroviruses were isolated only during the June to September period. Additionally, 13 of 21 enteric virus isolations were made from samples yielding no coliphage, though whether this applied to all samples examined or those of shellfish only is not clear from the paper. Variation in the choice of host *E. coli* strains produced different isolation rates of coliphage but no strain of the 3 tested yielded consistently better results. Similar limitations on the accurate assessment of coliphage in aquatic systems have been noted by Hilton & Stotsky (1973), and more detailed studies are needed.

My own work on the uptake of coliphage MS2 and *E. coli* by oysters has included field comparison of coliphage/*E. coli* content of oysters and laboratory studies on uptake and elimination. Table 2 illustrates some typical results obtained from the examination of samples of shellfish dredged at various sites in a local river. The stations 1 to 8 listed are at varying distances below a source of sewage pollution, station 1 being the most remote and station 8 almost at the site of the discharge. Estimates of *E. coli* and coliform bacteria show a decline with increasing distance from the outfall, but this is not reflected in the estimates of coliphage where the highest value obtained (12 $PFU.ml^{-1}$) corresponded with the lowest *E. coli*. Greater survival of coliphage may possibly account for its presence at some distance from the source of pollution but its apparent absence at source is less easily explained. Differences between the uptake of coliphage and *E. coli* have been noted by Canzonier (1971) and confirmed by my own laboratory studies. *Escherichia coli* are rapidly accumulated and eliminated provided that conditions are such that physiological activity

Table 2

Coliphage, Escherichia coli *and coliform bacteria in oysters* (Ostrea edulis) *from a sewage polluted estuary*

Station number	Coliphage ($PFU.ml^{-1}$ tissue)	*E. coli* per ml tissue	coliforms per ml tissue
1	12	<0.2	9.2
2	3	0.4	3.4
3	6	1.0	2.0
4	0	1.2	4.8
5	0	5.6	12.6
6	3	6.6	12.4
7	0	32.0	34.2
8	6	48.8	54.4

Water temperature at time of sampling 16°C.

of the shellfish may proceed. In contrast, the accumulation and elimination of coliphage MS2 is much slower and phage particles may persist for at least five days after all *E. coli* cells have been eliminated.

6. Summary

Coliphages are commonly encountered in sewage and, although numbers and types vary considerably, there would appear to be no seasonal pattern in their distribution. They are reduced by sewage treatment processes in a manner comparable to bacteria such as *E. coli* and certain enteroviruses, but do not appear to be numerically related to either. The available evidence suggests that coliphages are not an important factor contributing to the observed reduction of coliform bacteria in sewage treatment, although this may occur. The primary mechanism for removal of phage from sewage during treatment is by adsorption onto particulates, emphasizing the marked efficiency of activated sludge treatment over other forms of sewage treatment. Coliphages such as f2 have been successfully applied as virus models in studies on treatment processes, particularly with regard to disinfection.

In marine waters coliphages persist considerably longer than bacteria of faecal origin and some studies suggest that they may also be more persistent than enteroviruses. A major factor in the decline of coliphage numbers in sea water is adsorption, though this is probably also linked with a complex array of chemical and biological factors. Few studies specifically orientated towards coliphage in sediments have been made, but from related work on phage–virus removal in soils, adsorption is considered to be the main mechanism of removal. Differences in the behaviour of coliphages and enteroviruses in soils suggests that phage would not be acceptable as a model, though clearly some work on marine sediments is necessary to determine possible differences in a saline environment.

Experimentally the use of coliphage as a virus model in studies on the uptake and elimination of viruses by shellfish has yielded some useful though often conflicting results. The possibility of long term retention of enteroviruses in a manner which does not permit complete removal by the physiological activity of the animal poses potential problems in that very low levels of virus may constitute an infective dose.

The promise shown by a coliphage indicator system for viruses in sewage and treatment systems cannot, on existing evidence, be extended to sewage contaminated waters and shellfish, though further work on direct comparisons between bacterial indicators such as *E. coli*, coliphages and enteroviruses would undoubtedly be desirable.

7. References

ADAMS, M. R. 1959 *Bacteriophages.* New York: Interscience Publishers.

AKIN, E. W., HILL Jr, W. F., CLINE, G. B. & BENTON, W. H. 1976 The loss of poliovirus I infectivity in marine waters. *Water Research* **10**, 59–63.

BECKWITH, T. D. & ROSE, E. J. 1930 The bacteriophage content of sewage and its action upon bacterial organisms. *Journal of Bacteriology* **20**, 151–159.

BERG, G. 1973 Removal of viruses from sewage effluents and water. I. A review. *Bulletin of the World Health Organization* **49**, 451–460.

BERRY, S. A. & NOTON, B. G. 1976 Survival of bacteriophages in seawater. *Water Research* **10**, 323–327.

BITTON, G. & MITCHELL, R. 1974 Effect of colloids on the survival of bacteriophages in seawater. *Water Research* **8**, 227–229.

BLOOM, H. H., MACK, W. N., KRUEGER, B. J. & MALLMAN, W. L. 1959 Identification of enteroviruses in sewage. *Journal of Infectious Diseases* **105**, 61–68.

CALABRO, J. F., COSENZA, B. J. & KOLEGA, J. J. 1972 Bacteriophages recovered from sewage. *Journal of the Water Pollution Control Federation* **44**, 2355–2358.

CANZONIER, W. J. 1971 Accumulation and elimination of coliphage S-13 by the hard clam, *Mercenaria mercenaria. Applied Microbiology* **21**, 1024–1031.

CARLSON, G. F., WOODARD, F. E., WENTWORTH, D. F. & SPROUL, O. J. 1968 Virus inactivation on clay particles in natural waters. *Journal Water Pollution Control Federation* 40.R89.

CARLUCCI, A. F. & PRAMER, D. 1960*a* An evaluation of factors affecting the survival of *Escherichia coli* in seawater. II Salinity, pH and Nutrients. *Applied Microbiology* 8, 247–250.

CARLUCCI, A. F. & PRAMER, D. 1960*b* An evaluation of factors affecting the survival of *Escherichia coli* in seawater. IV Bacteriophages. *Applied Microbiology* 8, 254–256.

CLARKE, N. A., BERG, G., KABLER, P. W. & CHANG, S. L. 1964 Human enteric viruses in water: Source, survival and removability. Proceedings of an International Conference, London, 1962, Vol. 2. In *Advances in Water Pollution Research.* London: Pergamon Press.

COOKSON, Jr J. T. & NORTH, W. J. 1967 Adsorption of viruses on activated carbon: equilibria and kinetics of the attachment of *E. coli* bacteriophage T4 on activated carbon. *Environmantal Science and Technology* **1**, 46–52.

CRAMER, W. N., KAWATA, K. & KRUSE, C. W. 1976 Chlorination and iodination of poliovirus and f2. *Journal of the Water Pollution Control Federation* **48**, 61–76.

DE MICHELLE, E. 1974 Water re-use, virus removal and public health. In *Virus Survival in Water and Wastewater Systems*, eds Malina, J. R., Jr & Sagik, B. P. Texas, USA: Centre for Research in Water Resources, University of Texas.

DHILLON, T. S., CHAN, Y. S., SUN, S. M. & CHAU, W. S. 1970 Distribution of coliphages in Hong Kong sewage. *Applied Microbiology* **20**, 187–191.

DOUGHERTY, W. & ALTMAN, R. 1962 Viral hepatitis in New Jersey 1960–1961. *American Journal of Medicine* **32**, 704–716.

ENGLAND, B., LEACH, R. E., ADAME, B. & SHIOSAKI, R. 1967 Virologic assessment of sewage treatment at Santee, California. In *Transmission of Viruses by the Water Route*, ed. Berg, G. New York: Interscience Publishers.

GERBA, C. P. & SCHAIBERGER, G. E. 1975 The effects of particulates on virus survival in seawater. *Journal of the Water Pollution Control Federation* **41**, 93–103.

GERBA, C. P., WALLIS, C. & MELNICK, J. L. 1975 Fate of wastewater bacteria and viruses in soil. *Journal of the Irrigation and Drainage Division of the American Society of Civil Engineers* **101**, IR3 157–173.

GILCREAS, F. W. & KELLY, S. M. 1955 Relation of coliform-organism test to enteric-virus pollution. *Journal of the American Water Works Association* **47**, 683–694.

GUNDERSEN, K., BRANDBERG, Å., MAGNUSSON, S. & LYCKE, E. 1967 Characterization of a marine bacterium associated with virus inactivating capacity. *Acta pathologica et microbiologica Scandinavica* **71**, 281–286

HEDSTRÖM, C. E. & LYCKE, E. 1964 An experimental study on oysters as virus carriers. *American Journal of Hygiene* **79**, 134–142.

HILTON, M. C. & STOTZKY, G. 1973 Use of coliphages as indicators of water pollution. *Canadian Journal of Microbiology* **19**, 747–751.

HOFF, J. C. & BECKER, R. C. 1968 The accumulation and elimination of crude and clarified poliovirus suspensions by shellfish. *American Journal of Epidemiology* **90**, 53–61.

JONES, G. E. 1967 Precipitates from autoclaved seawater. *Limnology and Oceanography* **12**, 165–167.

KELLY, S. & SANDERSON, W. W. 1959 The effect of sewage treatment on viruses. *Sewage and Industrial Wastes* **31**, 683–689.

KOLLINS, S. A. 1966 The presence of human enteric viruses in sewage and their removal by conventional sewage treatment methods. In *Advances in Applied Microbiology*, ed. Umbreit, W. W. New York & London: Academic Press.

KOTT, Y., BEN ARI, H. & BURAS, N. 1969 The fate of viruses in a marine environment. Advances in Water Pollution Research, 14th International Conference on Water Pollution Research. Prague. pp. 823–835.

LEFLER, E. & KOTT, Y. 1974 Virus retention and survival in sand. In *Virus Survival in Water and Wastewater Systems*, eds Malina, Jr F. & Sagik, B. P. Texas, USA: Centre for Research in Water Resources, University of Texas.

LIU, O. C., SERAICHEKAS, H. R. & MURPHY, B. L. 1966 Fate of poliovirus in northern quahaugs. *Proceedings of the Society for Experimental Biology and Medicine* **121**, 601.

LUND, E. 1969 Discussion of paper by SPROUL *et al.* 'Virus removal by adsorption in wastewater treatment processes. Advances in Water Pollution Research, 14th International Conference on Water Pollution Research. Prague.

MCDERMOTT, J. H. 1975 Virus problems in water supplies. *Water and Sewage Works,* May 1975.

MACKOWIAK, P. A., CARAWAY, C. I. & PORTNOY, B. L. 1976 Oyster associated hepatitis: lessons from the Louisiana experience. *American Journal of Epidemiology* **103**, 181–191.

MAGNUSSON, S., HEDSTRÖM, C. E. & LYCKE, E. 1966 The virus inactivating capacity of seawater. *Acta pathologica et microbiologica scandinavica* **66**, 551–559.

MALINA, J. F., Jr, RANGANATHAN, K. R., MOORE, B. E. D. & SAGIK, B. P. 1974 Poliovirus inactivation by activated sludge. In *Virus Survival in Water and Wastewater Systems*, eds Malina, J. F., Jr & Sagik, B. P. Texas, USA: Centre for Research in Water Resources, University of Texas.

MASON, J. O. & MCLEAN, W. R. 1962 Infectious hepatitis traced to the consumption of raw oysters. *American Journal of Hygiene* **75**, 90–111.
MATOSSIAN, A. M. & GARABEDIAN, G. A. 1967 Virucidal action of seawater. *American Journal of Epidemiology* **85**, 1–858.
METCALF, T. G. & STILES, W. C. 1967 Survival of enteric viruses in estuary waters and shellfish. In *Transmission of Viruses by the Water Route*, ed. Berg, G. New York: Interscience Publishers.
METCALF, T. G. & STILES, W. C. 1968 Viral pollution of shellfish in estuary waters. *Journal of the Sanitary Engineering Division of the American Society of Civil Engineers.* **94**, SA4. 595–609.
METCALF, T. G., WALLIS, C. & MELNICK, J. C. 1974 Virus enumeration and public health assessments in polluted surface waters contributing to transmission of virus in nature. In *Virus Survival in Water and Wastewater Systems*, eds Malina Jr J. R., & Sagik, B. P. Texas, USA: Centre for Research in Water Resources, University of Texas.
MITCHELL, R. & JANNASCH, H. W. 1969 Processes controlling virus inactivation in seawater. *Environmental Science and Technology* **3**, 941–943.
MITCHELL, J. R., PRESNELL, E. W., AKIN, E. W., CUMMINS, J. M. & LIU, O. C. 1966 Accumulation and elimination of poliovirus by the eastern oyster. *American Journal of Epidemiology* **84**, 40–50.
ROOS, R. 1956 Hepatitis epidemic conveyed by oysters. *Svenska Lakartidningen* **53**, 989–1003.
ROPER, M. M. & MARSHALL, K. C. 1974 Modification of the interaction between *Escherichia coli* and bacteriophage in saline sediment. *Microbial Ecology* **1**, 1–13.
SAFFERMAN, R. S. & MORRIS, M. E. 1976 Assessment of virus removal by a multi-stage activated sludge process. *Water Research* **10**, 413–420.
SCHAUB, S. A., SORBER, C. A. & TAYLOR, G. W. 1974 The association of enteric viruses with natural turbidity in aquatic environments. In *Virus Survival in Water and Wastewater Systems*, eds Malina, Jr J. F. & Sagik, B. P. Texas, USA: Centre for Research in Water Resources, University of Texas.
SERAICHEKAS, H. R., BRASHEAR, D. A., BARNICK, J. A., CAREY, P. F. & LIU, O. C. 1868 Viral depuration by assaying individual shellfish. *Applied Microbiology* **16**, 1865–1871.
SHUVAL, H. I., THOMPSON, A., FATTAL, B., CYMBALISTA, S. & WIENER, Y. 1971 Natural virus inactivation processes in seawater. *Journal of the Sanitary Engineering Division of the American Society of Civil Engineers* **SA5**, 587–600.
SORBER, C. A., MALINA, Jr J. R. & SAGIK, B. P. 1972 Virus rejection by the reverse osmosis-ultra filtration processes. *Water Research* **6**, 1377–1388.
SPROUL, O. J., WARREN, M., LA ROCHELLE, L. R. & BRUNNER, D. R. 1969 Virus removal by adsorption in wastewater treatment processes. Advances in Water Pollution Research, 14th International Conference on Water Pollution Research, Prague.
TWORT, F. W. 1915 An investigation on the nature of ultramicroscopic viruses. *Lancet* **2**, 1241.
VAUGHN, J. M. & METCALF, T. G. 1975 Coliphages as indicators of enteric viruses in shellfish and shellfish raising waters. *Water Research* **9**, 613–616.
VAUGHN, J. M. & RYTHER, J. H. 1974 Bacteriophage survival patterns in a tertiary sewage treatment – aquaculture model system. *Aquaculture* **4**, 399–406.
WALLIS, C. & MELNICK, J. L. 1965 Photodynamic inactivation of enteroviruses *Journal of Bacteriology* **89**, 41–46.
WARE, G. C. & MELLON, M. A. 1956 Some observations on the coli/coliphage relationship in sewage. *Journal of Hygiene, Cambridge* **54**, 99–101.
WOOD, P. C. 1966 Lobster storage and shellfish purification. *Laboratory Leaflet, Fisheries Laboratory, Burnham-on-Crouch (N.S.) No. 13.*

Some Ecological Aspects of the Bacterial Fish Pathogen — *Aeromonas salmonicida*

D. H. McCarthy

Ministry of Agriculture, Fisheries and Food,
Fish Diseases Laboratory, The Nothe,
Weymouth, Dorset, England

CONTENTS

1. Introduction

THE MARKED UPSURGE of interest in commercial fish farming, which has occurred in the UK during the past 10 years, has served to highlight the prevalence and economic significance of communicable fish diseases such as furunculosis (causative organism, *Aeromonas salmonicida*). However, many of the epizootics of this disease reported from fish farms have in fact resulted from misuse of intensive rearing methods, usually by inexperienced personnel. It is, of course (or should be) well known that such abuse predisposes fish stocks to disease because it often leads to the potentially dangerous situation where severely stressed fish are held in suboptimal environmental conditions without the protection of basic hygienic procedures. What is not so well known, however, is precisely how diseases such as furunculosis are introduced into a fish farm environment and are maintained thereafter in the fish population. Various assertions concerning the route of spread of furunculosis are frequently advanced, but there is surprisingly little unequivocal experimental data to support them: there appears to be dispute, for example, over the basic question whether *A. salmonicida* is an obligate fish pathogen or capable of a free-living existence in fish pond water (Christensen 1972).

This paper describes experimental work with *A. salmonicida*, performed as part of a larger study designed to investigate the epizootiology of furunculosis with particular reference to the fish farming industry. Detailed historical reviews have been made (McCraw 1952; Herman 1968) and only certain important facets of the disease will be described and emphasized by way of introduction and background against which the work can be clearly outlined and discussed. The term ecological will be interpreted here in its widest sense in order to embrace all aspects of the life cycle of the bacterium including its existence in the fish host.

2. Furunculosis – the Disease

Furunculosis was first described from a fish farm in Germany in 1894 by Emmerich & Weibel and is generally thought to have been introduced to Europe in the 1880s with first imports of the rainbow trout (*Salmo gairdneri*). However, others think that it originated in Europe and was brought to North America with the brown trout (*Salmo trutta*). Wherever its origin, the disease was first reported in the UK in 1911 and, following a recommendation of the Government appointed Furunculosis Committee (Mackie *et al.* 1930, 1933, 1935), it was made notifiable under the Diseases of Fish Act, 1937. Although this act has undoubtedly prevented subsequent importation of certain other highly destructive fish diseases, it had little effect on the spread of furunculosis which is now enzootic in the UK, being present in both cultured and feral fish. Furthermore, the disease now has a worldwide distribution and, with the possible exceptions of Australia and New Zealand, is a significant problem wherever salmonid culture is practised.

Furunculosis has an extensive host range, affecting all age-groups, but fish of the Salmonida family are by far the most susceptible, particularly the brook trout (*Salvelinus fontinalis*) and the Atlantic salmon (*Salmo salar*), with the rainbow trout being relatively resistant. Furunculosis is primarily a disease of fresh water fish, but there have been several recent reports of epizootics among salmonid fish held, or being cultured in sea water. Confirmation of the disease in a strictly marine host – the sable fish (*Anoplopoma fimbria*) – by Evelyn (1971), is of considerable epizootiological importance.

Furunculosis is a communicable septicaemic disease appearing in two main recognizable forms: the first, the acute type, characterized by the appearance of few or no external signs, is of sudden onset and causes high mortality; the second, a subacute form, usually manifested by the presence of prominent, focal muscle lesions or 'furuncles' (hence 'furunculosis'), is of more gradual onset and causes relatively lower mortality.

The causative organism, *A. salmonicida*, is a chemo-organotrophic, non-motile, Gram negative rod which usually produces a characteristic brown, water-soluble pigment when grown on appropriate culture media. Its precise pathogenic mechanisms have not yet been fully elucidated.

Outbreaks of furunculosis commonly accompany rising water temperatures and epizootics are therefore more frequently encountered during summer months. Although epizootics may be controlled by chemotherapy provided it is instituted promptly, they frequently re-occur, and therefore emphasis is quite properly placed on prevention of the disease (Ghittino 1972).

The precise route of transmission of furunculosis is unknown but contact with infected fish or contaminated water and fish farm materials, and transovarian transmission, have all been stated as probable routes of infection. Furthermore, carrier fish, which show no overt signs of disease, yet harbour the bacterium in their tissues, may also act as a reservoir for furunculosis in addition to being involved in its transmission. It is with this particular facet of the disease – its transmission – that the present work is primarily concerned, consisting of a study of the organism's survival in certain fish farm habitats and an investigation of some possible routes of infection.

3. Source and Viability of *A. salmonicida*

The possibility of a free-living life cycle for *A. salmonicida* in the fish farm environment has been considered by many workers since its existence would render disease prevention and control methods virtually inoperable. In this section a number of fish farm habitats have been considered as possible sources of the organism and its survival in them studied.

(a) *In fish ponds*

Possible sources of the bacterium in commercial fish farm ponds are: (a) fish (see below); (b) other animals; (c) pond water; (d) mud in earth ponds or detritus in concrete or fibreglass tanks.

The possibility that animals, other than fish, harbour and thereby possibly transmit the bacterium either directly or indirectly seems unlikely in view of the report of Cornick *et al.* (1969) who examined unsuccessfully a total of 2954 vertebrates and invertebrates (including leeches), taken from fish ponds during an epizootic of furunculosis, for the presence of *A. salmonicida.* Williamson (1928) also failed to find the organism in water snails under similar conditions.

Most of the published studies on the existence and survival of *A. salmonicida* in water are of scientific interest but of little practical value since they are concerned with the viability of the organism in distilled water or pure tap water. Similarly, other workers have limited the usefulness of their results by prior sterilization of their river water or sea water test media. Such highly artificial environments are far removed from the fish pond situation where *A. salmonicida* must compete in a rich and heterogeneous microbial community. However, from the technical viewpoint prior sterilization of water samples is perhaps understandable since *A. salmonicida* is notoriously difficult to isolate from mixed microbial populations, being quickly overgrown by most common water bacteria. Moreover, close proximity of colonies of other bacteria to those of *A. salmonicida* inhibits production of its brown pigment which is commonly used as presumptive evidence of presence of the organism on primary agar plates.

In the present study an attempt was made to simulate fish pond conditions as closely as possible so that results might be of the maximum practical relevance. Survival of *A. salmonicida* was therefore studied in mixed culture in open circulation fish tanks fully stocked with salmonid fish. However, because of the technical problems referred to above concerning recovery of the organism from mixed microbial populations and in the absence of a satisfactory selective medium for *A. salmonicida*, it was decided to use a streptomycin-resistant (1000 $\mu g.ml^{-1}$) mutant strain for these studies. While it is possible that this mutant, which was obtained by serial passage on culture-media containing increasing concentrations of streptomycin until the required level of resistance was obtained, differs from its precursor strain in characteristics other than antibiotic resistance, no difference in growth rate or virulence to fish was detected in the present study. Furthermore, Danso *et al.* (1973) also found no difference in pathogenicity of their streptomycin-resistant strains of *Xanthomonas campestris* and *Erwinia carotovora.*

An 850 l fish tank, containing either fresh water, brackish water (average salinity 2.34% (w/v) or full sea water (average salinity 3.46%) fully stocked with rainbow trout, was used for each experiment, the stocking and feeding rates and water flow being adjusted to produce a water clarity and quality consistent with that of an average commercial fish farm pond. During the course of the experiments water temperature varied within the range 11 to 13°C. After a period of acclimatization, 12 sterile 'Viskring' (Gallenkamp Ltd, London) dialysis bags were filled with *ca.* 100 ml of the appropriate tank water and then inoculated with the streptomycin-resistant mutant to a final approximate concentration of 10^7 viable $cells.ml^{-1}$. The open end of each bag was knotted, and the closed bags suspended below the water surface. Dialysis bags were used in order to allow free movement of metabolites to and from the test bacteria and yet, at the same time, retain the inoculum for convenient sampling. One bag was removed from each tank at *ca.* three-day intervals, the contents thoroughly mixed and the viable cell count of *A. salmonicida* assessed by the agar-surface method using duplicate culture plates of Tryptone Soya Agar (TSA, Oxoid Ltd) and TSA containing 1000 $\mu g.ml^{-1}$ of streptomycin, colonies being counted after four days incubation at 22°C.

Figure 1 clearly demonstrates the ability of *A. salmonicida* cells to survive for a considerable time under all the fish tank conditions tested. Although survival was most prolonged in the brackish water environment (24 days) and the organism survived for eight days in full sea water, the NaCl tolerance of the isolate used (2.2 g%) did not differ from that of other *A. salmonicida* strains. The times for survival of *A. salmonicida* in unsterilized river and brackish water (*ca.* 1–2 days) published by Williamson (1928) and Lund (1967) were not corroborated by the present results which were closer to those reported by Smith (1962): i.e. river water, 7–19 days; brackish water, 16–25 days. It should be noted, however, that at least for some of her experiments, Smith employed a selective medium. However, the present result for survival in sea water (8 days) was closer to that reported by Lund (1967): i.e. 4–6 days, than the result found by I. W. Smith (pers. comm.): i.e. 21–26 days. The discrepancy between Williamson's (1928) and Lund's (1967) results and those found in the present study for survival in fresh water and brackish water was probably due to the technical difficulty of isolating *A. salmonicida* from mixed cultures; in the present study the TSA plate without streptomycin was consistently overgrown by common water bacteria and its use was discontinued very early in the study.

Although the present data clearly demonstrate the organism's ability to survive for a significant length of time under fish farm conditions, it should be

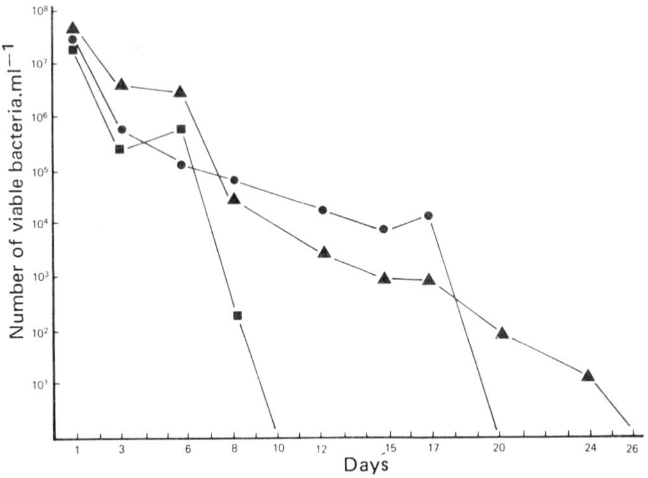

Fig. 1. Fate of *A. salmonicida* in: brackish water (▲); fresh water (●); sea water (■) at 11–13°C.

noted that several workers, including Cornick *et al.* (1969) and Kimura (1970) failed to isolate *A. salmonicida* from fish pond water during epizootics of furunculosis. It is possible of course that the organisms may have been present, perhaps in low numbers, but for technical reasons not detected; clearly, an efficient selective medium for *A. salmonicida* would greatly facilitate such studies.

During an epizootic of furunculosis, *A. salmonicida* is present in large numbers in the faeces of moribund fish and consequently fish pond detritus and mud is likely to be contaminated with the bacterium. Experiments designed to assess the viability of *A. salmonicida* in mud and detritus were carried out in fresh water only and in a manner essentially similar to those described above for survival in pond water. Two sets of dialysis bags, one containing pond water and the other mud and detritus, both previously seeded with *ca.* 10^9 viable cells.ml^{-1} of the streptomycin-resistant strain, were buried in the mud of an earth-bottomed fish tank. When both sets of experiments were terminated prematurely (because of decomposition of the bags presumably due to the action of cellulolytic bacteria) *ca.* 10^5 viable cells of *A. salmonicida* were still present in the pond water bags (after 29 days) and 10^6 remained in the soil bags (after 10 days). These experiments, again, demonstrate a protracted time of survival for *A. salmonicida* in the fish farm environment. Furthermore, the large number of viable cells surviving after 29 days is highly significant since it can be seen from Fig. 1 that the survival time for 10^5 viable cells released into fresh water, is about 14 days.

(b) *On fish nets*

During a furunculosis epizootic most fish farm implements may become contaminated with *A. salmonicida*, but the fish net, being used both to remove dead diseased fish and also to move healthy fish during procedures such as grading, is an obvious choice for study, particularly where there appears to be no published data on the survival of *A. salmonicida* on fish nets or similar implements.

Dry nylon fish net was cut into 3 cm squares and soaked for 30 min in a mixture of fish slime, fish faeces and mud containing *ca.* 10^8 viable cells.ml^{-1} of the streptomycin-resistant strain. Half of the squares were then dried at 10°C for 24 h and the remainder tested wet.

In addition, three compounds, 1% hypocholorite solution (Boots Farm Sales Ltd, Nottingham); 1 : 4000 acriflavine (BDH Ltd, Poole, Dorset) and 0.1% Teepol in 1% NaOH (a compound recommended by the Fish Disease Laboratory, Weymouth, for use as a virus disinfectant on fish farms), were tested for their efficacy in disinfecting contaminated fish netting. Both wet and dry contaminated netting, prepared as described previously, were dipped into the respective disinfectant solution for several seconds and then tested for the survival of *A. salmonicida* as described below.

Both wet and dry contaminated nets and the disinfected nets were incubated at 10°C and tested daily for six days for the presence of *A. salmonicida.* Duplicate test squares of each test group were dropped into 1% Glucose Tryptone Soya Broth (Oxoid Ltd, London) containing 1000 μm.ml^{-1} of streptomycin and, after two days incubation, broths were subcultured on TSA plates also containing 1000 μg.ml^{-1} of streptomycin.

The test organism was found to survive on both wet and dry contaminated nets for the full six days duration of the experiment. The acriflavine and Teepol–NaOH solutions disinfected both wet and dry nets; however, although the hypochlorite solution successfully disinfected wet netting it failed to disinfect dry nets. These results clearly demonstrate the potential danger of using both contaminated and improperly disinfected nets with healthy stock since it is well known that even the most careful netting abraids fish to some extent and such abrasions undoubtedly facilitate invasion by pathogenic bacteria such as *A. salmonicida.*

(c) *In dead fish*

Carcasses of fish which have succumbed to furunculosis are heavily contaminated with *A. salmonicida*: material from a furuncle, for example, may contain as many as 10^8 viable cells.ml^{-1} of necrotic tissue. Consequently, in fish ponds or large fish 'stews' where dead fish cannot be removed conveniently they constitute a significant source of contamination or, alternatively, they may be taken by piscivorous birds and perhaps dropped in uninfected waters.

To assess the viability of *A. salmonicida* in dead fish and the degree to which such diseased fish contaminate pond water, rainbow trout were infected experimentally with an intramuscular injection of 10 times the minimum lethal dose of the streptomycin-resistant strain. After the injections were given, the tank water was tested daily for the presence of *A. salmonicida*, as described above, until the first fish deaths occurred. Thereafter, both diseased fish tissue and tank water were tested for the presence of the organism every four days.

The majority of fish died four days after injection and all survivors were then removed. It can be seen from Fig. 2 that, although *A. salmonicida* was not detected in tank water prior to the death of test fish, 10^3 organisms were present on the fourth day when deaths first occurred, the corresponding infected muscle-tissue sample containing *ca.* 10^8 viable organisms. This result indicates, that at least under experimental conditions, detectable water contamination arises from terminally ill fish. The number of viable bacteria in diseased fish tissue then gradually decreased until by the twenty-fourth day they were fewer in number than those present in water, which now began to increase possibly

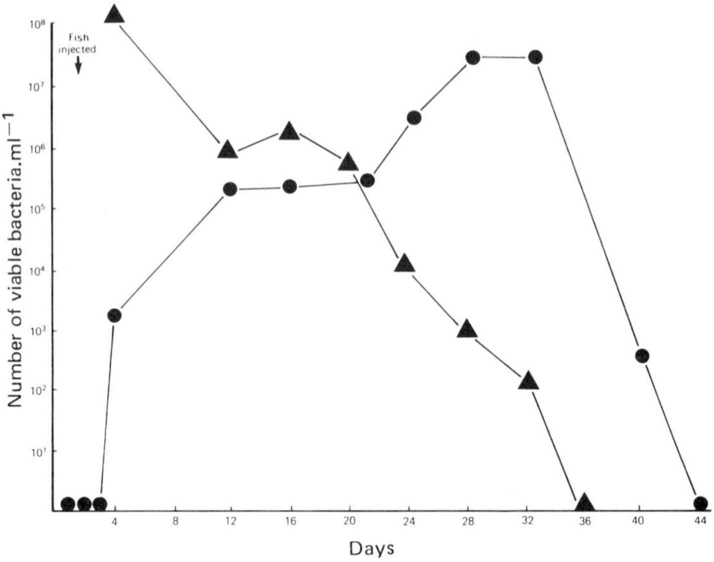

Fig. 2. Survival of injected cells of *A. salmonicida* in dead, experimentally infected, fish (▲) and the fate of shed bacterial cells (●) in fresh water at 11–13°C.

due to disintegration of the test fish. *Aeromonas salmonicida* survived in fish tissue for 32 days and was still present in the tank water for a further eight days despite an open circulation of water through the tank. It is presumed that the test organism must have contaminated the tank sides heavily and was gradually being released from that site.

Prolonged survival of *A. salmonicida* in dead infected fish, in addition to presenting a possible source of infection for healthy fish, adequately demonstrates the danger of feeding scrap fish which might be affected with a chronic form of furunculosis. This is of particular importance in view of the report by Cornick *et al.* (1969) who demonstrated that the organism could survive for up to 49 days in infected trout tissue when stored at −10°C.

(d) *Carrier fish*

The existence of apparently healthy carriers of *A. salmonicida* is well known and such fish have long been suspected of playing an important role in the spread of furunculosis. The main technical problem which has prevented research being carried out in this important area is the difficulty involved in detecting carriers with any certainty; for example, it is generally accepted that culture-isolation methods are at present far too insensitive. However, during this study it was found that intramuscular injection of *ca.* 20 mg.kg^{-1} of a corticosteroid, prednisolone acetate (PA; 'Deltastab', Boots Ltd, Nottingham) accompanied by thermal stress at 18°C, activated latent infections in carrier fish, all injected carriers dying of clinical furunculosis in 4–10 (mean 5) days. Similar work was reported by Bullock & Stuckey (1975) with 'Kenalog' (triamcinolone acetonide); however, a comparative evaluation of this compound obtained as 'Adcortyl-A' from E. R. Squibb & Sons, Liverpool with PA and another corticosteroid, dexamethazone sodium phosphate ('Decadron'; Merck, Sharp & Dohme, Hoddesdon, Herts), with known carrier brown trout, clearly demonstrated PA to be more effective in activating latent furunculosis infections.

Various experiments were then carried out with PA in order to assess the prevalence and significance of carriers in fish populations. In the first, representative fish from four populations of one-year old brown trout, taken from four different commercial fish farms, were found to have surprisingly high carrier rate (40–80%) while similar populations of rainbow trout also tested had either a very low incidence (<5%), or were apparently free from carriers. The usefulness of the PA technique is not confined to salmonid fish since it has also been used successfully with a 'coarse' fish, the silver bream (*Blicca bjoerkna*). A clinical outbreak of furunculosis, caused by an aberrant variety of *A. salmonicida*, was reported (McCarthy 1975*a*) in this fish in 1973 and, although no re-occurrence of the disease was reported during 1975, nine out of a sample of 12 apparently healthy bream tested in that year with PA died of clinical furunculosis, the same non-pigmented aberrant strain being isolated from all dead fish. Two further experiments were carried out with brown trout from a fish farm with a high carrier rate; in the first, a sample was taken from all age-groups on the farm and tested with PA and the following percentage population carrier rates obtained: 1 + (fish > one year old), 96%; 2 + (fish > two years old), 100%; 3 + (fish > three years old), 50%. The second experiment consisted of testing samples of the 1 + population at different times throughout a 12 month period and it was found that the carrier rate did not differ significantly throughout the year. It is indeed surprising that this particular farm has never reported clinical furunculosis, nor has it been observed there on frequent official inspections.

The site of 'carriage' of *A. salmonicida* in carrier fish is not precisely known but most workers believe that the kidney is involved while others have suggested the intestinal tract as a site of harbourage. In the present study known carrier fish were injected with PA and the kidney, spleen, liver, intestinal tract and heart blood were sampled bacteriologically at 24, 48, 72 and 96 h intervals after injection. The results of this experiment indicated that the kidney was the primary site of carriage, although the organism could also be isolated from the intestinal tract of PA-inoculated (48 h after injection), but not control (saline-injected), carrier fish. A great deal of further research is required both into the mechanism of carriage of *A. salmonicida* and the development of indirect non-lethal techniques for reliable carrier detection. One attractive and convenient technique – detection of specific circulating anti-*A. salmonicida* agglutinins – was investigated since it is well known that fish surviving a furunculosis epizootic (and therefore possibly carrier fish) may have demonstrable anti-*A. salmonicida* serum agglutinins (Kimura 1970). A population of 30 brown trout from a known carrier population were bled, individually tagged, and then injected with PA. However, when serum antibody titres for each fish were assessed no consistent significant difference was detected between fish which were shown to be carriers and the non-carrier fish.

Because of the possible existence of vertical transmission of furunculosis, a treatment was sought which might free brood stock fish from the carrier state. Sixty brown trout were taken from a population with a known carrier rate of *ca.* 80%, and divided into groups of 10 fish which were then treated by the following regimes for a period of 10 days (water temperature 11–13°C): (a) fed a normal diet; (b) fed 75 $mg.kg^{-1}$ of furazolidone (Smith, Kline & French, Welwyn Garden City, Herts.); (c) fed 200 $mg.kg^{-1}$ of sulphamerazine (May & Baker Ltd, Dagenham, Essex); (d) fed 100 $mg.kg^{-1}$ of sulphamethylphenazole (Ciba Laboratories Ltd, Horsham, Sussex) and trimethoprim (Burroughs Wellcome Ltd, London) in a ratio of 5 : 1; (e) fed normal food and given a single intramuscular injection of 50 $mg.kg^{-1}$ of gentimicin sulphate (British Schering Ltd, Slough, Bucks.); (f) fed normal food and given a single intramuscular injection of 50 $mg.kg^{-1}$ of tetracycline hydrochloride (Pfizer Ltd, Sandwich, Kent). Fourteen days were first allowed to elapse following medication for elimination of any drug residue, then all fish were injected with PA. Only fish receiving the tetracycline injection survived, all other groups sustaining mortalities ranging from 80% to 100%. In addition to demonstrating an apparently satisfactory treatment for economically important fish which can be injected conveniently, such as brood stock which can be treated several weeks prior to spawning and then segregated, this experiment also revealed that the recommended treatments for furunculosis (sulphamerazine, furazolidone and potentiated sulphonamide) are apparently not capable of removing the bacterium completely from the fish's tissues. These results provide a possible explanation for recrudescence of the disease. Moreover, it must be expected that fish populations treated for furunculosis will remain carriers, a possibility suspected during the use of sulphamerazine by Gutsell & Snieszko (1949) many years ago.

The existence of a high carrier rate in brown trout populations without a history of furunculosis emphasizes the necessity of research workers taking into account the possible pitfalls of using undetected carrier fish in experiments; for example, when carrying out virulence studies it is important to bear in mind the possibility that mortalities in test fish might be due, at least in part, to activation of a latent infection, perhaps as a result of an undetected stress factor. Furthermore, during animal-passage procedures the possibility of inadvertently contaminating or even perhaps replacing, a vaccine strain with the organism being carried, should be borne in mind. The use of biochemical and antibiotic resistance markers or phage sensitivity pattern can be used to distinguish test strains.

4. Lateral Transmission of Furunculosis in the Fish Farm Environment

It is generally accepted that furunculosis is spread by lateral transmission of *A. salmonicida* and more specifically, in the apparent absence of a known animal vector, by the water-borne route although mechanical transmission by infected nets may be important. Water contamination probably arises initially from moribund fish or possibly from apparently healthy carrier fish shedding the organism but, as shown by earlier experiments, the organism may then persist for a considerable length of time in the environment and the disease spread in this way.

Several experiments were carried out to investigate lateral transmission of *A. salmonicida* as part of the present study and these are described below under two general headings: (a) contact with contaminated water and infected fish; (b) infection via the gastro-intestinal tract (GIT). Both experiments are primarily concerned with the ability of *A. salmonicida* to survive on and penetrate into the host's tissues, i.e. its ability to invade its host: they are not so strictly concerned with its virulence, although these two facets are obviously related. It is important to remember, however, that although many workers have compared virulence of *A. salmonicida* (usually by injection of live cells) surprisingly few (if any) studies of the comparative infectivity of strains of this organism have been carried out. The existence of strains of widely differing ability to invade test fish by the water-borne route would provide a powerful tool for investigation of the precise route and mechanism of entry of the bacterium into the fish.

(a) *Contact with contaminated water and infected fish*

An experiment to examine transmission by contaminated water was performed with six brown trout from a population of non-carriers, previously tested over 12 months by the PA technique with consistently negative results. Test fish were held under similar environmental conditions to those described above, with an open water circulation, except that the water temperature was adjusted to 16°C, a temperature at which epizootics of furunculosis frequently occur. A saline suspension of *A. salmonicida* cells, calculated to contain sufficient viable cells to produce a final concentration of $10^6 .ml^{-1}$ of tank water was pumped in continuously with the incoming water for a total of 5 days; the purpose of this procedure being not only to expose fish to a continuous infection rather than a single and perhaps less natural challenge, but also to simulate conditions which existed in the earlier experiment when a similar number of bacteria were released from dead diseased fish (Fig. 2). Five of the six fish died of furunculosis (on days 6, 8(2), 9 and 10) and when the single survivor was later challenged with PA it also succumbed to the disease. These results are somewhat surprising in view of the very short incubation time and clearly demonstrate the facility with which the disease may be spread to susceptible non-carrier fish by the water-borne route. A further experiment was then carried out with a larger number (50) of fish from the same non-carrier population of brown trout by transferring them to a fish farm affected by an active summer epizootic of furunculosis and placing them in a pond receiving water from infected and dying fish. Mortalities among test fish commenced on the eighth day and by the time they had ceased 28 days later, 41 fish had died. The nine remaining fish were then brought back to the laboratory and challenged with PA; all succumbed. This experiment again demonstrates the apparent ease with which the disease is spread through water. Furthermore, it indicates that non-carrier brown trout are extremely susceptible to furunculosis and that survivors will in all probability become carriers.

Because of these results, an experiment was carried out to compare the relative susceptibility of carrier and non-carrier brown trout when placed in contact with a naturally infected population. Twenty-seven brown trout, taken from a carrier population suffering from a natural active outbreak of furunculosis, were placed in a fish tank at 16°C. When the first mortality occurred amongst these fish, 15 carrier (carrier rate in population, *ca.* 60%) and 15 non-carrier brown trout were introduced. Mortalities among the two populations ceased 15 days after their introduction at which time 13/15 non-carriers, but only 7/16 carrier fish, had died. Fourteen days later the 11 surviving test fish were injected with PA; both surviving non-carriers, but only 3/9 carrier fish, succumbed. It is difficult to interpret these results precisely, but they suggest that non-carrier fish are apparently more susceptible than carrier fish. Moreover, the six survivors from the carrier population probably survived because they were immune non-carrier fish: consequently, it seems possible that immune fish need not be carriers and non-carriers are not necessarily immune. This experiment indicates that, although non-carrier fish are preferable, the extra cost involved in raising such fish for stocking may not be justified insofar as an enzootic disease such as furunculosis is concerned: indeed, under certain conditions, such fish may be at a distinct disadvantage. However, MacDermott & Berst (1968) reported no evidence of transmission of furunculosis in a Canadian stream during a two-year period after stocking with wild populations of infected and non-infected brook trout.

I recently witnessed an interesting case of initial introduction of furunculosis to the specific pathogen-free fish farm from which FDL, Weymouth, received its non-carrier experimental brown trout. On receipt of each new batch of fish a number were challenged with PA in order to ensure that the population was still free from the disease and for several years all fish had been consistently negative. However, early in the summer of 1976 a positive result was unexpectedly obtained and a carrier rate of *ca.* 70% subsequently established in the population. The farm is well managed and supplied solely with spring water which is free from feral fish, but investigations at the farm revealed that the severe drought conditions had compelled the manager to supplement his much diminished spring water supply with water from a nearby stream and, subsequent electro-fishing of this stream yielded 40 feral brown trout of which two succumbed to furunculosis when challenged with PA. The only possible explanation of this phenomenon is that carrier fish in the stream had spread the disease to the farm, since live fish have never been imported to the farm and very strict hygienic measures are employed. It is intriguing that a population of apparently healthy fish with a very low carrier rate can be responsible for infection of a large fish farm population. No clinical furunculosis has been observed on the farm (at the time of writing) due probably to good management techniques and no doubt to the abundance of extra water. It is important to remember, therefore, that a history of freedom from clinical furunculosis cannot be taken as evidence of absence of the disease, and also that large populations of captive non-carrier fish may be readily infected without clinical evidence of disease by small numbers of carrier feral fish.

Rainbow trout are known to be more resistant to furunculosis than brown trout, a fact which may readily be observed when both species are present on a fish farm affected by furunculosis, and which was confirmed experimentally by Blake & Clark (1931) when they placed seven brown and eight rainbow trout in co-habitation with an infected brown trout and reported that all seven browns but none of the rainbows succumbed. In the present study, simple addition of bacteria to tank water, as described above, failed to infect rainbow trout; however, if the fish's body flanks were roughly abraded with sandpaper, under anaesthesia, infection and death from furunculosis followed. It was mentioned earlier that fish species differ markedly in their susceptibility to furunculosis. The result of the present infectivity experiment with rainbow trout might suggest that this apparent resistance was due, at least in the case of this fish, to the inability of the organism to survive on or penetrate the intact skin. However, differences in resistance also exist among fish when injected with *A. salmonicida* cells, clearly demonstrated by Table 1 which lists the minimum lethal dose (MLD) for an attenuated strain of *A. salmonicida* and its virulent precursor strain for four different fish species.

Table 1
Minimal lethal dose of a virulent and an attenuated strain of A. salmonicida

Fish	Minimum lethal dose of Virulent strain	Minimum lethal dose of Attenuated strain
Brown trout	6.8×10^{1} *	4.0×10^{6}
Rainbow trout	3.4×10^{3}	9.5×10^{9}
Roach (*Rutilus rutilus*)	3.8×10^{6}	1.0×10^{11}
Carp (*Carpio carpio*)	6.0×10^{9}	4.2×10^{9}

Water temperature 11–13°C.
* Viable cells.

(b) *Infection via the gastro-intestinal tract*

The possibility that fish might be infected by the gastro-intestinal tract (GIT) was investigated by feeding a commercial pelleted diet, previously infected with *A. salmonicida*, daily for a total of five days, to 30 brown trout taken from a non-carrier population and held in a fish tank at 16°C. The infected food was prepared by soaking it in a saline suspension containing *ca.* 10^8 viable cells.ml^{-1}, and the presence of viable *A. salmonicida* cells confirmed each day prior to feeding. No mortalities occurred within seven days of cessation of feeding infected food, nor did any of the test fish succumb to subsequent PA challenge. This result conflicts with reports by Plehn (1911) and Blake & Clark (1931) who both claimed success in infecting fish by feeding contaminated food. However, the present result is consistent with that reported by Krantz *et al.* (1964*b*) who also failed to infect brown trout by the GIT with food containing 10^8 viable cells.g^{-1}. It is possible, as suggested by Krantz *et al.* (1964*b*), that discrepancies in the results might be due to differences in the experimental and environmental conditions.

A further experiment was carried out to study the fate of *A. salmonicida* when introduced into the GIT of fish. Rainbow trout, held at 16°C, were each force-fed by a stomach tube, with *ca.* 10^7 viable cells of the streptomycin-resistant strain in fish food paste. Test fish were then removed at 5, 12, 24, 36 and 48 h after feeding and the stomach, foregut, hindgut and kidney examined for the presence of the bacterium by maceration, dilution with saline and subsequent inoculation on TSA plates containing 1000 μg.ml^{-1} of streptomycin. Figure 3 shows that test fish eliminated the bacterium within 48 h of feeding, but it is highly significant that the streptomycin-resistant strain was detected in kidney tissue from 5 h onwards, albeit in low numbers. Although no mortalities

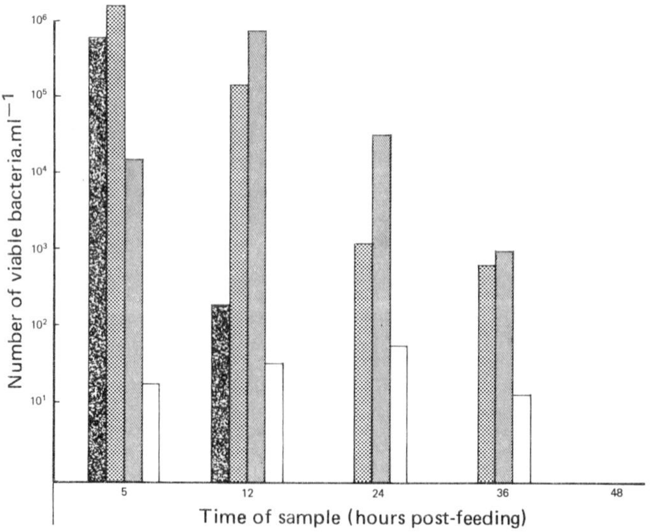

Fig. 3. Fate of an intragastric inoculum of *A. salmonicida* in rainbow trout (*Salmo gairdneri*) at 16°C. Number of viable cells in: stomach (■); foregut (▒); hindgut (▨); kidney (□).

occurred during this experiment, despite almost certain abrasion of the oesophagus and stomach tissues (through which the organism possibly gained access to the kidney), infection via the GIT must remain a possibility in view of the report of Klontz & Wood (1972) who observed clinical furunculosis in the sable fish, apparently resulting from ingestion of carrier Coho salmon (*Oncorhyncus kisutch*).

5. Vertical Transmission of Furunculosis in the Fish Farm Environment

Vertical or transovarian transmission of *A. salmonicida* is regarded as a possible route of infection (Snieszko 1974); however, published evidence is inconclusive; for example, Smith (1939; cited by McFaddon 1969) claimed that the organism was carried on the egg surface while Plehn (1911) and Mackie *et al.* (1930) considered that the bacterium was unable to infect fish eggs. Furthermore, although Mackie *et al.* (1930) and Lund (1967) both failed to isolate *A. salmonicida* from fertilized eggs obtained from experimentally infected fish, Lund (1967) described a positive isolation from the interior of fertile eggs from naturally infected mature fish. Moreover, the bacterium has also been isolated from the ovary and testes of infected mature fish (Lund 1967; McDermott & Berst 1968). In the present study, several experiments were carried out to investigate the possible existence of vertical transmission of furunculosis.

(a) *Carrier brood stock*

It was noted earlier that as carrier fish are present on a fish farm throughout the year there is the possibility of such fish transmitting the bacterium with their sex products.

Eight fully mature brood stock brown trout (four male and four female), taken from a known carrier population, were stripped and the eggs fertilized in the usual way. A 30 ml volume of fertilized eggs ('green' eggs) was thoroughly crushed with an equal volume of saline in a 'stomacher' (A. J. Seward & Co Ltd, London) and the viable count assessed in the usual way after growth on duplicate TSA plates incubated for seven days at 22°C. After stripping, the eight fish were then challenged with PA and although five of the eight (three male and two female) proved to be carriers, *A. salmonicida* was not isolated from the fertilized egg sample.

A further eight brood stock fish from the same carrier population were first injected with PA and, after a post-injection incubation period of four days to allow for activation of the latent disease, were stripped and the eggs fertilized and tested as described above. All test fish were then sacrificed and swabs taken from the kidneys for bacteriological examination. Although *A. salmonicida* was isolated from five of the eight kidney swabs (two male and three female) again, the organism was not recovered from the egg sample. Some difficulty was experienced in stripping PA-injected fish, presumably as a result of their incubation at 18°C; however, sufficient fertilized eggs were obtained for experimentation purposes.

(b) *Artificially infected brood stock*

Although active epizootics of furunculosis have been reported during winter (Klontz & Wood 1972) they are, at least in the UK, relatively rare and unfortunately it proved impossible to obtain natural actively infected brood stock fish for experimentation. Instead, eight carrier brown trout (four male and four female) were inoculated by the intramuscular route with 10 times the minimal lethal dose of the streptomycin-resistant strain and held at 10°C. After the first mortality occurred (day 6) the remaining seven fish, all by now exhibiting well-developed signs of furunculosis, were stripped and the eggs fertilized and tested as described above with the sole exception that duplicate plates containing 1000 μg of streptomycin were also inoculated. All test fish were then autopsied and samples taken for bacteriological examination from the injection site, kidney, spleen, heart blood and retained ova or milt. All were positive for *A. salmonicida*. The fertilized eggs were also positive (Fig. 4) and, although *ca.* 10^6 viable cells.ml^{-1} of eggs were present when first tested, numbers of *A. salmonicida* rapidly declined thereafter, not being detected when

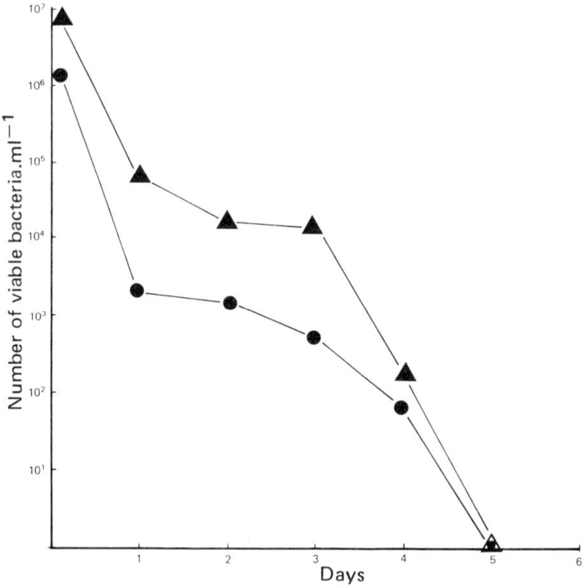

Fig. 4. Survival of *A. salmonicida* on fertilized fish ova: ova infected *in vitro* (▲); ova produced by experimentally infected fish (●).

the eggs were tested after five days incubation in the hatchery water and also when tested after a further two days. It is of interest that when the green eggs were tested immediately after fertilization they were, except for the presence of *A. salmonicida*, apparently sterile, a finding confirmed to some extent by the scanning electron microscopy (Fig. 5a). However, after incubation in hatchery water *A. salmonicida* appeared to be quantitatively displaced by organisms (myxobacteria and pseudomonads) similar to those described as commensal inhabitants of the salmonid egg surface by Bell *et al.* (1971) and Trust (1972).

These experiments indicate that vertical transmission of *A. salmonicida* is not a significant route of transmission for furunculosis. Moreover, in the unlikely event that heavily infected fish are used for egg production, and if fertilized eggs are accidentally contaminated, the organism is unlikely to survive to the eyed-egg stage at which they are sold. It is important to remember, however, that the stress imposed by spawning and the attendant netting may provoke furunculosis among a carrier population.

(c) *Infection experiments with fish ova* in vitro

Attempts to infect eyed eggs by simple shaking with *A. salmonicida* cells as described by Gee & Sarles (1942) were not successful. Instead, eggs were

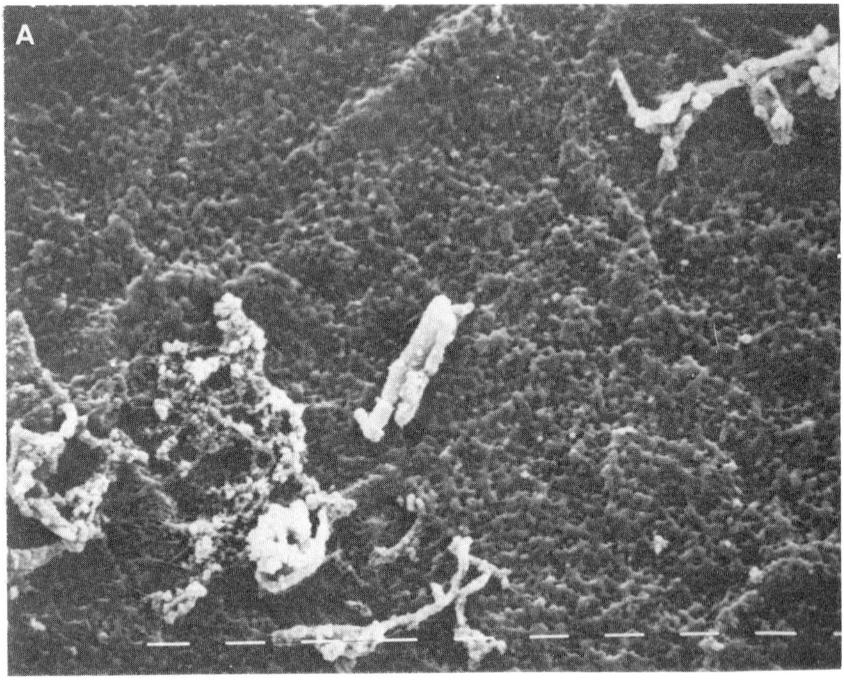

Fig. 5. Scanning electron micrograph of the surface of fertilized brown trout (*Salmo trutta*) ova (each scale unit: 1 μm). (a) Contaminated 'green' ova produced by experimentally infected fish showing cells of *Aeromonas salmonicida*. (b) Normal eyed ova from hatchery water illustrating mixed microbial flora.

incubated for 16 h at 4°C with an equal volume of saline containing *ca.* 10^8 viable cells of the streptomycin-resistant strain. After incubation an 'immediate' sample was taken and the viable count of *A. salmonicida* assessed as described above, the remaining eggs being placed in the hatchery water and re-tested daily for up to seven days. The results obtained (Fig. 4) were essentially similar to those obtained for 'green' eggs; that is, when tested immediately after a 16 h incubation period, *A. salmonicida* was isolated on ordinary TSA, virtually in pure culture but in later samples *A. salmonicida* was gradually replaced by bacteria from the hatchery water and inspection of the streptomycin agar plates became essential for detection of the organism. It is obvious from Fig. 5(b) that the bacterial flora of normal uninfected eyed eggs is very rich and varied.

A further batch of eyed eggs was infected artificially, late in their incubation period, as described above and then replaced in the hatchery water. The free-swimming sac fry which emerged nine days after artificial infection were tested for the presence of *A. salmonicida* two days after emergence and were found to be positive. It is presumably possible therefore that if eyed eggs are

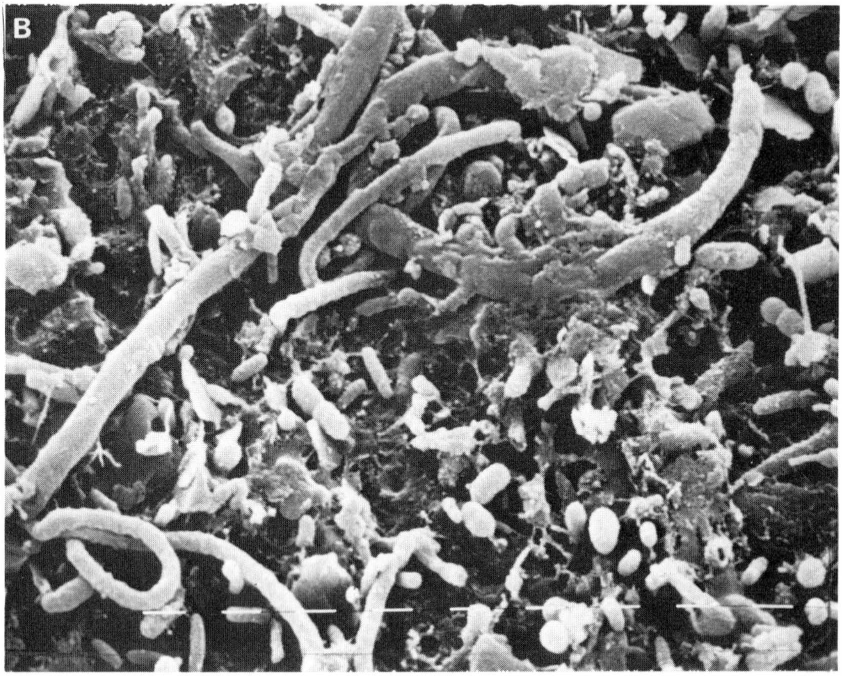

constantly in contact with very large numbers of *A. salmonicida* cells, then contaminated sac fry may be produced. The significance of this finding is at present being investigated.

(d) *Disinfection of fish ova*

Herman (1972) advised routine disinfection of fish ova because of the possibility of vertical transmission of furunculosis. Blake (1930) recommended acriflavine for disinfection of fish eggs, but McFadden (1969) asserted that an iodophor compound, Povidone iodine, was more reliable than acriflavine for disinfection of salmonid eggs contaminated with *A. liquefaciens* (syn. *A. hydrophila*). In 1972 Ross & Smith demonstrated that two iodophors, Wescodyne and Betadine, rapidly killed saline suspensions of *A. salmonicida* and certain other fish pathogenic bacteria. Since then, the use of iodophors has been generally recommended for routine disinfection of salmonid ova possibly contaminated with *A. salmonicida* (Frantsi & Withey 1972; Snieszko 1974).

In the present study an iodophor, Wescodyne (Ciba-Geigy, Cambridge, England) and acriflavine (BDH Ltd, Poole, Dorset), were compared for their ability to disinfect (a) infected 'green' eggs produced by artificially infected brood stock and (b) eyed eggs artificially infected *in vitro* with *A. salmonicida*. Ova for disinfection were suspended in 10 times their volume of hatchery water (pH 7.2) containing the relevant concentration of disinfectant. Following frequent agitation for 10 min (Wescodyne) or 20 min (acriflavine) eggs were removed, placed in running hatchery water for 30 min and then tested, as described above, for the presence of the streptomycin-resistant strain of *A. salmonicida* for up to 10 days. Both types of untreated infected eggs were tested in parallel with disinfected eggs to act as a positive control. The results obtained (Table 2) do not support recommendations that Wescodyne should be used to

Table 2

Results of disinfection of infected eyed and 'green' eggs with Wescodyne and Acriflavine

Disinfectant	($mg.l^{-1}$)	'Green' eggs	Eyed eggs
Wescodyne	50	+*	+
	100	+	+
	150	+	+
	200	–	–
Acriflavine	1000	–	–
	500	+	NT†

* *A. salmonicida* present.
† Not tested.

disinfect eggs infected with *A. salmonicida* since both the usually recommended levels (50 and 100 $mg.l^{-1}$) were inactive with contaminated eyed and 'green' eggs. Furthermore, although treatment with 200 $mg.l^{-1}$ of Wescodyne successfully disinfected eggs, this concentration is above the safe non-toxic limit (100 $mg.l^{-1}$) recommended for iodophor compounds (Amend 1974). It was also found that a stronger acriflavine solution (1/1000), than recommended previously (1/2000), was necessary for successful disinfection of contaminated ova. However, this compound is non-toxic to fish ova and can safely be used at the stronger concentration.

It may now be argued that, in the light of the negative vertical transmission experiments and the demonstrated short survival of *A. salmonicida* on fish eggs, routine disinfection of eyed eggs is unnecessary. However, eggs may be infected by continuous exposure to contaminated hatchery water, so if very large numbers of *A. salmonicida* are present, then it is probably best to disinfect eyed eggs before they are removed from, and when they are received on, a fish farm.

The present work did not corroborate the report of Lund (1967) who claimed to have isolated *A. salmonicida* from the interior of fish ova, since in the disinfection experiment with infected 'green' eggs acriflavine could only be expected to disinfect the egg surface, and if the bacterium had in fact been present within the egg, as speculated, its presence would certainly have been revealed by the maceration techniques used. However, it was not recovered.

In order to assess the practical significance of these experiments it was decided to follow 'bacteriologically' the progress of a batch of fertilized eggs, taken from known carrier brood stock, 'through' a fish farm with a known carrier rate among its fish. Samples of this population examined to date for the presence of *A. salmonicida* are: eyed eggs (three and six weeks); sac fry (two and six weeks); first-feeding fry (10, 14 and 20 weeks): all the foregoing test fish and eggs were held in spring water and segregated from both feral and adult captive fish. Eggs were examined for the presence of *A. salmonicida* as described above and fish were macerated (Silverson Machines Ltd, London) in twice their volume of 0.9% NaCl and decreasing serial decimal dilutions of macerate in 0.9% NaCl streaked on TSA plates. All test samples, judged by the absence of colonies of *A. salmonicida* on agar plates, proved negative. However, it must be borne in mind that *A. salmonicida* may have been inhibited or overgrown by commensal bacteria, an observation made previously in this study. A further sample of feeding fry was tested at 30 weeks, four weeks after having been removed from spring water and held in a fish pond supplied with water inhabited by adult fish, and found to be negative. This last sample was tested at the same time as the 1+, 2+ and 3+ age-groups (described in Section 3.d) which had a high carrier rate of furunculosis. However, it must again be pointed out that the feeding fry were examined by simple maceration, being too small for injection with PA. This experiment will continue throughout the year and when fish are large enough they will be challenged with PA.

6. General Discussion

The present work is part of a larger study still in progress and is therefore incomplete; nevertheless, it has provided sufficient information to propose certain hypotheses concerning the epizootiology of furunculosis which can be tested by future experiments. Some of these hypotheses are speculative but if they provoke argument or, better still further work, they will have achieved their aim.

The first hypothesis is that furunculosis is introduced to a fish farm by importation, not of contaminated fish ova but of healthy carrier fish: alternatively, non-carrier resident fish may be infected from a water supply contaminated by feral or imported captive carrier fish. Thus, continual segregation of a population from carrier fish is all that is required to produce, and also to maintain, a specific pathogen-free stock. If, however, stock must be introduced to a farm or other water body free from contact with feral fish, then it is recommended that they should be tested by injection with prednisolone acetate, preferably during the summer months, at a laboratory close to their origin. It is not, of course, envisaged that this procedure would be carried out on a fish farm, but if there is strong objection to allowing clinical furunculosis to develop among test fish then they may be sacrificed four days after injection and their kidneys tested for the presence of *A. salmonicida*.

Secondly, the mechanism by which carriers spread the disease to non-carrier fish depends mainly upon environmental conditions at the fish farm in question. In the absence of severe stress conditions resident fish can become carriers without exhibiting clinical evidence of the disease: moreover, under such conditions a high carrier rate can be maintained on a farm, without clinical evidence of furunculosis. This is not consistent, however, with the observation of Mackie & Menzies (1938) who were of the opinion that carrier fish die "sooner or later" of clinical furunculosis. Furthermore, in the absence of stress conditions and also vertical transmission, the disease must be perpetuated on a fish farm by annual summer infection of the 0+ population and since this age-group is relatively more resistant to furunculosis (Plehn 1911; Krantz *et al.* 1964; McCarthy, unpublished observations) they may be expected successfully to combat a challenge with the bacterium, perhaps developing a subclinical infection and becoming carriers to the disease. This hypothesis can be disputed on the grounds that at the present time, probably because of technical problems, it has not been possible to isolate the bacterium from the faeces of healthy carrier fish or detect it in their water supply. However, the sudden appearance of large numbers of *A. salmonicida* cells in the gastro-intestinal tract (48 h post-injection) of carrier fish injected with prednisolone acetate, is consistent with multiplication of the organism *in situ.* Moreover, although it could not be substantiated by the present worker, Klontz (1968) reported detection of the organism by immunofluorescence in the gut wall of carrier fish. It is likely, therefore, that in the absence of severe stress conditions healthy carrier fish infect non-carriers by shedding bacteria in their faeces, perhaps in greatly increased numbers when mildly stressed. However, in the presence of severe stress conditions introduction of carrier fish will result in severe mortalities among the non-carrier resident populations, most survivors becoming carrier fish and the disease being established on the farm.

Whatever the mechanism of their creation, once carrier fish are present on a fish farm they are responsible for initiation of epizootics. Adverse environmental conditions, such as low oxygen levels (Kingsbury 1961) exert considerable stress on the fish's limited homeostatic mechanisms (Wedeymeyer 1974), probably overwhelming the protection afforded by the carrier state. Simple activation of latent infections among the four brown trout populations examined in the present study would result in mortalities in the region of 40–80% without infection occurring. However, infection does occur by the water-borne route, perhaps aided by the use of contaminated nets, and the organism does survive for a significant length of time outside its host. Although it is not suggested that the bacterium enjoys a free-living life cycle in pond water, the present study has demonstrated that large numbers of the bacterium will be present in water during, and for a long time after, an epizootic. Furthermore, the demonstration that the commonly employed chemotherapeutics do not entirely remove the bacterium from infected fish, together with the organism's proved longevity in water, explains common recrudescence of the disease, particularly if it has not been possible to restore a favourable environment.

Epizootics among salmonid fish held in sea water are probably also initiated by stressed carrier fish. Moreover, the natural susceptibility of the marine sable fish (Evelyn 1971) together with successful experimental infection of the plaice (*Pleuronectes platessa*) by I. W. Smith (pers. comm.) and also possible transmission of the disease in sea water (Scott 1968) are of considerable epizootiological importance to those comtemplating mixed culture of salmonids and marine fish. Furthermore, the possibility that scavenger fish such as saithe, *Gadus* (*Pollachius*) *virens*, which abound near fish culture sea cages, attracted by excess food, acting as carriers of the disease must be seriously considered since they will undoubtedly eat moribund or dead carrier salmonid fish; Klontz & Wood (1972) have demonstrated that the marine sable fish can be infected in this way. Although non-carrier salmonids may be at a disadvantage when used to stock open fresh waters, because these may contain or be subsequently stocked with carrier fish with perhaps disastrous consequences, they have considerable advantages for use in marine fish farming.

The third hypothesis is that the carrier state is probably a synergistic relationship, the organism providing a measure of protection to the fish in return for shelter from the aquatic environment. This phenomenon would explain the marked difference in susceptibility to furunculosis noted in the present study between non-carrier and carrier fish and also the sudden outbreaks of furunculosis among 0+ fish since they are also non-carriers in the presence of a carrier population and, if held under stress conditions, perhaps insufficient to activate latent disease in older fish, yet sufficient to overwhelm their innate resistance, they will succumb. More research into the significance of the carrier state is required and one particular area that would repay investigation concerns the effect of injecting or feeding vaccines in order to protect fish populations against a disease such as furunculosis when as many as 80% of its individuals are already carrying the causative organism.

In conclusion, if carriers are present on a fish farm, or the fish farm receives water from a source containing feral fish then a satisfactory environment must be maintained at all times if the farm is to remain free from clinical disease. If the disease occurs then treatment must be correctly applied (McCarthy 1975*b*) to prevent, as far as is possible, recrudesence. If vaccination of a population is contemplated its aim must be to produce an immune non-carrier stock by immunization of the 0+ age-group fish.

7. Acknowledgement

I am grateful to Mr W. F. Sheldon of Pye Unicam Ltd, Cambridge, England, for undertaking the scanning electron microscopy.

8. References

AMEND, D. F. 1974 Comparative toxicity of two iodophers to rainbow trout eggs. *Transactions of the American Fisheries Society* **103**, 73–78.

BELL, G. R., HOSKINS, G. E. & HODGKISS, W. 1971 Aspects of the characterisation, identification, and ecology of the bacterial flora associated with the surface of stream-incubating Pacific salmon (*Oncorhynchus*) eggs. *Journal of the Fisheries Research Board of Canada* **28**, 1511–1525.

BLAKE, I. 1930 *The External Disinfection of Fish Ova with Reference to the Prophylaxis of Furunculosis.* Fishery Board for Scotland. Salmon Fisheries No. 2. Edinburgh: H.M.S.O.

BLAKE, I. & CLARK, J. C. 1931 Observations on experimental infection of trout by *B. salmonicida*: with particular reference to "carriers" of furunculosis and to certain factors influencing susceptibility. *Fishery Board for Scotland. Salmon Fisheries* **7**, 1–13.

BULLOCK, G. L. & STUCKEY, H. M. 1975 *Aeromonas salmonicida*: detection of asymptomatically infected trout. *Progressive Fish Culturist* **37**, 237–239.

CHRISTENSEN, N. O. 1972 In *Diseases of Fish*, ed. Mawdesley-Thomas, L. E. Symposia of the Zoological Society of London No. 30. London & New York: Academic Press.

CORNICK, J. W., CHUDYK, R. V. & MCDERMOTT, L. A. 1969 Habitat and viability studies on *Aeromonas salmonicida*, causative agent of furunculosis. *Progressive Fish Culturist* **31**, 90–93.

DANSO, S. K. A., HABTE, M. & ALEXANDER, M. 1973 Estimating the density of individual bacterial populations introduced into natural ecosystems. *Canadian Journal of Microbiology* **19**, 1450–1451.

EMMERICH, R. & WEIBEL, E. 1894 Uber eine durch Bakterien erzeugte Seuche unter den Forellen. *Archiv für Hygiene und Bakteriologie* **21**, 1–21.

EVELYN, T. P. T. 1971 An aberrant strain of the bacterial fish pathogen *Aeromonas salmonicida* isolated from a marine host, the sablefish (*Anoplopoma fimbria*), and from two species of cultured Pacific salmon. *Journal of the Fisheries Research Board of Canada* **28**, 1629–1634.

FRANTSI, C. & WITHEY, K. G. 1972 *A procedure for disinfecting Atlantic salmon (*Salmo salar*) eggs using a polyvinylpyrrolidone-idodine (PVP-I) solution.* Fisheries Service, Department of the Environment Progress Report No. 6, pp. 1–8.

GEE, L. L. & SARLES, W. B. 1942 The disinfection of trout eggs contaminated with *Bacterium salmonicida. Journal of Bacteriology* **44**, 111–126.

GHITTINO, P. 1972 In *Diseases of Fish*, ed. Mawdesley-Thomas, L. E. Symposia of the Zoological Society of London No. 30. London & New York: Academic Press, pp. 25–38.

GUTSELL, J. S. & SNIESZKO, S. F. 1949 Response of brook, rainbow and brown trout to various dosages of sulfamerazine. *Transactions of the American Fisheries Society* 77 (1947), 93–101.

HERMAN, R. L. 1968 Fish furunculosis 1952–1966. *Transactions of the American Fisheries Society* **97**, 221–230.

HERMAN, R. L. 1972 A review of the prevention and treatment of furunculosis. *FI:EIFAC 72/SC II* Symposium 19, 1–6.

KIMURA, T. 1970 Studies on a bacterial disease of adult "Sakuramasu" (*Oncorhynchus masou*) and pink salmon (*O. gorbuscha*) reared for maturity. *Scientific Report of the Hokkaido Salmon Hatchery* **24**, 9–100.

KINGSBURY, O. R. 1961 A possible control of furunculosis. *Progressive Fish Culturist* **23**, 136–137.

KLONTZ, G. W. 1968 Immunopathology. In *Progress in Sport Fishery Research, 1970.* Resource Publication of the Bureau of Sport Fisheries and Wildlife.

KLONTZ, G. W. & WOOD, J. W. 1972 Observations on the epidemiology of furunculosis disease in juvenile Coho salmon (*Oncorhynchus kitsutch*). *FI:EIFAC 72/SC II* – Symposium **27**, 1–8.

KRANTZ, G. E., REDDECLIFF, J. M. & HEIST, C. E. 1964*a* Immune response of trout to *Aeromonas salmonicida* Part 1 Development of agglutinating antibodies and protective immunity. *Progressive Fish Culturist* **26**, 3–10.

KRANTZ, G. E., REDDECLIFF, J. M. & HEIST, C. E. 1964*b* Immune response of trout to *Aeromonas salmonicida* Part 2 Evaluation of feeding techniques. *Progressive Fish Culturist* **26**, 65–69.

LUND, M. 1967 *A study of the biology of* Aeromonas salmonicida. MSc Thesis, Department of Agriculture, University of Newcastle-upon-Tyne.

MCCARTHY, D. H. 1975*a* Fish furunculosis caused by *Aeromonas salmonicida* var. *achromogenes. Journal of Wildlife Diseases* **11**, 489–493.

MCCARTHY, D. H. 1975*b* Fish furunculosis. *Journal of the Institute for Fisheries Management* **6**, 13–17.

MCCRAW, B. M. 1952 Furunculosis of fish. *United States Fish and Wildlife Service Special Scientific Report* **84**, pp. 87.

MCDERMOTT, L. A. & BERST, A. H. 1968 Experimental plantings of brook trout (*Salvelinus fontinalis*) from furunculosis-infected stock. *Journal of the Fisheries Research Board of Canada* **25**, 2643–2649.

MCFADDEN, T. W. 1969 Effective disinfection of trout eggs to prevent egg transmission of *Aeromonas liquefaciens. Journal of the Fisheries Research Board of Canada* **26**, 2311–2318.

MACKIE, T. J. & MENZIES, W. J. M. 1938 Investigations in Great Britain of furunculosis of the *Salmonidae. Journal of Comparative Pathology and Therapeutics* **51**, 225–234.

MACKIE, T. J., ARKWRIGHT, J. A., PRYCE-TANNATT, T. E., MOTTRAM, J. C., JOHNSTON, W. D. & MENZIES, W. J. M. 1930 *Interim Report of the Furunculosis Committee.* Edinburgh: H.M.S.O.

MACKIE, T. J., ARKWRIGHT, J. A., PRYCE-TANNATT, T. E., MOTTRAM, J. C., JOHNSTON, W. D. & MENZIES, W. J. M. 1933 *Second Interim Report of the Furunculosis Committee.* Edinburgh: H.M.S.O.

MACKIE, T. J., ARKWRIGHT, J. A., PRYCE-TANNATT, T. E., MOTTRAM, J. C., JOHNSTON, W. D. & MENZIES, W. J. M. 1935 *Final Report of the Furunculosis Committee.* Edinburgh: H.M.S.O.

PLEHN, M. 1911 Die Furunkulose der Salmoniden. *Zentralblatt für Bakteriologie, Parasitenkunde, Infetionskrankheiten und Hygiene, Abt. I* **60**, 609–624.

ROSS, A. J. & SMITH, C. A. 1972 Effect of two iodophors on bacterial and fungal fish pathogens. *Journal of the Fisheries Research Board of Canada* **29**, 1359–1361.

SCOTT, M. 1968 The pathogenicity of *Aeromonas salmonicida* (Griffin) in sea and brackish waters. *Journal of General Microbiology* **50**, 321–327.

SMITH, I. W. 1962 *Furunculosis in kelts.* Department of Agriculture & Fisheries for Scotland, Freshwater & Salmon Fisheries Research No. 27.

SNIESZKO, S. F. 1974 *Fish furunculosis. EIFAC* Symposium, FID:CFD/74/Inf, 197–199.

TRUST, T. J. 1972 The bacterial population in vertical flow tray hatcheries during incubation of salmonid eggs. *Journal of the Fisheries Research Board of Canada* **29**, 567–571.

WEDEMEYER, G. A. 1974 *Stress as a predisposing factor in fish diseases.* United States Department of the Interior, Fish and Wildlife Service Leaflet FDL-38.

WILLIAMSON, I. J. F. 1928 *Furunculosis of the Salmonidae.* Fishery Board for Scotland, Salmon Fisheries No. 5. Edinburgh: H.M.S.O.

Current Problems in the Study of the Biology of Infectious Pancreatic Necrosis Virus and the Management of the Disease it Causes in Cultivated Salmonid Fish

A. L. S. MUNRO AND I. B. DUNCAN

Department of Agriculture and Fisheries for Scotland, Marine Laboratory, Victoria Road, Aberdeen, Scotland

CONTENTS

1. Introduction

INFECTIOUS PANCREATIC NECROSIS VIRUS (IPNV) causes a well-characterized acute disease (Wolf *et al.* 1960; Vestergaard Jørgensen 1974) in hatchery reared rainbow trout (*Salmo gairdneri*) and brook trout (*Salvelinus fontinalis*) during the 3–5 months after fry start to feed. A disease closely resembling IPN was first described by McGonigle (1940) in Canada. That IPN is due to infection with a virus was first proposed by Wood *et al.* (1955) and substantiated by Wolf *et al.* (1960); since then several serotypes of the virus have been identified by neutralization with specific antisera (Vestergaard Jørgensen & Grauballe 1971; Vestergaard Jørgensen & Kehlet 1971; Wolf & Quimby 1971; McMichael *et al.* 1975) as well as by complement fixation (Finlay & Hill 1975).

2. Nature of Infectious Pancreatic Necrosis Virus

The inefficacy of certain metabolic inhibitors on the growth of IPNV (Malsberger & Cerini 1965) identified its genetic material ('genome') as ribonucleic acid (RNA): this has been confirmed more directly by the incorporation of radioactive uridine into the genome (Kelly & Loh 1972; Cohen *et al.* 1973; Cohen & Scherrer 1974; Alayse *et al.* 1975). In early attempts to gauge the size of the free particle ('virion') of IPNV, it was found that infectivity was not retained by ultrafilters, even at a porosity of 50 nm (Wolf 1964) and examination of negatively stained preparations in the electron microscope revealed roughly spherical moieties of *ca.* 20 nm diam. Initially, then, IPNV was placed among the picornaviruses.

Later publications (Kelly & Loh 1972; Cohen *et al.* 1973; Cohen & Scherrer 1974) showed a clearly icosahedral virion, but quoted a much larger diameter (of the order of 60–70 nm); for this reason, and because of the overall appearance of its growth process in the cytoplasm of infected cells (Moss & Gravell 1969; cf. Gomatos *et al.* 1972), IPNV came to be compared with the reoviruses, a large group of RNA-containing pathogens of the plant and animal kingdoms. Even this revised classification, however, warrants further inspection, as the following brief summary is intended to show.

First of all, the double-stranded RNA genome of reoviruses is divided into 10 or more different-sized pieces (Verwoerd 1970; Joklik 1974). There are two main opinions regarding the structure of IPNV RNA; one advocates a double-stranded genome (Argot & Malsberger 1972; Cohen *et al.* 1973; Cohen 1975), the other a fundamentally single-stranded molecule (Wolf 1964; Nicholson 1971*a*; Kelly & Loh 1972, 1975). A recent and particularly careful study by P. Dobos (1976*a*, pers. comm.) indicates that the first alternative is likely to be true, but that the RNA contained within the IPNV virion is of only two size classes. Secondly, the protein ('capsid') which encloses the genome of IPNV consists of a single shell (Kelly & Loh 1972; Cohen *et al.* 1973; Cohen & Scherrer 1974) whereas reoviruses characteristically have two concentric shells (Joklik 1974); this structural difference is accompanied by a difference in the protein composition of the capsids (Dobos 1976*b*) and in the buoyant density of the virions (Kelly & Loh 1972; Cohen *et al.* 1973; Cohen & Scherrer 1974; Joklik 1974). Finally, the early part of the IPNV growth cycle in tissue culture displays a far greater sensitivity to the antimetabolite actinomycin D than does that of the reoviruses (Malsberger & Cerini 1965; Nicholson 1971*b*; Kelly & Loh 1975), and indeed the processes of nucleic acid (Alayse *et al*. 1975; Cohen 1975) and protein metabolism (Dobos 1976*b*) in the two cases seem to follow entirely different pathways. IPNV resembles the reoviruses, then, to the extent that its virion is icosahedral and of appropriate size, and contains double-stranded RNA. However, the points of difference outlined above tend to prevent the inclusion of IPNV in this important group of infectious agents. Where then, should it lie?

Infectious bursal disease causes a severe inflammation of the bursa of Fabricius – a lymphoid organ situated next to the hindgut in chickens. The RNA virus (IBDV) which gives rise to this condition is strikingly similar to IPNV in several respects; first, in its size and shape (Almeida & Morris 1973; Pattison *et al.* 1975; Harkness *et al.* 1975; Nick *et al.* 1976); second, in its having only one protein shell (Almeida & Morris 1973; Nick *et al.* 1976); third, in its low buoyant density (Pattison *et al.* 1975; Nick *et al.* 1976) and fourth in the microscopic appearance of its development in infected cells (Lunger & Maddux 1972). The overall impression gained is that IPNV and IBDV may be more closely related to each other than to the known reoviruses. Furthermore, Hill (1976) has isolated an IPN-like virus from the marine bivalve *Tellina tenuis*, and it is conceivable that this virus, together with IPNV and IBDV, might be the founder members of a new taxonomic group.

Finally, it is interesting to note that an auxiliary, smaller particle of *ca.* 20 nm diam. has been seen in preparations of IBDV (Almeida & Morris 1973; Pattison *et al.* 1975; Harkness *et al.* 1975). It has been suggested that this small particle could be a 'satellite' virus, (Almeida & Morris 1973) though it could be a virion breakdown product instead (Almeida & Morris 1973; Harkness *et al.* 1975). By invoking a similar situation for IPNV (although as yet there is no clear evidence), one could attempt to explain the early electron microscopic observation of 20 nm particles in this system, rather than admit to the somewhat embarrassing alternative that IPN has been attributed consecutively to at least two distinct viruses!

3. Growth of Infectious Pancreatic Necrosis Virus

IPN virus will grow in a variety of cultured fish cells in the laboratory, at an optimal temperature of *ca.* 20°C (Scherrer *et al.* 1974). Infected cultures release virus steadily into the surrounding medium, though an appreciable amount of infectivity remains cell associated (Malsberger & Cerini 1965; Moss & Gravell 1969; Piper *et al.* 1973); yields of the order of 10^8 infectious units.ml^{-1} of tissue culture fluid are regularly obtainable.

Apparently not all serotypes of IPNV are able to grow equally in any given cell line (Vestergaard Jørgensen & Kehlet 1971). At the same time, there are indications that on regular passage, the virus will adapt rapidly to the type of tissue culture in use in preference to others (Scherrer & Cohen 1975); such adaptability would have an obvious bearing on the potential host range of IPNV in its natural environment.

It has been recorded (Malsberger & Cerini 1963; Nicholson & Dunn 1974) that serial passage of IPNV in tissue culture at consistently high infecting doses of virus per cell can lead to the eventual depression of infectious yields. This effect, known as 'homologous interference' or 'autointerference' has been observed in many other virus systems (Huang 1973) and may in fact be general. It is caused by the creation, from infectious (or 'standard') virus, of so-called 'defective interfering' or 'DI' particles (Huang & Baltimore 1970) which depend for their growth and survival on the assistance of standard virus, in the course of which they inhibit the growth of the latter. The common features of such DI particles, irrespective of virus type, are: (a) that they contain only a portion of the normal genome, and are consequently non-infectious; (b) that they contain the normal structural proteins, and are therefore serologically identical to the standard virus from which they are derived; (c) that they interfere *specifically* with the original, standard virus, and that this interference is not mediated through interferon.

The DI particles of IPNV have yet to be properly isolated and characterized; however it can be said at least that they are specific in their interference effect and are inactivated by immune serum (Nicholson & Dunn 1974).

One may ask whether DI particles have a particular biological role to play? Certainly it would be of no evolutionary advantage for a pathogen to be too rapidly destructive in its encounters with the susceptible hosts on which its survival depends, since this would limit its chance of spreading through the population, and any mechanism which could be depended upon to regulate the growth of virus might therefore be of overall benefit. The outcome of natural infections, then, could be controlled as much by viruses themselves as by, say, ambient temperature and the age, immunological competence and hormonal activity of their hosts.

In tissue culture, the interaction between DI particles and standard virus, if it strikes a certain balance, can result in a protracted infection; this is apparently as true for IPNV (Nicholson & Dexter 1975; B. L. Nicholson pers. comm.) as it is for some other virus systems (e.g. Huang 1973; Palma & Huang 1974). It is relevant to note that isolates of IPNV from asymptomatic, 'carrier' fish (see below) showed signs of a comparatively high level of interfering activity (Nicholson & Dexter 1975), and it has been proposed (Huang & Baltimore 1970) that DI particles effect the course of many viral infections in nature: one may construct models to account for self-imposed termination of infection, for recurrence of disease and for various forms of persistence.

The interfering activity of DI particles, together with their non-infectiousness and their serological identity with standard virus, might lead one to suppose that they could serve very usefully as antiviral agents. However, present evidence points to the hazards underlying such an approach, for DI particles are able to convert a potentially straightforward, acute infection to one which is more persistent or otherwise less amenable to diagnosis and treatment, so increasing the problem of disease control.

4. Pathology of Infectious Pancreatic Necrosis Disease

A clinically acute disease condition occurs in very young fish, the only consistent histopathological feature being a focal or generalized destruction of exocrine pancreatic tissue (Wood *et al.* 1955; Yasutake *et al.* 1965; Yasutake 1970). The extent of this damage varies considerably and often intact acini can be found which are contiguous with areas of complete necrosis. The necrotic changes involve rupture of infected cells with release of zymogen granules, or a slower degeneration with shrinkage and basophilia of cytoplasm, coupled with pyknosis of the nucleus, often associated with a perinuclear halo. These, and also cytoplasmic debris have been seen in light microscopy preparations and described as inclusion bodies (Wood *et al.* 1955). Electron microscopic examination of necrotic pancreatic cells has revealed crystalline arrays of virus-like particles (Ball *et al.* 1971) (Fig. 1), whose relation to the inclusions described by Wood *et al.* (1955) remains to be ascertained.

Of other organs, only the intestine commonly carries microscopic lesions. Epithelial cells of the pyloric caeca and upper intestine show a very distinctive necrotic pattern involving swelling and rounding-up of individual cells, with fragmentation of the nucleus and marked eosinophilia of the cytoplasm. These cells slough into the intestinal lumen, contributing to a distinctive catarrhal exudate. The complete sloughing of the intestinal mucosa is associated with cases of acute disease and sudden death (McKnight & Roberts 1976). When intestinal epithelial involvement is minimal or absent, and tissue damage is restricted to a slow but often extensive and sometimes even total destruction of the pancreatic acini extending into the peripancreatic adipose tissue, the disease is best described as being in a chronic phase. Many fish die over extended periods in this phase, commonly doing so during or just after a period of stress such as after grading or other handling procedures. Fish which survive the chronic phase develop fibrous tissue in the acinar areas which have been destroyed by virus. If the areas of missing acinar or replacement fibrous tissue are extensive (I. M. McKnight pers. comm.), poor growth, and probably poor food conversion, subsequently results.

The onset of disease in young infected fish is not obligatory but apparently depends on external circumstances which are considered to stress the fish. Amongst these, water quality (e.g. with low oxygen tension) and poor husbandry (e.g. high stocking density) may play a significant role. Increasing water temperatures are also implicated (Frantsi & Savan 1971) as a factor inducing clinical disease in infected fish. Some 3–5 months after first feeding, infected fish seldom show clinical signs of disease. In Europe the Sp serotype of the virus is much more likely to cause significant mortality than the Ab serotype, but it should be remembered that the virulence of this virus in general is low and neither serotype need cause clinical disease. The same conclusion, namely low virulence of all known isolates, has been made also in North America.

5. Infection and Persistence

Infection with IPNV can apparently take place at any age in susceptible species (Wolf & Quimby 1967, 1969; Vestergaard Jørgensen 1974). Only in young fish

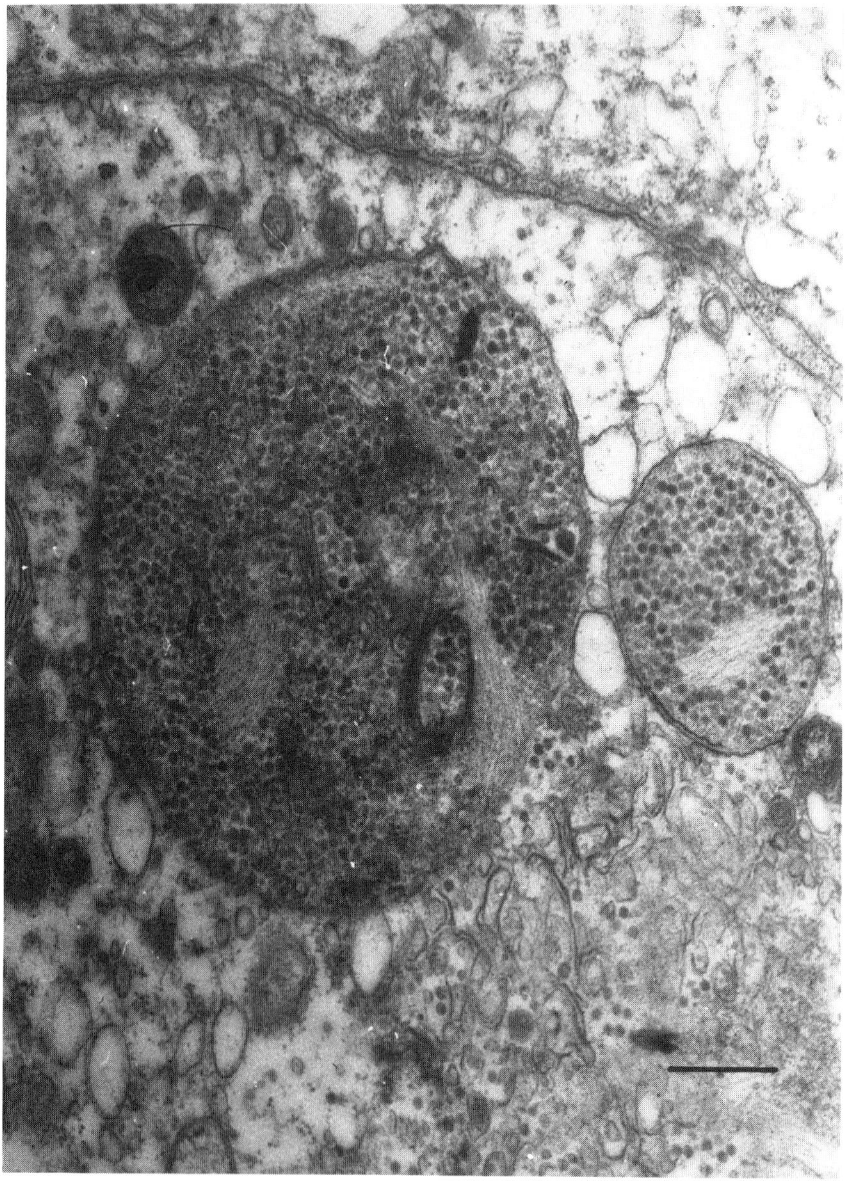

Fig. 1. Electron micrograph showing membrane-bound accumulations of IPN virus particles in a pancreatic acinar cell of a morbid rainbow trout fly; bar: 0.4 μm.

is the occurrence of clinical disease common and only then if associated with certain environmental factors such as crowding or poor water quality which stress the fish and in turn are believed to depress protective mechanisms. These qualificiations are important because it implies that adoption of husbandry techniques which ensure specified environmental conditions will prevent disease even when the virus is present in fish. To what extent husbandry techniques can be relied upon to maintain non-stressing conditions over a 3–5 month period has not been proven in the field although some initial attempts are promising (Poophard pers. comm.).

A feature of IPN virus infection is that regardless of whether disease follows infection or not a lifelong chronic persistent infection (Billi & Wolf 1969; Fenner *et al.* 1974) is established in young cultivated brook and rainbow trout. These fish become asymptomatic 'carriers' of virus for life and it can be isolated from many organs (Yamamoto 1975*a, b*); the kidney (containing most of the leuco- and haemopoietic cells of the fish) has the highest titres but spleen, liver, reproductive organs, pancreas and intestine also yield virus periodically. Virus is often shed in the faeces and with the sex products. It is believed to be carried in the egg (Wolf *et al.* 1963).

The cellular sites of persistent infection are unknown, but the consistent recovery of virus from kidney tissues of asymptomatic 'carrier' fish suggests one of the leuco- or haemopoietic cell types. The virus is immunogenic and high titres of neutralizing antibody are produced in older fish (Wolf & Quimby 1969). Yamamoto (1976*b*) has demonstrated a correlation between increasing serum neutralization titre and decreasing number of fish yielding virus in a study of two-year-old hatchery-reared rainbow trout which survived IPN disease early in life. It is likely that at the persistent stage of infection virus is present in intracellular sites protected from antibody. In older fish first exposed to virus there is some evidence (Wolf & Quimby 1969) that after an acute infection involving an immune response virus is eliminated. Such a conclusion would support the findings of Munro *et al.* (1976) that the prevalence of virus found in wild fish populations exposed to virus-containing wastes and asymptomatic 'carrier' escapees from a fish farm has not increased over a 3 year investigation period.

Fish can mount a rapid immune response detectable within days of antigenic stimulation (Avtalion 1969) although production of humoral antibody commonly takes 3–12 weeks to reach significant titres. The susceptibility of young fish to IPN disease is unlikely to be due to a poorly developed state of immune responsiveness because lymphocytes, characterized by the presence of immunoglobulin on their surface, appear in great number during the 7–14 days after first feeding in rainbow trout fry (Ellis, unpublished). The suggestion that stress is a disease-determining factor implies a possible hormonal effect depressing immune responsiveness.

Although IPNV has not yet been shown to stimulate interferon production in fish it does so in fish cell tissue cultures (De Sena & Rio 1975). Interferon production is therefore probably also a major factor in limiting acute infection or containing it until other defence mechanisms are mobilized.

6. Epizootiology

In addition to causing infection and disease in brook and rainbow trout the virus has been isolated from, and on occasion caused disease in many other species of salmonid fish kept in culture, namely Atlantic salmon (*Salmo salar*) (MacKelvie & Artsob 1969), brown trout (*S. trutta*) (Rasmussen 1965), cutthroat trout (*S. clarki*) (Parisot *et al.* 1963), coho salmon (*Oncorhynchus kisutch*) (Wolf & Pettijohn 1970), Amago trout (*O. rhodurus macrostomus*) (Sano & Yamazaki 1973), and arctic char (*Salvelinus alpinus*) (Ljungberg & Vestergaard Jørgensen 1972). Where the description is sufficient the pattern of disease is similar to that in brook and rainbow trout. The virus has been isolated from non-salmonids, such as the white sucker (*Catastromus commersoni*) (Sonstegard *et al.* 1972), perch (*Perca fluviatilis*), minnow (*Phoxinus phoxinus*) and lamprey (*Lampetra fluviatilus*) (Munro *et al.* 1976) in the vicinity of salmonid fish farms carrying stock infected with IPN virus. The disease has been reported from many parts of North America, Europe (including Scandinavia and the British Isles) and Japan. However there are no reports of the *disease* occurring in any species of *wild fish* although Desautels & MacKelvie (1975) report that the virus is enzootic in fish in the Maritime Provinces of Canada and in all probability this holds for some parts of the eastern and western USA as well. Undoubtedly many hatchery-reared fish carrying IPN virus as a persistent infection have been released to the wild or to sport fisheries in North America and Europe but to what extent these fish perpetuate themselves and spread IPN virus to their progeny and other wild fish is unknown. In a 3 year study (Munro *et al.* 1976) of the spread of IPN virus from a farm with infected stock to wild fish (mainly adult and juvenile brown trout and juvenile Atlantic salmon), a low prevalence and a distribution limited to the area close to the farm was found in a nearby loch. It showed that the virus (the Sp strain and possibly all strains) is absent from native fish in this area of Scotland and that the presence of virus is due to fish farming activity. Although the time scale was short no increase in the prevalence was found in successive years suggesting that persistent infections were not established in the majority of wild fish. In contrast, high numbers of asymptomatic 'carrier' rainbow trout were found in the escaped fish from the farm. The number of very young wild salmonids found carrying virus was too small and too close to the farm to give positive evidence of vertical transmission. A much longer period of examination will be necessary to establish if transmission between wild fish, particularly vertical transmission, is occurring.

The major sources of infection leaving the farm were identified as escaped fish, free virus in the waste water during epizootic outbreaks of disease and particulate materials especially washings from a fish gutting plant in the farm. Examination of wild Atlantic salmon from the river system draining this loch and from six other separate major Scottish rivers (Munro & Liversidge, unpublished observations) has not led to the isolation of virus or to the demonstration of anti-IPNV serum titres, thus indicating no present or past exposure to IPNV.

7. Control Measures

Should IPN virus be established in truly wild stocks of salmonids the current evidence suggests it is unlikely to pose a significant health hazard to them but may threaten any salmonid cultivation project using the same water. Current views on IPN virus as a serious hazard in fish cultivation vary. It is classed as a major communicable disease by the Food & Agriculture Organization and the International Office of Epizootics Inter-Government Consultation Committee on the control of the spread of the major communicable fish diseases. Countries signing the declaration of the committee agree to classify the disease status of fish moving across international boundaries and the disease status of their farm of origin. However, legislation within countries is quite varied: for example, Norway and Sweden require total slaughter of infected fish, the UK restricts movement of infected fish and in Denmark and France, specific areas, sometimes all farms on a river or watershed or sometimes individual farms, are cleared of all fish and only re-stocked from sources free of virus. No Federal laws restrict the movement of IPN virus infected fish in the USA.

Where there is no legislation demanding slaughter of IPN-infected fish, and as there are no therapeutic measures to eliminate virus from infected fish, farmers have the choice of tolerating the virus or avoiding it. A precondition of avoiding it is an existing virus-free stock and virus-free water supplies, the latter depending on the absence of infection in any other fish (wild or cultured) in this supply. In the light of the limited current knowledge regarding the control of IPN disease it would be folly to introduce the virus to any virus-free farm-rearing young fish or holding brood stock. Where the disease does exist it is likely that certain management proceedures must be carried out, at some additional cost, either to reduce the risk of susceptible fish meeting virus, or if this is inevitable, to ensure that disease does not ensue from infection.

Other methods of control may be possible and although some are actively being researched none are on offer to the practising culturist. Selection of avirulent strains of virus would be the basis for an attenuated vaccine. Use of killed vaccines introduced by novel methods (Amend & Fender 1976) might offer another approach. The selection of fish with a greater degree of inherited resistance is another possibility. However, assessment of resistance will be a demanding exercise in the absence of a sure knowledge of factors or conditions precipitating disease. A more speculative aspect is that as most cultivated rainbow trout brood stocks are cited as IPN virus free there is a suggestion that their progeny may have an increased or increasing susceptibility to IPN disease.

8. Conclusions

The present evidence does not indicate IPNV as a danger to the health of wild fish in the British Isles unless they are already under environmental stress, in which case endemic diseases already present a hazard. Where the virus exists in a fish farm it is likely that wild fish living in waters receiving the discharge from it will be exposed to virus with the result that some wild fish may become infected. Whether the virus will spread among wild fish outside the infective area of the fish farm's effluent is uncertain; current evidence suggests that if the virus does spread it will be a slow process. The present information does not indicate that IPNV is enzootic in native fresh water fish, rather that the virus is present in some trout farms and in the wild fish in the vicinity of these farms. Whether the virus is enzootic in marine shellfish living on the coasts of the British Isles is the subject of continuing investigation.

The virus may cause a serious disease in young rainbow trout and possibly also in brown trout and salmon in certain conditions of culture; additionally some survivors of disease have a reduced growth rate making them of dubious merit for commercial culture. The extent of losses from IPN disease in farms with infected stock is not known on a national scale but with individual farms losses are seldom above 30–40% and are often nil. Definition of environmental and husbandry conditions which do not provoke disease in spite of presence of the virus is not possible but if current investigations in this area result in the description of a satisfactory process for preventing disease it is likely that some additional cost in the rearing process must be expected.

With a distribution of virus limited to some fish farms and with a growing fish farming industry in the hands of many who are new to fish culture techniques, it is the authors' view that in the UK the use of the existing statutory controls to restrict the spread of virus is justified. Looking to the future, if prophylactic measures which clearly prevent IPN disease in spite of the presence of virus can be defined there may well be a case for relaxing restrictions on the spread of virus, in which case the fish culturist can choose, as he cannot at present, between avoiding the virus and accepting its presence but avoiding the disease.

9. Acknowledgement

We wish to acknowledge the technical assistance of W. Hodgkiss in preparing the electron micrograph.

10. References

ALAYSE, A. M., COHEN, J. & SCHERRER, R. 1975 Étude de la synthèse des ARN viraux dans les cellules FHM et RTG-2 infectées par le virus de la necrose pancréatique infectieuse (NPI). *Annales de Microbiologie* **126 B**, 471–483.

ALMEIDA, J. D. & MORRIS, R. 1973 Antigenically-related viruses associated with infectious bursal disease. *Journal of General Virology* **20**, 369–375.

AMEND, D. F. & FENDER, D. C. 1976 Uptake of bovine serum albumin by rainbow trout from hyperosmotic solutions: A model for vaccinating fish. *Science, New York* **192**, 793–794.

ARGOT, J. & MALSBERGER, R. G. 1972 Intracellular replication of infectious pancreatic necrosis virus. *Canadian Journal of Microbiology* **18**, 865–867.

AVTALION, R. R. 1969 Temperature effects on antibody production and immunological memory in carp (*Cyprinus carpio*) immunised against bovine serum albumin. *Immunology* **17**, 927–931.

BALL, H. J., MUNRO, A. L. S., ELLIS, A., ELSON, K. G. R., HODGKISS, W. & MCFARLANE, I. S. 1971 Infectious pancreatic necrosis in rainbow trout in Scotland. *Nature, London* **234**, 417–418.

BILLI, J. & WOLF, K. 1969 Quantitative comparison of peritoneal washes and faeces for detecting infectious pancreatic necrosis (IPN) virus in carrier brook trout. *Journal of the Fisheries Research Board of Canada* **26**, 1459–1465.

COHEN, J. 1975 Ribonucleic acid polymerase activity in purified infectious pancreatic necrosis virus of trout. *Biochemical and Biophysical Research Communications* **62**, 689–695.

COHEN, J. & SCHERRER, R. 1974 Le virus de la nécrose pancréatique infectieuse. I Purification et structure de la capside. *Annales de Recherches Vétérinaires* **5**, 87–100.

COHEN, J., POINSARD, A. & SCHERRER, R. 1973 Physico-chemical and morphological features of infectious pancreatic necrosis virus. *Journal of General Virology* **21**, 485–498.

DESAUTELS, D. & MACKELVIE, R. M. 1975 Practical aspects of survival and destruction of infectious pancreatic necrosis virus. *Journal of the Fisheries Research Board of Canada* **32**, 523–531.

DOBOS, P. 1976*a* Size and structure of the genome of infectious pancreatic necrosis virus. *Nucleic Acid Research* (in press).

DOBOS, P. 1976*b* Virus-specific protein synthesis in cells infected by infectious pancreatic necrosis virus. *Journal of Virology* (in press).

FENNER, F., MCAUSLAN, B. R., MIMS, C. A., SAMBROOK, K. & WHITE, D. O. 1974 In *The Biology of Animal Viruses*, 2nd Edn. London & New York: Academic Press.

FINLAY, J. & HILL, B. J. 1975 The use of the complement fixation test for rapid typing of infectious pancreatic necrosis virus. *Aquaculture* **5**, 305–310.

FRANTSI, C. & SAVAN, M. 1971 Infectious pancreatic necrosis virus – temperature and age factors in mortality. *Journal of Wildlife Diseases* 7, 249–255.

GOMATOS, P. J., TAMM, I., DALES, S. & FRANKLIN, R. M. 1972 Reovirus 3: Physical characteristics and interaction with L cells. *Virology* **17**, 441–454.

HARKNESS, J. W., ALEXANDER, D. J., PATTISON, M. & SCOTT, A. C. 1975 Infectious bursal disease agent: morphology by negative stain electron microscopy. *Archives of Virology* **48**, 63–73.

HILL, B. J. 1976 Properties of a virus isolated from the bivalve *Tellina tenuis* (Da Costa). In *Proceedings of the Third International Wild Life Disease Symposium*, 1975. New York & London: Plenum Press.

HUANG, A. S. 1973 Defective interfering viruses. *Annual Review of Microbiology* **27**, 101–117.

HUANG, A. S. & BALTIMORE, D. 1970 Defective viral particles and viral disease processes. *Nature, London* **226**, 325–327.

JOKLIK, W. K. 1974 Reproduction of Reoviridae. In *Comprehensive Virology*, Vol. 2, eds Fraenkel-Conrat, H. & Wagner, R. R. New York & London: Plenum Press.

KELLY, R. K. & LOH, P. C. 1972 Electron microscopical and biochemical characterisation of infectious pancreatic necrosis virus. *Journal of Virology* **10**, 824–834.

KELLY, R. K. & LOH, P. C. 1975 Replication of IPN virus: a cytochemical and biochemical study in SWT cells. *Proceedings of the Society for Experimental Biology and Medicine* **148**, 688–693.

LJUNGBERG, O. & VESTERGAARD JØRGENSEN, P. E. 1972 Infectious pancreatic necrosis of salmonids in Swedish Fish Farms. *Report No. 14. FAO/EIFAC, 725C, II Symposium.*

LUNGER, P. D. & MADDUX, T. C. 1972 Fine-structure studies of the avian infectious bursal agent. I *In vivo* viral morphogenesis. *Avian Diseases* **16**, 874–893.

MACKELVIE, R. M. & ARTSOB, H. 1969 Infectious pancreatic necrosis virus in young salmonids of the Canadian Maritime Provinces. *Journal of the Fisheries Research Board of Canada* **26**, 3259–3262.

MCGONIGLE, R. H. 1940 Acute catarrhal enteritis of salmonid fingerlings. *Transactions of the American Fisheries Society* 297.

MCKNIGHT, I. M. & ROBERTS, R. J. 1976 The pathology of infectious pancreatic necrosis. I The sequential histopathology of the naturally occurring condition. *British Veterinary Journal* **132**, 76–85.

MCMICHAEL, J., FRYER, J. L. & PILCHER, K. S. 1975 An antigenic comparison of three strains of infectious pancreatic necrosis virus of salmonid fishes. *Aquaculture* **6**, 203–210.

MALSBERGER, R. G. & CERINI, C. P. 1963 Characteristics of infectious pancreatic necrosis virus. *Journal of Bacteriology* **86**, 1282–1287.

MALSBERGER, R. G. & CERINI, C. P. 1965 Multiplication of infectious pancreatic necrosis virus. *Annals of the New York Academy of Sciences* **126**, 320–327.

MOSS, L. H. & GRAVELL, M. 1969 Ultrastructure and sequential development of infectious pancreatic necrosis virus. *Journal of Virology* **3**, 52–58.

MUNRO, A. L. S., LIVERSIDGE, J. & ELSON, K. G. R. 1976 The distribution and prevalence of infectious pancreatic necrosis virus in wild fish in Loch Awe. *Proceedings of the Royal Society of Edinburgh* (in press).

NICHOLSON, B. L. 1971*a* Macromolecule synthesis in RTG-2 cells following infection with infectious pancreatic necrosis (IPN) virus. *Journal of General Virology* **13**, 369–372.

NICHOLSON, B. L. 1971*b* Effect of actinomycin D on the multiplication of the infectious pancreatic necrosis virus of trout. *Experientia* **27**, 1362–1363.

NICHOLSON, B. L. & DEXTER, R. 1975 Possible interference in the isolation of IPN virus from carrier fish. *Journal of the Fisheries Research Board of Canada* **32**, 1437–1439.

NICHOLSON, B. L. & DUNN, J. 1974 Homologous viral interference in trout and Atlantic salmon cell cultures infected with infectious pancreatic necrosis virus. *Journal of Virology* **14**, 180–182.

NICK, H., CURSIEFEN, D. & BECHT, H. 1976 Structural and growth characteristics of infectious pancreatic necrosis virus from white suckers (*Catastromus commersoni*)

PALMA, E. I. & HUANG, A. S. 1974 Cyclic production of vesicular stomatitis virus caused by defective interfering particles. *Journal of Infectious Diseases* **129**, 402–410.

PATTISON, M., ALEXANDER, D. J. & HARKNESS, J. W. 1975 Purification and preliminary characterisation of a pathogenic strain of infectious bursal disease virus. *Avian Pathology* **4**, 175–187.

PIPER, D., NICHOLSON, B. L. & DUNN, J. 1973 Immunofluorescent study of the replication of infectious pancreatic necrosis virus in trout and atlantic salmon cell cultures. *Infection and Immunity* **8**, 249–254.

PARISOT, T. J., YASUTAKE, W. T. & BRESSLER, V. 1963 A new geographic and host record for infectious pancreatic necrosis. *Transactions of the American Fisheries Society* **92**, 63–66.

RASMUSSEN, C. J. 1965 A biological study of Egtved disease. *Annals of the New York Academy of Sciences* **126**, 427–460.

SANO, T. & YAMAZAKI, T. 1973 Studies of viral disease of Japanese fishes. V. Infectious pancreatic necrosis of Amago trout. *Bulletin of the Japanese Society of Scientific Fisheries* **35**, 477–480.

SCHERRER, R. & COHEN, J. 1975 Studies on infectious pancreatic necrosis virus interactions with RTG-2 and FHM cells: selection of a variant virus type in FHM cells. *Journal of General Virology* **28**, 9–20.

SCHERRER, R., BIC, E. & COHEN, J. 1974 Le virus de la nécrose pancréatique infectieuse: étude de la réplication et de l'induction de la synthèse d'inteféron en function de l'hôte et de la température. *Annales de Microbiologie* **125 A**, 455–467.

DE SENA, J. & RIO, G. J. 1975 Partial purification and characterisation of RTG-2 fish cell interferon. *Infection and Immunity* **11**, 815–832.

SONSTEGARD, R. A., MCDERMOTT, L. A. & SONSTEGARD, K. S. 1972 Isolation of infectious pancreatic necrosis virus from white suckers (*Catastromus commersoni*) *Nature, London* **236**, 174–175.

VERWOERD, D. W. 1970 Diplornaviruses: a newly recognised group of double-stranded RNA viruses. *Progress in Medical Virology* **12**, 192–210.

VESTERGAARD JØRGENSEN, P. E. 1974 *A Study of Viral Diseases in Danish Rainbow Trout, their Diagnosis and Control.* København: A/S Carl Fr. Mortensen.

VESTERGAARD JØRGENSEN, P. E. & GRAUBALLE, P. C. 1971 Problems in the serological typing of IPN virus. *Acta veterinaria scandinavica* **12**, 145–147.

VESTERGAARD JØRGENSEN, P. E. & KEHLET, N. P. 1971 Infectious pancreatic necrosis (IPN) viruses in Danish rainbow trout: their serological and pathogenic properties. *Nordisk veterinaermedicin* **23**, 568–575.

WOLF, K. 1964 Characteristics of viruses found in fishes. *Developments in Industrial Microbiology* **5**, 140–148.

WOLF, K. & PETTIJOHN, L. L. 1970 Infectious pancreatic necrosis virus isolated from coho salmon fingerlings. *Progressive Fish Culturist* **32**, 17–18.

WOLF, K. & QUIMBY, M. C. 1967 Infectious pancreatic necrosis (IPN): its diagnosis, identification, detection and control. *Revista Italiana di Piscicoltura e Ittiopatologica* **2**, 76–80.

WOLF, K. & QUIMBY, M. C. 1969 Infectious pancreatic necrosis: clinical and immune response of adult trouts to inoculation with live virus. *Journal of the Fisheries Research Board of Canada* **26**, 2511–2516.

WOLF, K. & QUIMBY, M. C. 1971 Salmonid viruses: infectious pancreatic necrosis virus. Morphology, pathology and serology of first European isolations. *Archiv für die gesamte Virusforschung* **34**, 144–156.

WOLF, K., SNIESZKO, S. F., DUNBAR, C. E. & PYLE, E. 1960 Virus nature of infectious pancreatic necrosis in trout. *Proceedings of the Society for Experimental Biology* **104**, 105–108.

WOLF, K., QUIMBY, M. C. & BRADFORD, A. D. 1963 Egg-associated transmission of IPN virus of trouts. *Virology* **21**, 317–321.

WOOD, E. M., SNIESZKO, S. F. & YASUTAKE, U. T. 1955 Infectious pancreatic necrosis in brook trout. *A.M.A. Archives of Pathology* **60**, 26–28.

YAMAMOTO, Y. 1975*a* Frequency of detection and survival of infectious pancreatic necrosis virus in a carrier population of brook trout in a lake. *Journal of the Fisheries Research Board of Canada* **32**, 568–570.

YAMAMOTO, Y. 1975*b* Infectious pancreatic necrosis (IPN) virus carriers and antibody production in a population of rainbow trout (*Salmo gairdneri*) *Canadian Journal of Microbiology* **21**, 1343–1347.

YASUTAKE, W. T. 1970 Comparative histopathology of epizootic salmonid virus diseases. In *A Symposium on Diseases of Fishes and Shellfishes*, Part 1, Disease in fish. American Fisheries Society Washington D.C.

YASUTAKE, W. T., PARISOT, T. J. & KLONTZ, G. W. 1965 Virus diseases of the salmonidae in western United States. II Aspects of pathogenesis. *Annals of the New York Academy of Sciences* **126**, 520–530.

Selected Abstracts Presented at the Annual General Meeting and Summer Conference

Session I

Numerical Taxonomy of Heterotrophic Bacteria in a Canadian River

M. A. HOLDER-FRANKLIN, T. KANEKO and M. FRANKLIN
University of New Brunswick, Fredericton, New Brunswick, Canada

Taxometric methods were employed to characterize the species diversity of heterotrophic bacteria which predominate in the Saint John River as a result of seasonal changes of industrial effluent discharged. Samples were obtained above, below and at the effluent from a large food processing plant. The bacterial flora at the effluent outflow reflected the waste disposal of the plant. Greater similarity in species was observed in the samples above and below the outflow. The methodology of numerical taxonomy is being refined to accommodate the numbers of organisms and the species diversity created by extremes in environmental conditions.

A modified Blattner Column method has been utilized to analyse the bacterial DNA. The phenotypic characteristics of the organisms are compared with the DNA analysis. The development of taxometric methods for the characterization of bacterial species obtained from aquatic sources was discussed.

The Taxonomy of Some Oxidase-negative Vibrio-like Aquatic Bacteria

J. V. LEE, T. J. DONOVAN and A. L. FURNISS
Public Health Laboratory, Preston Hall, Maidstone ME20 7NH, Kent, England

It has become increasingly evident to us that laboratories frequently isolate oxidase-negative organisms that otherwise resemble vibrios. Such organisms have been isolated throughout Europe from sea water, divers crustacea, marine (coastal) sediments, river water, sewage and animal intestines. They do not appear to be pathogenic for man.

The characteristics of the group are: Gram negative; usually curved rods; motile by means of a single polar flagellum; oxidase negative; do not reduce nitrate to nitrite; metabolize glucose fermentatively without gas production; the DNA base composition is in the range 45.4 to 46.9 mol% GC, i.e. within the range for vibrios; sensitive to the vibriostatic compound 2,4-diamino-6,7-di-

iso-propyl pteridine phosphate; require NaCl for growth but grow optimally in the range 0.5 to 5% NaCl; grow at 37°C and some grow at 43°C; grow well on ordinary culture media; usually possess arginine dihydrolase and/or lysine decarboxylase but never ornithine decarboxylase; the majority are sucrose fermenters that grow on thiosulphate-citrate-bile salt-agar (TCBS) as yellow colonies, but sucrose-negative strains often do not grow on TCBS. They appear very similar to the strains deposited in various culture collections as *V. metschnikovii* (Gamaleia) and *Vibrio proteus* (Buchner). The descriptions of these species are inadequate.

We have carried out a taxonomic analysis on 83 organisms including representatives of all the recognized *Vibrio* spp., aeromonads; certain other possibly related organisms and 40 of the oxidase-negative organisms from a wide range of habitats. They have been tested for their abilities to grow on 49 compounds as the sole source of carbon; their fermentation of carbohydrates; their production of extracellular enzymes and various physiological tests. The results are being analysed by computer and the final results of the analysis will be discussed.

A Rapid Screening Method for *Vibrio alginolyticus* in Materials of Marine Origin

T. J. DONOVAN, J. V. LEE and A. L. FURNISS
Public Health Laboratory, Preston Hall, Maidstone ME20 7NH, Kent, England

Vibrio alginolyticus often forms the majority of the vibrio population in marine samples. It is not considered to cause enteritis in man.

A number of laboratories throughout the world are now examining samples from the marine environment for vibrios pathogenic for humans: *V. parahaemolyticus*, non-cholera vibrios (NCVs) and *V. cholerae*. After suitable enrichment procedures thiosulphate-citrate-bile salt-agar (TCBS) (Kobayashi *et al.* 1963 *Japanese Journal of Bacteriology* **18**, 10) is commonly used for the isolation of these vibrios. Non-cholera vibrios, *V. cholerae, V. alginolyticus* and other marine vibrios all produce similar yellow (sucrose-positive) colonies on TCBS.

In order, therefore, to recognize sucrose-positive vibrios other than *V. alginolyticus*, a quick method of screening for *V. alginolyticus* is needed. After attempts to provide a single test for such screening had been unsuccessful the following system using three simple tests was developed.

Yellow colonies are picked from TCBS plates and the following tests inoculated.

1. Cystine lactose electrolyte deficient medium – CLED – (Bevis 1968 *Journal of Medical Laboratory Technology* **25**, 38) for growth in the absence of electrolytes (or sodium chloride).

2. Voges-Proskauer using the semi-solid medium and test method of Method 3 in Cowan & Steel. (*Manual for the Identification of Medical Bacteria*, 2nd Edn, p. 169.)
3. Swarming on freshly prepared Marine Agar (Difco Marine Agar 2216 product no. 0979).

Typical strains of *V. alginolyticus* give the following reactions after overnight incubation.

CLED —No growth
VP —Positive
Swarming—Positive

Organisms giving other combinations of results need further tests for identification.

We have found all strains recognized as *V. alginolyticus* by this method have been confirmed by full identification. On the other hand, 2% of *V. alginolyticus* strains would not be recognized by this method alone.

Surface Water Vibrios – a Mixed Bunch!

R. W. A. PARK and SANUSI BIN JANGI
Microbiology Department, Reading University, London Road, Reading RG1 5AQ, Berks, England

Curved, rod-shaped bacteria occur in a variety of habitats and are members of several genera. The trivial name 'vibrio' is often applied to them. Vibrioid morphology is of some diagnostic value when present and taken with other characters, but reports of its absence from some authentic members of *Vibrio*, of its loss on repeated subculture, and of its occurrence in apparently unrelated bacteria all suggest that the character cannot be relied upon. Vibrios are frequently seen in surface waters and in organic enrichments using such waters yet there is little information about these organisms. We decided to see if these water vibrios had any distinctive characters.

Direct microscopic observation revealed that vibrios were common in scums of surface waters and isolation of 30 strains of Gram negative curved rods was achieved following extensive examination of plates inoculated using various organic enrichments. Electron microscopy and 65 physiological tests were used to characterize the strains but identification was not possible. All had polar flagella but only three produced the spheres characteristic of old cultures of *Vibrio* spp. (see Baker & Park 1975 *Journal of General Microbiology* **86**, 12). None fermented carbohydrates or was sensitive to the vibriostat 0/129. Most strains were lipolytic: only seven showed any use of carbohydrates. Base ratio analysis of DNA revealed mol% G + C values for the 11 strains studied as 28.3,

28.5, 29.5, 55.1, 59.8, 65.0, 65.1, 65.4, 65.4, 65.4, respectively indicating that a wide range of types was represented. The utility of vibriod morphology as a diagnostic feature is considered in the light of this work.

The Delineation of a Point Source Plume Characterized by Anisotropic Propagation of Bacterial Populations

S. S. RAO and R. P. BUKATA
Microbiology Laboratories Section. Applied Research Division, Canada Centre for Inland Waters, 867 Lakeshore Road, P.O. Box 5050, Burlington, Ontario L7R 4A6, Canada

Recent point source microbiological studies, using zonal grid sampling stations, have indicated the potential of using microbiological biotypes and densities in the delineation of mixing zones associated with nutrient input. Such zonal grid approaches have been effective in identifying the plumes under conditions of basically isotropic propagation of bacteria, i.e. for the case of plumes diffusing under quasi-static lake conditions.

In this communication, the discharge from the Niagara River into Lake Ontario is considered. Such a discharge displays spatial and temporal variations resulting from the dynamic interaction between lake and river velocities, and the bacterial populations become characterized by anisotropic propagation following the point source discharge. For such examples of dynamic bacterial movements, it is shown that the data from the zonal grid network must be analysed in an anisotropic mode, i.e. one which takes direct account of the preferential directions of movement of the microbiological populations. Under such an analysis scheme, the microbiological data clearly delineate the main plume area (impact zone), zone of minor influence of the plume, and the non-plume lake waters.

The microbiological observations were found to be in direct agreement with synoptic thermal observations of the plume area made by remote sensing scanning techniques.

Some Statistical Methods Useful in Bacteriological Water Examinations

G. J. BONDE
Institute of Hygiene, University of Aarhus DK-8000, Denmark

Authors working with distribution functions pertaining to counts of micro-organisms in water agree that these belong to the exponential family, such as the simple Poisson, the negative binomial, the Polya, or the log normal distribution (e.g. Armitage 1957 *Journal of Hygiene, Cambridge* **55**, 564; Bliss

1956 *Proceedings 10th International Congress on Entomology* **2**, 1015; Bonde 1963 *Bacterial Indicators of Water Pollution.* Thesis. Teknisk Forlag, Copenhagen; Roberts & Coote 1965 *Biometrics* **21**, 600).

From almost 15,000 counts of plates and deep-agar, Bonde (1963, Thesis) found the variance to mean relationship $\sigma^2 \approx \gamma\mu^\beta$, both γ and β being close to 1, and thus not excluding the application of the Poisson law as an approximation.

The possibilities to overcome computational problems are different in research work, where planning is possible, and in routine examinations where a verdict must be given under less favourable conditions. In fact, no sample can be replaced.

Application of the Central Limit Theorem implies that the distribution of the sum of n independent random variables tends to the normal distribution for $n \to \infty$ under fairly general conditions. In the special case where the variables all have the same distribution, this theorem is valid if only the mean and the variance of this common distribution exists. This must be envisaged in planning and sampling.

Non-Parameter Methods are less powerful, and will too often lead to acceptance of null hypotheses. Useful tests in this group are (a) the sign test (Bonde 1963 Thesis) and (b) group correlation.

Transformations. The Poisson distribution as an approximation has convenient properties for practical work such as additivity and application of the square root transformation.

Comparing two counts in routine work, whether from same or different dilutions is a frequent and puzzling problem, which may be solved by Poisson reasoning (Bonde 1963 Thesis). The two volumes which can be varying in magnitude are designed n_1 and n_2 with sum: $n_1 + n_2 = n$. The hypothesis to be tested (Hs) is that the probability of colony growth in unit volume is supposed to be the same in both counts, say θ, and the corresponding number of colonies will be:

$$n_1\theta = \lambda_1 \qquad \text{and} \qquad n_2\theta = \lambda_2.$$

The observed numbers of colonies a_1 and a_2 and their sum $a_1 + a_2 = a$ all follow the Poisson law with parameters λ as above

$$\lambda_1 + \lambda_2 = \lambda.$$

The conditional probability

$$\mathrm{p}\{a_1 | a\} = (a/a_1)(\lambda_1/\lambda)^{a_1}(\lambda_2/\lambda)^{a_2}$$

is thus expressed by a binomial law with known parameters, the null hypothesis being a common θ in both counts in which case the test ratio a_1/a can be accepted as an estimate of the known ratio n_1/n.

The Marine *Bacillus*

G. J. BONDE
Institute of Hygiene, University of Aarhus DK-8000, Denmark

Bacillus strains occur often but with varying frequency in marine areas; however, the differences between water and sediment are even greater. In water *Bacillus* amount to 0–16% of the total flora of heterotrophs; in sediments much higher numbers may be found (1,500,000 g^{-1}), amounting to 7–30% of the heterotrophs, and on greater depths up to 100% (Bonde 1968 *Revue Internationale d' Océanographie Médicale* **IX**, 17; Rüger 1975 *Veröffentlichungen des Instituts für Meeresforschung* **15**, 227). The *Bacillus* counts on sea water media may be 10–50 times higher than on plain nutrient media and cells are as frequent as spores.

Much less has been recorded regarding species distribution; however, this seems to be a promising ecologic character (*cf.* Bonde 1975 *Danish Medical Bulletin* **22**, 41). In polluted sea water *B. licheniformis* (64%) dominates, followed by *B. subtilis* (15%), and *B. polymyxa* (10%); in more virgin areas *B. subtilis* and *B. pumilus* are more frequent (36% and 21% respectively). In sediments of polluted areas the soil species: *B. cereus* (18%), *B. sphaericus* (17%), and *B. brevis* (10%) dominate, the virgin sediments showing a higher diversity.

A 'specific' marine type might be found among 20–25% non-identified strains, which have many common features (75 characters). They grow in smooth butyrous yellow, ochre, or pink colonies. Rods are Gram variable or Gram negative, of medium size and bulging and non-bulging spores are found in the same culture. VP is negative, nitrates not reduced; the catalase reaction is weakly positive; sugars are not or weakly fermented. These strains cluster into one phenetic group and are not unlike *B. firmus* (and *B. coagulans*).

The Effect of Virus Particle Concentration on the Inactivation of Bacteriophage MS2 in Sea Water

JANICE M. TYLER and FELICITY M. BESWICK
Department of Botany and Microbiology, University College, Singleton Park, Swansea SA2 8PP, Wales

Bacteriophage MS2 is a small, spherical virus containing single stranded RNA. Since the coat contains only two polypeptides the virus provides a simple model for demonstrating mechanisms of virus inactivation. At temperatures ranging from 10 to 60°C we have shown that the rate of inactivation of phage MS2 in normal and aged sea water varies with the initial concentration of active virus. As concentrations are increased from 10^2 plaque forming units PFU.ml^{-1} to between 10^4 and 10^5 PFU.ml^{-1} the rate of inactivation decreases progressively.

Similar results were obtained with polio virus in sea water (Plissier *et al.* 1962 *Annales de L'Institut Pasteur* **103**, 665). Further increase in the initial concentration of virus reverses this trend; that is, the rate of inactivation increases progressively at concentrations from 10^5 to 10^8 PFU.ml^{-1}.

It has been suggested that the main virus inactivating capacity of sea water is caused by compounds produced by the marine microflora. (Lycke *et al.* 1965 *Archiv für die Gesamte Virusforschung* **17**, 409; Shuval *et al.* 1971 *Journal of the Sanitary Engineering Division, A.S.C.E.* **97**, 587). In our experiments however, comparable results were obtained in sea water at 10°C, and at 52–60°C, in aged sea water, and it seems unlikely that a biological compound is responsible for the effect.

Our results suggest that the loss of phage titre is due to a combination of irreversible inactivation, and reversible aggregation of the virus particles. The various inactivation curves obtained are consistent with the theoretical ones postulated by Chang (1965, In *Transmission of Viruses by the Water Route*, ed. Berg, G. Interscience Publishers) and Berg (1965, In *Transmission of Viruses by the Water Route*, ed. Berg. G. Interscience Publishers) for situations where aggregation of viruses occurs. Although it is not immediately apparent why the initial concentrations of viable virus should alter the inactivation curves obtained, we are doing experiments to try to establish whether aggregation is occurring with phage MS2 in sea water. The variation observed does provide an explanation for instances where different laboratories have reported different rates of inactivation of viruses under largely similar conditions.

Session II

The Seasonal Variations of Selected Bacterial Populations in Estuarine Sediments

R. J. Parkes and N. J. Poole

Department of Microbiology, Marischal College, University of Aberdeen, Aberdeen, Scotland

The seasonal variations of selected bacterial populations were investigated in the estuary of the River Don, Aberdeenshire, using their growth on selective media for enumeration and characterization. Aerobic and anaerobic heterotrophic bacteria, aerobic and anaerobic cellulolytic bacteria, and sulphate-reducing bacteria together with certain physical parameters (e.g. sulphide, pH, E_h and temperature) were monitored approximately every three weeks between April 1975 and May 1976. The results obtained indicate the following.

1. That the characteristic 'hydrogen sulphide' smell which develops during the summer months was due to the *physical* release of sulphide as a result of the

marked increase in sediment temperature during the summer. The increased temperature probably results in a decrease in the solubility of sulphide and hence facilitates the release of H_2S gas.
2. The levels of sulphide in the sediment were a result of the activity of sulphate-reducing bacteria.
3. The sulphide released during the summer was probably produced during the spring bacterial 'bloom', as during the period of H_2S gas release the levels of sulphate-reducing bacteria were at a minimum.
4. That during winter the sulphate-reducing bacteria seem to tolerate a higher level of sulphide than in summer and hence temperature may not only control the amount of sulphide a sediment can hold, but the toxicity of that sulphide to bacterial populations.
5. The observed decrease in all bacterial populations during late winter was probably due to the attainment of minimum winter temperatures (approximately 3°C).

All bacterial groups studied showed similar patterns of variation to those of the sulphate-reducing bacteria, hence indicating that they may be controlled by common factors. In the Don Estuary it seems that temperature is the main controlling factor, limiting growth during late winter due to extremely low temperatures and during summer indirectly determining the extent of the 'spring bloom' via its effect on the toxicity of sulphide to bacterial populations.

The Influence of Water on the Properties of Faecal Indicators

I. DAUBNER
Institute of Experimental Biology and Ecology of the Slovak Academy of Sciences, Limnology Section, Bratislava, Czechoslovakia

In a series of laboratory experiments the stability of some properties of the main indicators of faecal pollution (*Escherichia, Citrobacter, Enterobacter*) in water was examined. Presterilized distilled water, surface waters and high mineralized medium were employed paying special attention to utilization of citrate as the sole C source.

Changes in biochemical activities of *E. coli, C. freundii* and *E. aerogenes* stored in waters were observed including fermentation of sucrose, lactose, mannitol, dulcitol, splitting of urea, H_2S and indol formation, methyl red test, gelatine liquefaction, utilization of citrate and gas production from glucose at 43°C. The fermentation of glucose at 37°C and at 43°C and the Voges-Proskauer reaction was never changed.

In 22 of 35 experiments citrate utilization was observed after 18–24 h. This property remained to the end of the experiments (1–6 weeks) but, after further cultivation under optimal conditions, almost 50% of strains ceased utilizing citrate. Strains isolated from stock culture preserved under laboratory conditions have never begun to utilize the citrate. This change appeared only after storing cells in waters. Conversely in citrate utilizing genera (*Citrobacter* and *Enterobacter*) this property never changed after survival in water. After detailed inoculation and purification of changed strains, as a rule two types of colonies were present: (a) those with typical metallic sheen which contained unchanged *E. coli*; (b) pink colonies which utilized citrate etc. Therefore we suppose that the observed citrate utilization was due to the accompanying flora, which are mostly bacteria of the family Pseudomonadaceae.

The Effect of Hydrostatic Pressure on Bacteria in Sea water Intended for Injection into Oil Formations

B. N. HERBERT
Shell Research Limited, Woodstock Laboratory, Sittingbourne ME9 8AG, Kent, England

Reports on the effects of hydrostatic pressure on marine bacteria (ZoBell & Oppenheimer 1950 *Journal of Bacteriology* **60**, 771) and the microscopic appearance of bacteria grown *in situ* in the ocean depths (Kriss 1963 *Marine Microbiology*, Edinburgh: Oliver and Boyd) indicate that although many marine bacteria can survive and multiply over a wide range of hydrostatic pressures their morphological appearance can undergo changes. Typically long filamentous forms can be seen.

The injection of sea water into North Sea reservoirs for the supplementary recovery of oil will provide high pressure environments which can range from 200 to 400 atmospheres.

A sulphate-reducing culture has been isolated from sea water in the Brent Field area and grown in a liquid medium (Postgate 1963 *Applied Microbiology* **11**, 265) prepared in 75% aged sea water. Subjection of this culture to a range of hydrostatic pressures (Morita 1970 In *Methods in Microbiology*, Vol. 2, eds Norris, J. R. & Ribbons, D. W. London & New York: Academic Press) at 25°C has indicated that it can grow and reduce sulphate at pressures as high as 400 atmospheres. The tolerance of this culture to pressure was greater than that of *Desulfovibrio selaxigens* NCIB 8364 and when grown at pressures near its threshold of tolerance, morphological changes were observed.

These results indicate that pressures likely to be encountered in sea water injection systems will probably not offer a barrier to the growth of sulphate-reducing bacteria that could be introduced from the sea water. If the potentially damaging effect of their growth is to be prevented, studies in the laboratory on potential control measures will need to take high pressure effects into account.

Prosthecate Bacteria from Fresh Water Environments

A. LAWRENCE and C. S. DOW
Department of Biological Sciences, University of Warwick, Coventry CV4 7AL, Warks, England

Although prosthecate bacteria have been observed in fresh water environments over several years (Henrici & Johnson 1935 *Journal of Bacteriology* **30**, 61; Staley 1968 *Journal of Bacteriology* **95**, 1921; Schmidt 1971 *Annual Review of Microbiology* **25**, 93; Hirsch 1974 *Annual Review of Microbiology* **28**, 1644), questions concerning the function or survival value of integral cellular extensions are unresolved. However, it is now becoming apparent that this group of micro-organisms forms part of a highly specialized indigenous population.

An electron microscopic survey of the bacterial populations of the Cumbrian Lakes has shown that *Hyphomicrobium* and *Caulobacter* spp. constitute up to 20% of such populations, the percentage being correlated with the nutrient status of the system. The multi-appendaged genera, at best, approach 1% of the total microbial population. 'Enrichment' systems to which no nutrients have been added select for the latter. Several isolates of multi-appendaged bacteria have been obtained from such systems. Studies on one of these has shown that gross phenotypic variation can be brought about by varying the nutrient status of the culture medium. The point of importance is that the prosthecae can be induced or repressed by environmental stimuli. This phenotypic variation consequently makes the estimation, simple on morphological grounds, of the incidence of multi-appendaged bacteria in the environment difficult. It almost certainly leads to gross underestimates of numbers.

These observations also highlight a fundamental split concerning prosthecae function: (a) where prosthecae formation is obligate and intimately involved in the cell cycle e.g. *Caulobacter* and *Hyphomicrobium*; (b) where the prosthecae are non-obligate and environmentally induced.

Stalk Synthesis and the *Caulobacter* Cell Cycle

U. K. RETNASABAPATHY and C. S. DOW
Department of Biological Sciences, University of Warwick, Coventry CV4 7AL, Warks, England

In recent years the dimorphic life cycle of *Caulobacter* spp. has been exploited as a model system in the field of cellular differentiation (Kurn & Shapiro 1975 *Current Topics in Cellular Regulation* **9**, 41). However, the question as to the involvement and function of the prosthecae or stalk in the cell cycle is unanswered.

Using a modified version of the selective filtration system devised by Maruyama & Yangita (1956 *Journal of Bacteriology* **71**, 542) we have obtained synchronized *Caulobacter* swarm cell populations. These we have used to study stalk synthesis and crossband formation with respect to the cell cycle.

It has been reported recently that the crossbands, characteristic of *Caulobacter* stalks (Jones & Schmidt 1973 *Journal of Bacteriology* **116**, 466) serve as generation markers (Staley & Jordan 1973 *Nature, London* **246**, 155), a crossband being synthesized during each turn of the mother cell cycle. To examine this system in detail and to study the influence of environmental changes on the overall sequence of events, e.g. stalk length and inter-crossband distances, a modified Helmstetter–Cummings apparatus (Helmstetter & Cummings 1963 *Proceedings of the National Academy of Sciences USA* **50**, 767) was used.

Synchronized swarm cells attached by the holdfast to the inverted membrane differentiate into stalked mother cells. Subsequent Coulter Counter monitoring of the effluent shows sequential waves of swarm cell production. These give the precise number of swarm cells produced per mother cell and the concomitant generation times. Electron microscopic examination of the mother cells from the membrane, after a known number of generations, has shown that there is no correlation between early swarm cell production and crossband formation, i.e. crossbands do not appear in the stalk until after the third or fourth generation.

The time of stalk synthesis with respect to the sequential events occurring in the mother cell has been established by studying penicillin induced spheroplast formation.

The Detection of False Negatives in Coliform Testing of Marine and Elevated Temperature Water Samples

BETTY H. OLSON, N. MADDOCKS and JANICE PRATTE
Program in Social Ecology, University of California, Irvine, California 92717 USA

False negatives have been detected in the examination of sea water and elevated temperature waters by the Most Probable Number Method (MPN). Sea water samples were examined from September to April and hot spring and stream samples for a six-week period. A false negative was classified as a tube which showed growth, but no gas in the presumptive coliform test, and growth with gas production in the confirmed and faecal coliform tests. Bacteria from the positive Eijkman tubes were streaked onto eosin-methylene blue (EMB) agar. A typical bacterial colony was selected and then restreaked onto EMB agar, and subsequently transferred after 24 h to nutrient agar slants. The bacterial cultures were screened with the IMVIC test, and the results showed that the majority of isolates belonged to the genus *Escherichia.* Tryptone glucose extract (TGE) agar was used for total plate counts. Plates were incubated at 20 and 35°C.

False negatives have occurred in sea water samples consistently over the eight months of this study and occurred at each sampling time in both the hot springs and the stream samples. In all the samples tested the increase in coliform numbers due to the presence of false negatives was substantial. In the elevated temperature water samples the effect of false negatives was on the average at least a ten-fold increase. In sea water samples the increase was more variable with increased coliform counts due to false negatives ranging from less than one log to a maximum of three logs. In sea water samples total plate counts were usually less than 500 bacteria.ml^{-1} and in elevated temperature water the counts ranged from 260 to 350 bacteria.ml^{-1}.

Session III

Experiments in Deep Sea Microbiology

H. W. JANNASCH
Woods Hole Oceanographic Institution, Woods Hole, Massachusetts 02543, USA

The Distribution and Characterization of Bacteria on the Surfaces of Some River Macrophytes

JUDITH C. HOSSELL and J. H. BAKER
Freshwater Biological Association, River Laboratory, East Stoke, Wareham BH20 6BB, Dorset, England

The periphytic bacterial populations found on three species of higher plants in a chalk stream have been studied. The plants, which are representatives of the submerged, emergent and floating vegetation respectively, were *Ranunculus penicillatus* var. *calcareus* (water crowfoot), *Rorippa nasturtium aquaticum* (watercress) and *Lemna minor.* The size of the bacterial population was estimated both by direct and viable counting techniques. On all three species the population, by direct counts, was approximately 5×10^6 bacteria cm^{-2} and all the samples were apparently healthy. The viable counts were lower than the direct counts and the ratio of the two varied from 1 to 100.

On all three species bacteria were seen to have accumulated in the grooves which occur above the epidermal cell walls. On *Ranunculus* accumulations were also observed in the depressions afforded by the dichotomies of the leaf filaments and in grooves on the stems not associated with cell boundaries. Experiments using plastic ribbon which has a grooved surface and microscope slides which were scratched with a diamond, suggest that the physical form of the surface to be colonized is important when there is a shearing force caused by the water flow. A second phenomenon common to all the plants studied was that the young tissue had fewer bacteria associated with it than mature material from the same plant. With respect to *Ranunculus* the experiments with plastic ribbon indicate that this second phenomenon is not linked with current velocity. Hence other factors such as time of exposure and antibiotic production may be important.

Characterization of bacteria taken from *Ranunculus* leaves revealed that they were predominantly Gram negative rods as are the bacteria in the water column. The most prevalent types were pseudomonads and *Flavobacterium*-like organisms which each constituted about 20% of the population.

Quantitative Bacteriology of *Mytilus edulis* on a Sewage Polluted Shore

D. L. WEBBER and D. R. TROLLOPE
Department of Botany and Microbiology, University College of Swansea, Wales

The quantitative bacteriology of macerated tissue from mussels (*Mytilus edulis*) native to or introduced to a sewage-polluted shore has been studied. The high degree of correlation between counts on MacConkey agar at 44, 37 and 26°C indicated that these bacteria were sewage derived. Coliform numbers varied with seasonal and climatic factors, peaking during the winter and during the presumed functioning of nearby storm overflows. With bacterial counts on ZoBell marine agar these relationships were not evident. These data will be discussed with respect to coliform and marine bacterial numbers in the sea.

Mussels returned to the site after laboratory or aquarium cleansing, or introduced from a lightly polluted site, were analysed bacteriologically and compared with an indigenous sample. The bacterial numbers were influenced by the particular cleansing treatment. Information on the use of such mussels to monitor coliform numbers in the sea will be presented.

The Role of Flagella in Adhesion of Bacteria: an Ecological Hypothesis

W. A. SCHEFFERS, the late WILLEMINA E. DE BOER
and ATTY M. LOOYAARD
Laboratory of Microbiology, Delft University of Technology, Delft, The Netherlands

Flagella may be aids in holding bacteria to a solid substrate and thus may play an ecological role besides their function in locomotion. This hypothesis will be discussed on the basis of the following considerations.

1. Adhering bacteria may profit from nutrients provided by particulate matter *per se* (plankton, organic detritus) or from nutrients adsorbed from surrounding liquid onto inert solid substrate.
2. Adhesion of bacteria to particles may be enhanced by presence of flagella, extending the cells' capacity for intermolecular interaction with the surface.

This effect will be most pronounced in cells with many flagella, especially when the flagella are long and have a small wavelength. This type of adhesion would not exclude motility: during locomotion of the cells, interaction would continually be maintained at the momentary points of contact between flagella and solid substrate. Slime excretions may further enhance adhesion between flagella and surface.

3. The capacity to form numerous flagella may be a selective advantage in habitats where adhesion of bacteria to solid particles is important. It is noteworthy that peritrichous bacteria are more characteristic of systems rich in particulate matter (e.g. soil, intestine), whereas many aquatic habitats show a preponderance of monotrichous bacteria.
4. Variable flagella organization is frequently encountered among marine bacteria. This is illustrated by strains of *Vibrio parahaemolyticus* biotype 2 (*alginolyticus*) from Dutch coastal water (Golten & Scheffers 1975 *Netherlands Journal of Sea Research* **9**, 351) which in liquid media form cells with a single, sheathed, polar flagellum only, whereas on agar media and other solid supprts cells with numerous unsheathed, lateral flagella of shorter wavelength develop. The two types of flagella also differ by their response to various physical and chemical factors (de Boer *et al.* 1975 *Netherlands Journal of Sea Research* **9**, 197, *Antonie van Leeuwenhoek* **41**, 385). We suggest that variable flagellar organization is of ecological importance in bacteria living alternately on solid support and freely in liquid surroundings.

Occurrence of Sessile Bacteria in a Pristine Subalpine Stream

G. G. GEESEY,* J. POMEROY and J. W. COSTERTON
Department of Biology, University of Calgary, Calgary, Alberta, Canada

Natural sessile bacterial populations were studied in an unpolluted mountain stream in Alberta, Canada. Preliminary results indicated that large numbers of bacteria were associated with a growth of slime which covered much of the submerged cobble substrate during the early and late summer months. Direct counts obtained by epifluorescence microscopy revealed as many as 5×10^9 bacteria.cm^{-2} rock surface during peak periods of slime growth. Also associated with the slime, but in much lower numbers, were algae, diatoms, and blue-green algae. The free flowing water contained from 10^3 to 10^6 bacteria.cm^{-2}. Thin-section preparations examined by electron microscopy revealed that the slime-associated bacteria were predominantly Gram negative. Ruthenium red stained preparations demonstrated that many of the bacteria were enmeshed in an 'acid polysaccharide-like' matrix which appeared to radiate directly from the

* Supported by Environment Canada.

bacterial cell wall. The extracellular matrix was occasionally observed to enclose microcolonies of morphologically similar bacteria. It is suggested that the sessile bacteria are an important community of the stream and that their activities contribute to the observed slime production on the submerged rock surfaces.

Physicochemical Aspects of Marine Bacterial Attachment to Solid Surfaces

MADILYN FLETCHER
Marine Science Laboratories, University College of North Wales, Menai Bridge, Gwynedd LL59 5EH, Wales

The attachment of a marine pseudomonad to polystyrene was investigated, and the number of attached cells was found to be dependent upon (a) culture concentration; (b) time allowed for attachment and (c) the growth phase of the culture. The number of attached cells was directly proportional to the $\log_{10}$ of culture concentration. With low cell concentrations (approximately $<10^9$ bacteria.ml^{-1}) attached cell numbers rose steeply with increase in cell concentration, whereas at greater cell concentrations the increase in attached cell numbers levelled off as the attachment surface approached saturation. Attached cell numbers also increased with the time allowed for attachment (until attachment surface saturation was reached), and the $\log_{10}$ of attached cells was directly proportional to the $\log_{10}$ of time. The proportion of cells attached and the rate of attachment was greatest with log phase cultures, and progressively decreased with stationary and death phase cultures.

The influence of cell concentration and time upon attachment demonstrates the importance of physicochemical factors in the attachment mechanism, whereas the influence of growth phase stresses biological factors. The results are compared with models describing molecular adsorption, and are shown to suggest that initial bacterial attachment is largely determined by physicochemical forces.

Pseudomonas aeruginosa in Sewage and Aquatic Environments

D. W. F. WHEATER, D. D. MARA and LOUZAN JAWAD
Department of Biological Sciences, The University, Dundee DD1 4HN, Scotland

Problems Associated with the Use of Membrane Filters for Enumerating *Thermoactinomyces* Endospores in Water Samples

BRIDGET A. UNSWORTH, LAMYA AL-DIWANY and T. CROSS
School of Biological Sciences, University of Bradford, Bradford BD7 1DP, Yorks, England

Membrane filtration is a common method used for the enumeration of bacteria in water. When this technique was employed to recover the viable endospores of the *Thermoactinomyces* (THA) spp. in streams, rivers and farm drains, significantly different counts were obtained on the membranes of different manufacturers.

Water samples, or spore suspensions in tap water from pure cultures grown on agar slopes, were filtered through 0.45 μm pore-size membranes. The membranes with retained spores were incubated on the surface of Czapek + yeast extract + casamino acids agar (CYC) at 50°C for 48 h. CYC agar containing novobiocin (25 μg.ml^{-1}) and cycloheximide (50 μg.ml^{-1}) was used for the water samples (Cross & Attwell 1974 In *Spore Research 1973*, eds Barker, A. N., Gould, G. W. and Wolf, J. London & New York: Academic Press).

When spore suspensions of pure THA cultures were filtered through membranes, highest recoveries were obtained on Millipore HAWG, Millipore HC, Gelman Green 6NGD and Gelman White GN-6 membranes, with significantly fewer colonies on membranes such as Nuclepore and Oxoid. The efficiency of the various membranes when used for water samples was different and with these samples Sartorius Black and Millipore HABG membranes gave highest recoveries of THA. The size and appearance of the colonies after incubation varied on the membranes used.

The pore opening diameter has been claimed to be the important factor determining the efficiency of membranes for recovering and counting coliform bacteria (Sladek *et al.* 1975 *Applied Microbiology* **30**, 685), and the membrane Millipore HC was developed for these organisms. This membrane gave high counts of THA presented as a pure spore suspension but was not as efficient as some other membranes when natural water samples containing a mixed bacterial flora were filtered.

Our results suggest that several membranes should be compared for their efficiency in recovering and growing a chosen bacterial species. The results obtained with pure cultures should not be used to predict the type of membrane to use with mixed cultures or natural populations.

Cycles in Fresh Water Microbial Ecology

N. C. B. Humphrey, T. C. Iles* and J. C. Fry
*Departments of Applied Biology and *Mathematics, University of Wales Institute of Science & Technology, Cathays Park, Cardiff CF1 3NU, Wales*

Regular seasonal changes are commonly observed in biological data, but objective time series analysis of the type used by Williamson (1974, in *Ecological Stability*, eds Usher, M. B. & Williamson, M. H. London: Chapman & Hall) are few. Such analysis assumes that cycles are regular sine waves and it is restricted to normally distributed data that have a large number of observations, each separated by an equal time interval. This paper describes a procedure for time series analysis which identifies the period and phase of cycles and distinguishes between genuine cycles and random observations. The procedure involves the following steps. (a) A power spectrum is calculated to approximately identify the period of the major cycle. To test if this cycle is due to random events a cumulative periodogram and a Kolmogorov–Smirnov test is used (Box & Jenkins *Time Series Analysis, Forecasting and Control*, Holden-Day). (b) The period and position of the maximum are then more accurately determined by correlation with a series of cosine curves varying in period and phase. (c) Multiple regression of the test variable is carried out with the optimal cosine curve to produce a series of residuals, which are the difference between the observed and the cosine predicted values of the test variable. (d) The presence of further cycles is tested by repeating steps one to three with the residuals, which should be random when all the cycles are explained. Using this procedure we have shown that many microbial and chemical parameters in a small pond show cyclical behaviour. For example, potential heterotrophic activity in water (Hobbie & Crawford 1969 *Limnology and Oceanography* **14**, 528) showed two significant cycles: the major cycle with a period of 53 weeks and a maximum in July–August; the minor cycle with a period of ten weeks with a maximum in April–May. Together these two cycles explained more than 75% of the variance in the data. In contrast the total number of bacteria in water counted by epifluorescence (Daley & Hobbie 1975 *Limnology and Oceanography* **20**, 875) showed no significant cyclic changes.

The Growth of *Enteromorpha* Germlings under Different Conditions

T. LOVEGROVE
International Marine Coatings, Newton Ferrers, Plymouth, Devon, England

The life cycle of the ship-fouling alga *Enteromorpha* is usually shown as a simple alternation between asexual generations producing zoospores and a morphologically similar sexual generation producing anisogametes which fuse and develop into the asexual generation. The cycle is not quite so simple since gametes of either type can produce viable filaments without fusion taking place.

Zoospores settle quickly but large gametes tend to settle more slowly and thus germination can be delayed, making settlement a continuous process since gametes can continue swimming for up to at least eight days before settling. Settling zoospores show a strong affinity for surface imperfections which is not shown by gametes, and the ageing of gametes does not induce the typical zoospore settling response to surface contact.

Enteromorpha from a normal shore area and from a quiet almost fresh water site provided settled germlings and both sets were cultured under static and aerated conditions as well as a cycle of immersion in nutrient for a period followed by exposure to the air. The results showed that the ecological background of the parent filament apparently determined the response of the offspring, since germlings from shore area plants responded markedly to aeration whilst the other material only gave a growth response with germlings produced from zoospores.

Tankers and other vessels create an artificial set of circumstances at their waterline and other areas where turbulence or aeration occur that *Enteromorpha* is naturally adapted to exploit.

Session IV

Estimation of Fruiting Myxobacteria in Water by a Modified MPN Procedure

E. R. BROCKMAN
Biology Department, Central Michigan University, Mt. Pleasant, Michigan 48859, USA

Myxobacteria do not typically form colonies on solid media and they are difficult to cultivate in liquid media. Thus, an accurate estimation of numbers of myxobacteria in the environment has been difficult, if not impossible, to achieve. A modification of the five-tube most probable number procedure, employing 0.45 μm membrane filters, a 25-fold increase in the normal sample volume, and a solid medium (Carlson & Pacha 1968 *Applied Microbiology* **16**, 795) has allowed the estimation of myxobacteria in the Chippewa River (Michigan, USA). The antibiotic cycloheximide was incorporated into the medium at a concentration of 25 mg.l^{-1} to inhibit fungal growth. Although the numbers of myxobacteria estimated in this river were low, less than 50 per 100 ml, the method could easily be modified by decreasing the volume filtered to accommodate larger populations of myxobacteria. In addition to obtaining an estimate of myxobacteria in the water the presence of different species could also be determined on the basis of fruiting body size, colour, and morphology and microcyst form and size.

Session V

Growth of the Marine Basidiomycete *Nia vibrissa* on Wood

L. E. LEIGHTLEY
Department of Biological Sciences, Portsmouth Polytechnic, King Henry I Street, Portsmouth PO1 2DY, Hants, England

Marine lignicolous fungi mostly belong to the Ascomycetes and Fungi Imperfecti and many are known to cause soft rot attack of wood. The Basidiomycetes have two known marine representatives. *Nia vibrissa* Moore and Meyers, a Gasteromycete, was isolated from wood submerged in the sea and a study of its growth and ability to degrade wood was made using the light and scanning electron microscopes.

Basidiocarps of *N. vibrissa* developed on wood and appeared as soft, globose, orange pink fruiting bodies 3–5 mm in diameter. These basidiocarps developed from white, spherical tufts of mycelium *ca.* 1 mm diameter. The surface of the basidiocarp was composed of hook-like bifurcate hyphae, which may aid in dispersal, especially since the basidiocarps were easily detached from the wood.

Mature basidiocarps contained basidiospores embedded in mucilage which were passively released on disintegration of the outer wall. Basidiospores had tetraradiate appendages with inflated tips.

Nia vibrissa cultured on beech (*Fagus sylvatica*), balsa (*Ochroma* sp.) and pine (*Pinus sylvestris*) caused a pattern of wood decay characteristic of the white rot type. Hyphal penetration from cell to cell was through borehole formation. Wood cell wall lysis occurred as the hyphae formed erosion channels in the inner cell wall. These channels caused a gradual 'thinning' of the wood cell walls, indicating that extracellular enzymes of this organism readily diffused from the hyphae and were active on the cell wall constituents.

Nia vibrissa exhibited the same pattern of wood attack as most terrestial Gasteromycetes. However, the hooked hyphae and more especially the tetraradiate basidiospores are characteristic of an organism adapted to the aquatic environment.

Factors Influencing the Biodegradation of Petroleum in Aquatic Environments

J. D. Walker
Environmental Technology Center, Martin Marietta Corporation, Corporate Research and Development, 1450 South Rolling Road, Baltimore, Maryland 21227 USA

The Growth of Micro-organisms on Materials Used in Plumbing in Contact with Treated Water and Techniques for Its Assessment

N. P. Burman and Jennifer Colbourne
Thames Water Authority, London EC1R 4TP, England

Water samples collected from plumbing installations have often been shown on analysis to be of poorer microbiological quality than samples collected from the local supply. This deterioration in water quality is often associated with a musty taste or visible turbidity in the water and a slime has sometimes been observed on the water fittings. The growth of micro-organisms in plumbing installations is encouraged by certain localized conditions, such as an increase in water temperature or a period of little or no flow, but the primary cause appears to be the presence of carbon donating materials within the system. The exact mechanism involved will depend upon the composition of the materials and the types of micro-organism present.

A technique to assess the ability of plumbing materials to support microbial growth is described, and the most common categories of materials capable of supporting such growth are identified. Micro-organisms of significance which have frequently been found to grow in the presence of unsuitable materials are coliform organisms, especially members of the genus *Citrobacter, Aeromonas hydrophila, Pseudomonas aeruginosa* and fungi. The growth can often be visible to the naked eye as a film on the surface of the material or as a turbidity in the surrounding water. The technique is currently being applied to all non-metallic materials, including paints and coatings, packing materials, jointing compounds, fluxes and lubricants, as well as components of fittings, in an attempt to minimize the use of those materials likely to deteriorate the water quality.

Evaluation of Potential Impacts of Silver Iodide Nucleating Agents on Aerobic and Anaerobic Aquatic Microbiological Processes

D. A. KLEIN and R. A. GIANGIORDANO
Department of Microbiology, Colorado State University, Fort Collins, Colorado 80523, USA

A concern in assessment of potential ecological impacts of silver iodide nucleating agents on natural ecosystems has been the possible effects of these agents on aerobic and anaerobic decomposition processes in lakes, muds and in sewage treatment plants where vital organic matter breakdown and methane production processes occur. To gain information on these points, comparisons of aerobic mineralization and microbial growth processes in soils and simulated aquatic environments, and of methane production in muds and anaerobic digester sludge were carried out using silver ion, silver iodide, silver iodide–sodium iodide mixtures, and silver iodide burn complexes prepared with sodium or ammonium iodide. Using muds which had these compounds present for a 1.5 year period, silver iodide at 10 $mg.l^{-1}$ showed a distinct inhibition of methane production, while silver present at the same level in the iodide complex showed no such effect. In anaerobic sludge which contained high populations of methane-producing organisms, a similar although less distinct pattern was observed when these silver forms were present for a two-week period, the usual retention time for sludges in an anaerobic digester. Related studies of aerobic processes of glucose mineralization and of silver-temperature interactions in relation to growth of a common appendaged aquatic bacterium *Hyphomicrobrium* suggest that under aerobic conditions the silver iodide–sodium iodide complex can cause more distinct biological effects than silver iodide alone. The unique complex of silver iodide with excess halide may thus have a lesser potential for causing changes in anaerobic, in comparison with aerobic, biological processes.

Bacteriology of a Marine Fish Tank

G. T. VALDIMARSSON
Department of Applied Microbiology, University of Strathclyde, Glasgow G1 1XW, Scotland

Samples were taken from the inlet and outlet of a tank stocked with 2- to 4-year-old saith. Surface plate counts were made on ZoBell's 2216 and Difco Marine Agar. Results showed counts ranging from 10^3 to 10^4 ml $^{-1}$ with the inlet water and 10^4 to 6 x 10^5 ml^{-1} with the outlet water over the period March 1974 to August 1975.

Randomly selected colonies (1050) were classified to group generic level. Percentages of isolates for the predominant groups in inlet and outlet water respectively were as follows: *Vibrio*/anaerogenic *Aeromonas* 25.6%, 33.3%; *Moraxella* 15.0%, 11.3%; *Pseudomonas* Group I and II/*Alcaligenes*/*Agrobacterium* 19.8%; *Pseudomonas* Groups III and IV/*Alcaligenes*/*Agrobacterium* 8.6%, 9.8%; *Photobacterium* 0.6%, 8.9%; 'Enterobacteriaceae' 2.2%, 10.5%. This latter group was frequently isolated and on one occasion constituted 32% of the outlet flora. These organisms are believed to be non-luminous *Photobacterium* spp. An inverse relationship between *Vibrio*/anaerogenic *Aeromonas* and *Moraxella* was observed with inlet water samples.

Interactions Between Micro-organisms and Invertebrates in Estuarine Sediments

P. S. MEADOWS,* J. G. ANDERSON, E. A. DEANS and D. C. SMITH
*Department of Applied Microbiology, University of Strathclyde, *Department of Zoology, University of Glasgow, Scotland*

We have been conducting a joint programme of research concerned with the extent to which microbial activity in intertidal estuarine mud sediments can influence the distribution of benthic invertebrates and also the extent to which invertebrate activity can influence the distribution and activities of the sediment microflora. Concerning the latter aspect we have examined the microbiological and chemical characteristics of sediment material from around the burrows of two common benthic invertebrates *Corophium volutator* and *Nereis diversicolor*. A comparison was made on samples collected from (a) the sediment surface (b) the subsurface sediment localized around the burrows (the burrow linings) and (c) the subsurface sediment at the same depth as (b) but where the burrowing had not occurred. Gross differences were assessed by counts of heterotrophic bacteria and sulphate reducing bacteria together with measurements of the levels of chlorophyll *a*, phaeopigments, carbon, nitrogen, sulphide and oxidation–reduction potential.

The results demonstrated that surface sediment and non-burrowed subsurface sediment differed markedly in their properties. The burrow lining sediment, however, had properties which were intermediate between these. When compared with the non-burrowed subsurface sediment the sediment from the burrow lining showed an increase in the number of heterotrophic bacteria and in the levels of chlorophyll *a*, phaeopigments and oxidation–reduction potential but a decrease in the level of sulphide. We suggest that these changes may be due to ventilation of the sediment by the penetration of surface water through the burrow channel.

The importance of the sediment–water interface as a zone for differential microbial activity and for rapid chemical and gaseous exchange is well recognized. The results of this study indicate that the burrows of sediment invertebrates may be particularly important in this respect since they represent a significant extension of the sediment–water interfacial area.

Subject Index